Springer Tracts in Electrical and Electronics Engineering

Series Editors

Brajesh Kumar Kaushik, Department of Electronics and Communication Engineering, Indian Institute of Technology Roorkee, Roorkee, Uttarakhand, India

Mohan Lal Kolhe, Faculty of Engineering and Sciences, University of Agder, Kristiansand, Norway

Springer Tracts in Electrical and Electronics Engineering (STEEE) publishes the latest developments in Electrical and Electronics Engineering - quickly, informally and with high quality. The intent is to cover all the main branches of electrical and electronics engineering, both theoretical and applied, including:

- Signal, Speech and Image Processing
- Speech and Audio Processing
- Image Processing
- Human-Machine Interfaces
- Digital and Analog Signal Processing
- Microwaves, RF Engineering and Optical Communications
- Electronics and Microelectronics, Instrumentation
- Electronic Circuits and Systems
- Embedded Systems
- Electronics Design and Verification
- Cyber-Physical Systems
- Electrical Power Engineering
- Power Electronics
- Photovoltaics
- Energy Grids and Networks
- Electrical Machines
- Control, Robotics, Automation
- Robotic Engineering
- Mechatronics
- Control and Systems Theory
- Automation
- Communications Engineering, Networks
- Wireless and Mobile Communication
- Internet of Things
- Computer Networks

Within the scope of the series are monographs, professional books or graduate textbooks, edited volumes as well as outstanding PhD theses and books purposely devoted to support education in electrical and electronics engineering at graduate and post-graduate levels.

Review Process

The proposal for each volume is reviewed by the main editor and/or the advisory board. The books of this series are reviewed in a single blind peer review process.

Ethics Statement for this series can be found in the Springer standard guidelines here https://www.springer.com/us/authors-editors/journal-author/journal-author-helpdesk/before-you-start/before-you-start/1330#c14214

Li Zhang · Shihao Yang · Qianqian Wang ·
Dongdong Jin

Magnetic Micro and Nanorobot Swarms: From Fundamentals to Applications

Li Zhang
Department of Mechanical and Automation
Engineering
The Chinese University of Hong Kong
Hong Kong SAR, China

Shihao Yang
Department of Mechanical and Automation
Engineering
The Chinese University of Hong Kong
Hong Kong SAR, China

Qianqian Wang
School of Mechanical Engineering
Southeast University
Nanjing, China

Dongdong Jin
School of Materials Science
and Engineering
Harbin Institute of Technology
Shenzhen, China

ISSN 2731-4200 ISSN 2731-4219 (electronic)
Springer Tracts in Electrical and Electronics Engineering
ISBN 978-981-99-3035-7 ISBN 978-981-99-3036-4 (eBook)
https://doi.org/10.1007/978-981-99-3036-4

© The Editor(s) (if applicable) and The Author(s), under exclusive license to Springer Nature Singapore Pte Ltd. 2023

This work is subject to copyright. All rights are solely and exclusively licensed by the Publisher, whether the whole or part of the material is concerned, specifically the rights of translation, reprinting, reuse of illustrations, recitation, broadcasting, reproduction on microfilms or in any other physical way, and transmission or information storage and retrieval, electronic adaptation, computer software, or by similar or dissimilar methodology now known or hereafter developed.
The use of general descriptive names, registered names, trademarks, service marks, etc. in this publication does not imply, even in the absence of a specific statement, that such names are exempt from the relevant protective laws and regulations and therefore free for general use.
The publisher, the authors, and the editors are safe to assume that the advice and information in this book are believed to be true and accurate at the date of publication. Neither the publisher nor the authors or the editors give a warranty, expressed or implied, with respect to the material contained herein or for any errors or omissions that may have been made. The publisher remains neutral with regard to jurisdictional claims in published maps and institutional affiliations.

This Springer imprint is published by the registered company Springer Nature Singapore Pte Ltd.
The registered company address is: 152 Beach Road, #21-01/04 Gateway East, Singapore 189721, Singapore

Preface

The recent rise of micro/nanorobots that can convert environmental energy into mechanical movement and/or forces has been a fascinating topic to academic and industrial communities, as well as the public. Featured with small size and active mobility, micro/nanorobots are envisioned to perform a variety of tasks that can hardly be achieved before, holding great potential in various fields, including environmental remediation, targeted delivery, biological diagnosis, and so on. Nevertheless, the capabilities possessed by a single micro/nanorobot are usually limited, and hence, a group of micro/nanorobots is needed to be actuated and manipulated in parallel, which derives the topic of this book, micro and nanorobot swarms.

In nature, swarm behaviors are ubiquitous yet striking phenomena presented by living organisms, which play an indispensable role in their vital activities. For example, the bird flocks fly in a V-formation to save energy during migration, the fish schools swim in a coordinated large group to forage and defend themselves from predators, and the insect colonies cooperatively construct structures with their bodies to traverse difficult terrains. These natural swarm behaviors exhibit group-level functionality that cannot be achieved by individuals and thus are inspiring for the development of micro/nanorobotic swarms whose interacting units are relatively simple and unsophisticated compared to the large-scale counterparts. With collective abilities emerged from the local interactions between individuals, the artificial swarm systems emerge as a feasible platform to compensate the limitation encountered by micro/nanorobots and contribute to the actual realization of their potential applications.

Such an encouraging prospect has stimulated substantial research efforts on the development of micro/nanorobot swarms. Among the developed systems, magnetic swarms that harvest energy from external magnetic fields have shown remarkable progress due to a series of unique advantages, e.g., the wireless and remote controllability, accurate, rapid yet simple modulation of magnetic field parameters, as well as the excellent compatibility of magnetic fields with many application scenarios. This exciting area of research is thus envisioned to promise major advances in diverse areas with new powerful tools and bring about revolutionary techniques that are currently beyond our reach to improve our quality of life.

The goal of this book is to convey the recent advances in the design and construction of magnetic micro/nanorobot swarm systems and to promote the filling of the gap between their theoretical investigation and practical applications. The book is suitable for scientists, engineers, and students who are engaged in the investigation of micro/nanorobots. We have tried to make the book a systematic introduction to the current state-of-the-art and future research directions of magnetic micro/nanorobot swarms, which may help the researchers have a better understanding, and further stimulate the development of such an exciting field.

The material of this book is composed of 13 chapters. Chapter 1 is devoted to briefing different kinds of swarming micro/nanorobots that have been developed in recent years. Chapters 2–3 present the principles for constructing the vortex-like and ribbon-like microswarms. Chapters 4–6 introduce heterogeneous structure, three-dimensional pattern, indent control, and pattern transformation rate control of micro/nanorobot swarms. Chapters 7–9 discuss the advanced manipulation of magnetic swarms, including the control and imaging strategies in biological fluids and dynamic environments. Chapters 10–12 are devoted to the applications of magnetic swarms in the fields of biomedicine, environmental remediation, and microelectronics. Finally, Chap. 13 summarizes the content of the book and discusses future opportunities and challenges of micro/nanorobot swarms.

The book is written with the kind help of many organizations and individuals. The authors would like to acknowledge support from the Hong Kong Research Grants Council with project Nos. RFS2122-4S03 (RGC Research Fellow Scheme), R4015-21 (RGC Research Impact Fund), E-CUHK401/20 (EU/RGC Research and Innovation Cooperation Co-funding Mechanism by RGC), 14300621 (RGC GRF), 14301122 (RGC GRF); the HKSAR Innovation and Technology Commission (ITC) with a project No. MRP/036/18X (ITC Midstream Research Program for Universities); the Croucher Foundation with project No. CAS20403; the Chinese University of Hong Kong (CUHK) internal grants; the SIAT-CUHK Joint Laboratory of Robotics and Intelligent Systems, the Multiscale Medical Robotics Center (MRC) at the Hong Kong Science Park, the CUHK Chow Yuk Ho Technology Centre for Innovative Medicine, and the CUHK T Stone Robotics Institute.

We are very grateful to Prof. Bradley Nelson from ETH Zurich, Switzerland; Prof. Joseph Sung Jao-yiu, Dean of Lee Kong Chian School of Medicine and Senior Vice President (Health and Life Sciences), at the Nanyang Technological University; Prof. Yangsheng Xu, President of CUHK (Shenzhen); Prof. Philip Chiu Wai Yan and Prof. Samuel Au Kwok Wai, Co-Directors of MRC, the other collaborators, as well as the colleagues from the Faculty of Engineering and Faculty of Medicine at CUHK, for their long-term guidance and/or generous support to our group's research on micro/nanorobot swarms. We appreciate all the laboratory ex- and current members, including Dr. Tiantian Xu, Dr. Xiaohui Yan, Dr. Jiangfan Yu, Dr. Yabin Zhang, Dr. Ben Wang, Dr. Tony Chan Kai Fung, Dr. Lidong Yang, Dr. Yue Dong, Dr. Mengmeng Sun, Dr. Xingzhou Du, Dr. Fengtong Ji, Dr. Jialin Jiang, and many other researchers for contributing their excellent research work to this book. We would also like to thank Haojin Yang, Zhaoyang Qi, Junjia Guo, Cheng Chak Kit, Yanfei Cao,

Preface

Yihang Jiang, and Bonan Sun in our group for their assistance in drawing figures and proofreading. The editorial and production staff of Springer Nature Press are gratefully acknowledged for the kind arrangement in writing this book that enables its publication. Thanks to the great and fascinating efforts achieved by numerous researchers all over the world, the continuous discoveries keep pushing the forefront of magnetic micro/nanorobot swarms, and we sincerely hope the prospect and imagination of micro/nanorobots presented in this book can come into reality in the near future.

Finally, we would like to use this chance to send our best greeting for celebrating the 60th anniversary of the Chinese University of Hong Kong in 2023.

Hong Kong SAR, China
December 2022

Li Zhang
(On behalf of all authors)

Contents

1 Introduction to Micro/Nanorobot Swarms 1
   1.1   Introduction 1
   1.2   Magnetic Field-Driven Microswarms 4
        1.2.1   Magnetic Field-Induced Microswarm 6
        1.2.2   Medium-Induced Microswarm 8
        1.2.3   Weakly Interacted Microswarm 10
   1.3   Light-Driven Microswarms 11
   1.4   Acoustic Field-Driven Microswarms 15
   1.5   Electric Field-Driven Microswarms 17
   1.6   Hybrid Field-Driven Microswarms 20
   References 25

2 Pattern Formation and Control of a Vortex-Like Paramagnetic Nanoparticle Swarm 31
   2.1   Introduction 31
   2.2   Swarm Formation 33
   2.3   Characterization of the Vortex-Like Swarm 36
   2.4   Pattern Transformation of the Vortex-Like Swarm 38
        2.4.1   Changes in Swarm Size 38
        2.4.2   Spread Swarm Pattern 40
        2.4.3   Inward Force 42
   2.5   Experimental Results 43
        2.5.1   Formation of the Vortex-Like Microswarm 43
        2.5.2   Characterization of the Vortex-Like Swarm 45
        2.5.3   Pattern Transformation 47
        2.5.4   Swarm Locomotion 51
   2.6   Elliptical Vortex-Like Swarm 56
        2.6.1   Elliptical Rotating Magnetic Field 56
        2.6.2   Formation of the Elliptical Vortex-Like Microswarm 58
        2.6.3   Experimental Results 64

		2.7	Conclusion	68
			References	69
3	**Ribbon-Like Magnetic Colloid Microswarm**			71
	3.1	Introduction		71
	3.2	Formation of the Ribbon-Like Swarm		73
		3.2.1	The Oscillating Field and Time-Varying Chain Length	73
		3.2.2	Chain-Chain Magnetic Interaction	76
		3.2.3	Pattern Formation Process	78
		3.2.4	Chain-Chain Hydrodynamic Interaction	80
		3.2.5	Pattern Formation Using Particles of Different Sizes and at Different Particle Concentrations	86
	3.3	Multimodal Reconfiguration of the Swarm Pattern		89
		3.3.1	Pattern Reconfiguration of the Ribbon-Like Swarm	89
		3.3.2	Mechanism of Pattern Reconfiguration	92
	3.4	Navigated Locomotion		95
		3.4.1	Two Locomotion Modes of the Ribbon-Like Swarm	95
		3.4.2	Pattern Stability and Controlled Locomotion	97
	3.5	Preliminary Application Demonstration		99
	3.6	Conclusion		102
		References		103
4	**Heterogeneous Colloidal Microswarm with Multifunction**			107
	4.1	Introduction		107
	4.2	Design and Preparation of Building Blocks as Drug Delivery Vehicles		111
	4.3	Magnetic Control of the Synchronized and Adaptive Microswarm		113
	4.4	Domino Reaction Encoded Cooperative Function of Heterogeneous Microswarm		116
	4.5	Adaptive and Heterogeneous Colloidal Swarm for Anticancer Targeted Drug Delivery		119
	4.6	Experimental Section		121
		4.6.1	Nanoparticle Preparation	121
		4.6.2	Cellular Experiments and Characterization Techniques	123
	4.7	Conclusion		124
		References		124
5	**Three-Dimensional Structure and Independent Control of Micro/Nanorobot Swarms**			127
	5.1	Introduction		127
	5.2	Two-Dimensional Swarm Under the Conical Rotating Field		131

		5.2.1	Building Blocks and the Magnetic and Optical Control System	131
		5.2.2	Formation of the 2D Swarm Pattern	132
	5.3	The Tornado-Like Swarm with a 3D Structure		136
		5.3.1	Light-Driven Hovering of the Swarm	136
		5.3.2	Reconfigurable Formation of the Tornado-Like Swarm	139
		5.3.3	Reaction Rate Control Using the Tornado-Like Swarm	142
	5.4	The Nickel Nanorod Swarm		144
		5.4.1	Modeling of the Magnetized Nickel Nanorod	144
		5.4.2	Formation Mechanism of the Nanorod Swarm	147
		5.4.3	Pattern Transformation of the Nanorod Swarm	149
	5.5	Independent Control of Micro/Nanorobot Swarms		153
		5.5.1	Comparison Between Nanorod and Nanoparticle Swarms	153
		5.5.2	Independent Pattern Formation and Transformation of Nanorod and Nanoparticle Swarms	157
	5.6	Conclusion		158
	References			160
6	**Pattern Transformation Rate Control of Magnetic Microswarms**			163
	6.1	Introduction		163
	6.2	Modeling of Chain-Chain Interactions		165
		6.2.1	Magnetic Interaction	165
		6.2.2	Hydrodynamic Interaction	168
		6.2.3	Driving Force of Pattern Transformation	170
	6.3	Pattern Transformation Rate Control Strategy		171
		6.3.1	Pattern Transformation Rate Control of the Ribbon-Like Microswarm	171
		6.3.2	Pattern Transformation Rate Control of the Vortex-Like Microswarm	173
	6.4	Pattern Transformation Rate Control in Various Environments		176
	6.5	Conclusion		180
	References			181
7	**Formation and Actuation of Micro/Nanorobot Swarms in Bio-Fluids**			185
	7.1	Introduction		185
	7.2	The Library of Magnetic Active Micro/Nanorobot Swarms		187
	7.3	Influence of Fluid Properties on Microswarms		190
		7.3.1	Main Physical Properties of Bio-Fluids	190

		7.3.2	Swarm Formation in Fluids with Controlled Physical Conditions	191
		7.3.3	Swarm Locomotion and Transformation in Ionic and Viscous Fluids	193
	7.4	Selection Strategy for Optimized Swarms in Bio-Fluids		194
	7.5	Swarm Formation and Navigation in Bio-Fluids		196
		7.5.1	Swarm Formation and Pattern Transformation in Bio-Fluids	196
		7.5.2	Swarm Navigation in Bio-Fluids	200
		7.5.3	Swarm in Whole Blood	201
		7.5.4	Swarm in Vitreous Humor and Bovine Eyeballs	203
	7.6	Conclusion		210
	References			210
8	**Localization of Microswarms Using Various Imaging Methods**			215
	8.1	Introduction		215
	8.2	Localization of a Rotating Colloidal Microswarm Under Ultrasound Guidance		218
		8.2.1	Mathematical Modeling and Simulations	218
		8.2.2	Localization of a Microswarm Using Ultrasound Feedback	221
	8.3	Magnetic Actuation of a Dynamically Reconfigurable Microswarm for Enhanced Ultrasound Imaging Contrast		224
		8.3.1	Estimation of the Imaging Contrast of a Rotating Microswarm	224
		8.3.2	Experiments of Ultrasound Contrast Under the Actuation of NRF	227
		8.3.3	Swarm Transformation and Ultrasound Contrast Under the Actuation of the PRF	228
		8.3.4	Swarm Navigation and Transformation in a Confined Environment Ex Vivo	231
	8.4	Fluorescence Imaging and Photoacoustic Imaging (PAI) of the Microswarm		232
	References			235
9	**Formation and Navigation of Microswarms in Dynamic Environments**			239
	9.1	Introduction		239
	9.2	Formation of a Magnetic Nanoparticle Microswarm in Whole Blood		241
	9.3	Rotating Microswarm Under Doppler Ultrasound Imaging		246
	9.4	Swarm Formation and Navigation in Flowing Blood		249
	9.5	Real-Time Swarm Formation and Navigation in Porcine Coronary Artery Ex Vivo		254
	9.6	Discussion and Conclusion		255
	References			258

Contents

10 Applications of Micro/Nanorobot Swarms in Biomedicine 261
    10.1 Introduction 261
    10.2 Enhance Local Hyperthermia 264
         10.2.1 Magnetic Local Hyperthermia 264
         10.2.2 Adjustment of the Nanoparticle Concentration 265
         10.2.3 Photothermal Effect of the Nanoparticle Swarm 268
         10.2.4 Enhanced Local Hyperthermia 269
    10.3 Biofilm Eradication 271
         10.3.1 Introduction to Biofilms 271
         10.3.2 The p-Fe_3O_4 Swarm with Catalytic and Antimicrobial Capabilities 274
         10.3.3 Biofilm Elimination by the p-Fe_3O_4 Swarm 279
         10.3.4 The MUCR@MLMD and Its Swarming Behavior 282
         10.3.5 Biofilm Eradication Using the MUCR@MLMD Swarm 285
    10.4 Accelerate Thrombolysis 291
         10.4.1 Introduction to Thrombolysis 291
         10.4.2 Swarm-Induced Fluid Flow and Shear Stress 292
         10.4.3 Accelerating TPA-Mediated Thrombolysis Using the Microswarm 295
    References 301

11 Biohybrid Microswarm for the Removal of Toxic Heavy Metals 307
    11.1 Introduction 307
    11.2 Synthesis and Characterization of PSFBA 308
    11.3 Controlled Locomotion of PSFBA Under the Guidance of Magnetic Field 311
    11.4 Enhanced Heavy Metal Removal Efficiency Enabled Flow Field Around PSFBAs Swarm 313
    11.5 Experimental Section 316
    11.6 Conclusion 316
    References 317

12 Ant Bridge-Mimicked Reconfigurable Microswarm for Electronic Application 321
    12.1 Introduction 321
    12.2 Design and Functionalization of Building Blocks 322
    12.3 Swarm Control of the Magnetic and Electrically Conductive Building Blocks 326
    12.4 Construction of Ant Bridge-Mimicked Magnetic Microswarm for Electrical Connection 330
    12.5 Applications of Microswarm in Electronic Field 331
    12.6 Experimental Section 333
    12.7 Conclusion 335
    References 335

13 Summary and Outlook 337
    13.1 Fundamentals 338
        13.1.1 Formation 338
        13.1.2 Transformation 339
        13.1.3 Locomotion 340
        13.1.4 Heterogeneous Microswarm 341
        13.1.5 3D Microswarms 342
        13.1.6 Independent Control 342
    13.2 Applications 343
    References 347

Chapter 1
Introduction to Micro/Nanorobot Swarms

Abstract Micro/nanorobot swarm inspired by collective behaviors in nature represents an emerging research field. This chapter provides a brief explanation of the concept of micro/nanorobot swarms, followed by some examples of representative work in this field. The magnetic field-driven micro/nanorobot swarms are first highlighted, which are divided into three types according to the difference in internal dominant interactions: magnetic field induced, medium induced, and weakly interacted. In addition, other types of external field-driven micro/nanorobot swarms and their general principles are also introduced, including light driven, acoustic field driven, electric field driven, and hybrid field driven.

Keywords Micro/nanorobot · Collective behavior · Microswarm · Magnetic field · External field driven

1.1 Introduction

The rapid development of micro/nanoscale science and technology in the last few decades enables the scale-down of remotely controlled machines. Micro/nanorobots, which mainly refer to micro/nanoscale machines capable of converting energy into force or motion, are able to complete tasks on demand in various environments (Kim and Julius 2012; Sitti 2017; Diller and Sitti 2013). Through careful design of materials and structures, micro/nanorobots can be actuated in many different ways, including external power sources (Sitti and Wiersma 2020; Kim et al. 2018; Chen et al. 2018a; Barbot et al. 2019; Yang et al. 2018; Kong et al. 2020), self-propulsion (Avila et al. 2020; Karshalev et al. 2018; Xu et al. 2020; Lin et al. 2016), and hybrid propulsion (Alapan et al. 2018; Bente et al. 2018; Ricotti et al. 2017; Medina-Sanchez et al. 2016). Micro/nanorobots have attracted wide interest due to their potential for applications in diverse fields (Sitti 2018; Field et al. 2019; Ou et al. 2020; Chen et al. 2018b), especially in biomedicine, such as targeted delivery (Nelson et al. 2010; Sitti et al. 2015; Yang et al. 2020; Li et al. 2018), biosensing (Li et al. 2017; Esteban-Fernández de Ávila et al. 2016; Molinero-Fernández et al. 2018; Zhang et al. 2019;

© The Author(s), under exclusive license to Springer Nature Singapore Pte Ltd. 2023
L. Zhang et al., *Magnetic Micro and Nanorobot Swarms: From Fundamentals to Applications*, Springer Tracts in Electrical and Electronics Engineering,
https://doi.org/10.1007/978-981-99-3036-4_1

Yang et al. 2019), micromanipulation (Steager et al. 2013; Pacchierotti et al. 2018; Jing et al. 2019), and minimally invasive surgery (Kummer et al. 2010; Bergeles and Yang 2014; Mahoney and Abbott 2016). Although the tiny size and wireless actuation endow micro/nanorobots with the ability to perform in hard-to-reach areas of traditional tools, their practical application faces many challenges. First, a single micro/nanorobot is limited in the force generating and cargo carrying due to its small size and volume and therefore may not be able to adequately influence the surrounding environment or deliver sufficient doses of drugs when serving as an end effector or drug carrier. Second, the morphology of micro/nanorobots is generally fixed or can only perform small-scale deformation, which lacks good adaptability to some complex environments, such as tiny and tortuous lumens in the human body. Third, the small propulsion force of micro/nanorobots makes them vulnerable to the interference of external environments, especially the complex liquid composition and fluid flow of the physiological environment, which may reduce the controllability and efficiency of locomotion. Moreover, the low imaging resolution and contrast of micro/nanorobots also challenge the real-time imaging and localization for in vivo applications. Overall, although significant progress has been made in developing advanced micro/nanorobots, the capability of an individual is always limited, and it is difficult for even the most elaborate single micro/nanorobots to complete various predetermined tasks in a complex environment.

The ubiquitous swarming phenomenon in nature has been observed long before the first micro/nanorobots were invented (Sumpter 2010; Ward and Webster 2016). As a common phenomenon in the biological world, collective behavior spans most of the biological scales ranging from bacterial colonies at the cellular level to insect swarms, bird flocks, fish schools, and even human crowds on the organismal scale (Parrish and Edelstein-Keshet 1999; Sumpter 2006). The animal swarm consists of hundreds, thousands, or even millions of individual organisms through local communication, which can migrate as an entity and transform or reconstruct in response to external stimuli (Vicsek and Zafeiris 2012; Nagy et al. 2010). The swarming behavior enables living organisms to perform tasks that are far beyond the ability of individuals (Fig. 1.1a). For example, birds fly together in a specific formation to save energy during long-range migrations (Weimerskirch et al. 2001; Usherwood et al. 2011; Mirzaeinia et al. 2020), fish in the ocean gather into schools to swim faster, escape predators, and obtain sufficient oxygen (Brierley and Cox 2010; Fish have it easy in schools; Hemelrijk et al. 2015), and ants link their bodies to form rafts to survive floods or bridges to across treacherous terrain (Mlot et al. 2011; Anderson et al. 2002; Szuba 2017). These amazing and spectacular natural phenomena provide inspiration for researchers in the field of robotics. At the macroscale, swarming artificial robotic systems have been reported (Fig. 1.1b) (Petersen et al. 2019), such as the particle robotics consisting of many loosely coupled components (Li et al. 2019), the thousand-robot swarm with programmable self-assemble two-dimensional patterns (Rubenstein et al. 2014a, b, 2012), and the robot swarm with autonomous task sequencing ability (Garattoni and Birattari 2018). At the micro/nanoscale, scientists propose the idea of breaking through the bottleneck in the development of micro/

1.1 Introduction

nanorobots by increasing the number and utilizing swarm intelligence, which leads to the concept of micro/nanorobot swarms (microswarm), as shown in Fig. 1.1c.

While retaining the inherent advantages of micro/nanorobots, the microswarm has enhanced capabilities to perform tasks that are hard to achieve using a single agent (Wang and Zhang 2018; Jin and Zhang 2022; Wang and Pumera 2020). First, the large number of agents improves the functions of the microswarm serving as delivery and micromanipulation tools, such as carrying more cargo, batch-manipulating multiple objects, and performing parallel operations. Second, microswarms are reconfigurable and have transformable patterns that significantly improve their environmental adaptability. The microswarms are able to change their morphology to adapt to a variety of

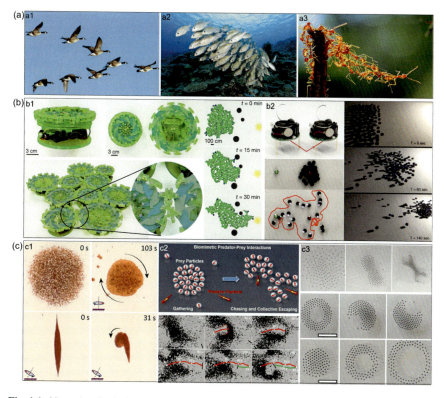

Fig. 1.1 Natural and robotic swarms. **a** Animal swarms in nature. **a1** V-shaped bird flock. Reproduced with permission from reference (Mirzaeinia et al. 2020). **a2** Fish school. Reproduced with permission from reference (Fish have it easy in schools). **a3** Ant bridge. Reproduced with permission from reference (Szuba 2017). **b** Macroscale robot swarms. **b1** The particle robot swarm. Reproduced with permission from reference (Li et al. 2019). **b2** Collective system of Kilobot. Reproduced with permission from reference (Rubenstein et al. 2014b). **c** Micro/nanorobot swarms. **c1** The swarm of magnetic microdroplets. Reproduced with permission from reference (Sun et al. 2021). **c2** The passive microparticles exhibit predator–prey-like swarming behaviors. Reproduced with permission from reference (Mou et al. 2019a). **c3** Pattern reconfiguration of ferromagnetic microrobots at the air–water interface. Reproduced with permission from reference (Dong and Sitti 2020)

confined environments, especially narrow lumens in the human body, such as blood vessels, bile ducts, and pancreatic ducts. Third, the coordinated behavior and agent-agent interactions in the microswarm can improve movement stability and enhance resistance to external disturbance. Fourth, the microswarm can significantly improve the quality of imaging and be observed more easily. The microswarm has higher contrast than a single micro/nanorobot in many imaging modalities (e.g., ultrasound imaging, magnetic resonance imaging, and fluoroscopic imaging), which is beneficial to real-time localization and control. In addition, the structure of microswarms does not require complicated design, and the desired functions can be obtained by simply functionalizing the building blocks or combining different types of agents.

Research related to the fundamentals and applications of microswarms has become a hot topic in the academic community in recent years. On the one hand, the microswarm is a typical active matter, i.e., a non-equilibrium system consisting of interacting agents that constantly dissipate and convert energy into mechanical forces or motion (Bowick et al. 2022; Fu et al. 2022; Marchetti et al. 2016). Studies on the fundamental physics of microswarms, including constructing theoretical models, exploring various types of microswarms, and analyzing internal interactions, can provide a new perspective for investigating active matter systems and deepen the understanding of collective behavior in nature. On the other hand, on the basis of fundamental research, microswarm systems with different functionalities have been developed. These versatile microswarms are expected to be revolutionary tools for traditional engineering and medical applications, and a variety of practical applications have been demonstrated. Microswarms can be driven in various different strategies, among which magnetic fields have the advantage of being easy to generate and program, exhibiting unique advantages in carefully customizing and elaborately modulating the interactions between micro/nanorobots, facilitating the study of a wide variety of collective behaviors. Besides, the magnetic field also has the advantages of good penetrability and is harmless to human tissue. Therefore, magnetically driven microswarms (magnetic microswarms) are considered to have great application potential in the biomedical field and have become a subfield of intense research in this field. In this chapter, we will first highlight some representative research work on magnetic microswarms. In addition, this chapter will also briefly introduce other field-driven microswarms, including light, acoustic field, electric field, and hybrid field-driven microswarms.

1.2 Magnetic Field-Driven Microswarms

Magnetically driven microswarms consist of ferro/paramagnetic building blocks (e.g., sphere, rod, helix, bacterium, etc.) with good magnetic responses and are energized by external dynamic magnetic fields. The paramagnetism or ferromagnetism of the building blocks can be qualitatively reflected by their magnetization behavior (Abbott et al. 2020). The magnetization curve hysteresis of ferromagnetic materials (e.g., Fe, Co, Ni) shows that they remain magnetized after magnetization and can

1.2 Magnetic Field-Driven Microswarms

therefore be regarded as tiny magnets. The magnetized ferromagnetic material can be demagnetized using a heat treatment at a temperature above the Curie temperature or a reverse magnetic field with a critical strength. For paramagnetic materials, their magnetism disappears when the magnetic field is removed. Therefore, the magnetic interactions between paramagnetic agents are easier to tune (Kei Cheang et al. 2014; Shapiro et al. 2014). In a magnetic field **B**, a magnetic dipole with moment **m** is subjected to the magnetic force **F** or magnetic torque τ, which are expressed as

$$\mathbf{F} = (\mathbf{m} \cdot \nabla)\mathbf{B} \tag{1.1}$$

$$\tau = \mathbf{m} \times \mathbf{B}. \tag{1.2}$$

The magnetic dipole moment is given by

$$\mathbf{m} = \frac{3V}{\mu_0}\left(\frac{\mu - \mu_0}{\mu + 2\mu_0}\right)\mathbf{B}, \tag{1.3}$$

where V is the volume of the magnetic agent, μ is the magnetic permeability, and μ_0 is the permeability of free space. In a magnetic microswarm, each individual agent is not only actuated by the external magnetic field, but also affected by the magnetic moments of other agents within the swarm. The interaction force between two magnetic dipoles can be expressed as

$$\mathbf{F}_{ij} = \frac{3\mu_0}{4\pi \|\mathbf{r}_{ij}\|^4}\left(\left(\hat{\mathbf{r}}_{ij}^T\mathbf{m}_j\right)\mathbf{m}_i + \left(\hat{\mathbf{r}}_{ij}^T\mathbf{m}_i\right)\mathbf{m}_j + \left(\mathbf{m}_i^T\mathbf{m}_j - 5\left(\hat{\mathbf{r}}_{ij}^T\mathbf{m}_i\right)\left(\hat{\mathbf{r}}_{ij}^T\mathbf{m}_j\right)\right)\hat{\mathbf{r}}_{ij}\right), \tag{1.4}$$

where $\mathbf{m}_i$ and $\mathbf{m}_j$ indicate the magnetic moment of the ith and jth dipoles, respectively. $\mathbf{r}_{ij}$ is the vector between the two dipoles. In addition to magnetic interactions, magnetic agents in the microswarm are also affected by the surrounding medium (usually liquid). For example, the field-driven motion of magnetic agents perturbs the surrounding fluid or deforms interfaces (e.g., liquid–air and liquid–liquid interfaces), affecting the motion behaviors of other magnetic agents. Both the magnetic and hydrodynamic agent-agent interactions play crucial roles in forming magnetic microswarm and coordinating the collective behavior of inner agents (Liljestrom et al. 2019; Aranson 2013; Shields and Velev 2017; Dobnikar et al. 2013). Therefore, we divide magnetic microswarms into three types according to the dominant inner interactions: magnetic field-induced microswarm, medium-induced microswarm, and weakly interacted microswarm (Yu et al. 2019).

1.2.1 Magnetic Field-Induced Microswarm

Various types of dynamic magnetic fields are capable of introducing agent-agent interactions in the magnetic field-induced microswarm. For example, in a horizontal rotating magnetic field, the magnetic interaction between magnetic agents changes with the field direction. It becomes maximally attractive and repulsive with $\boldsymbol{m}_i$ and $\boldsymbol{m}_j$ parallel and normal to $\boldsymbol{r}_{ij}$, respectively. From a time-average aspect, the paramagnetic colloids can assemble into a carpet-like microswarm since the interaction is attractive (Fig. 1.2a) (Massana-Cid et al. 2019). When removing the external field, the dipole–dipole interactions disappear, resulting in the disassociation of the assembled pattern. Non-magnetic particles can be endowed with permanent magnetic moments by coating a thin magnetic film, such as magnetic Janus particles (Yan et al. 2015a). Since the magnetic coating is asymmetric, the magnetic dipole shifts from the center of mass, and the dipole offset becomes larger when the layer is thinner. Actuated by the horizontal rotating magnetic field, the Janus particles perform a spinning motion, and their magnetic caps point perpendicular to the horizontal plane. The magnetic attraction between the particles leads to the formation of a hexagonal crystal-like pattern for most values of thickness (Fig. 1.2b). Furthermore, when the dipole offset is large enough, the Janus particles can stack in the vertical direction to minimize the distance between the dipoles. In this case, the swarm pattern is reconfigured into a body-centered tetragonal lattice. This demonstrates that designing the magnetic moments of the building blocks is an effective way to develop novel magnetic microswarm systems (Klapp 2016).

Besides the direct interactions-governed pattern formation, interactions between particle chains also play an important role in swarm formation. In the magnetic field, magnetic particles first assemble into chain-like structures through the dipole–dipole interactions of induced magnetic moments, which serve as the building blocks of the swarm. Actuated by a precessing magnetic field, superparamagnetic polystyrene microparticles form chains and then self-organize into cohesive swarms by balancing the magnetic dipolar attraction and multipolar repulsion, and they arrange themselves at steady-state distances from the neighboring chains (Fig. 1.2c) (Yigit et al. 2020). It is noted that the precession angle of the magnetic field essentially affects the magnetic dipolar interaction. The average interaction force between $\boldsymbol{m}_i$ and $\boldsymbol{m}_j$ over a rotating cycle is expressed as (Giovanazzi et al. 2002)

$$\mathbf{F}_m = -\frac{3\mu_0 \mathbf{m}_i \mathbf{m}_j}{4\pi \|\mathbf{r}_{ij}\|^4} \left(\frac{3\cos^2\varphi - 1}{2}\right), \tag{1.5}$$

where φ is the precession angle to the rotating axis. The above equation shows that the magnetic interaction is able to be modulated between repulsion and attraction when $0° < \varphi < 54.7°$ and $54.7° < \varphi < 90°$, respectively. φ is adjusted in 68°–72° to obtain chain-chain attraction where the pairwise distance converges to a dynamically steady range (Fig. 1.2c). Besides dipole–dipole interaction, dipole–hexapole interaction

1.2 Magnetic Field-Driven Microswarms

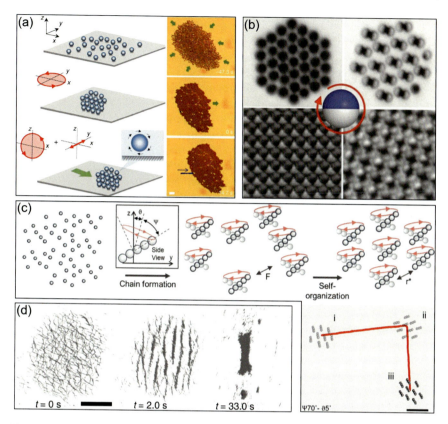

Fig. 1.2 Magnetic field-induced microswarms. **a** A carpet-like microswarm assembled by paramagnetic colloid particles. Reproduced with permission from reference (Massana-Cid et al. 2019). **b** The crystal-like pattern formed by magnetic Janus particles under a rotating field. Reproduced with permission from reference (Yan et al. 2015a). **c** Superparamagnetic polystyrene microparticles self-organize into cohesive swarms under a precessing magnetic field. Reproduced with permission from reference (Yigit et al. 2020). **d** The ribbon-like swarm formed by paramagnetic Fe$_3$O$_4$ nanoparticles under an oscillating magnetic field. Reproduced with permission from reference (Yu et al. 2018a)

dominates the repulsive multipolar interaction when $\varphi < 61.5°$, which leads to a short-range repulsion and long-range attraction (Tierno et al. 2009).

Paramagnetic Fe$_3$O$_4$ nanoparticles can also assemble into chains and form a ribbon-like microswarm under the actuation of an oscillating magnetic field, as shown in Fig. 1.2d (Yu et al. 2018a). The oscillating magnetic field consists of a sinusoidal alternating component and a constant component orthogonal to it and thus has a time-varying field strength and angular velocity, resulting in a time-varying nanoparticle chain length. When the magnetic field strength is maximum, the angular velocity is zero, while the angular velocity is maximum when the magnetic field strength is minimum. As a result, the nanoparticle chains are continuously broken into shorter ones and arranged in elongated patterns as they oscillate with the magnetic field. The attractive force between adjacent nanoparticle chains is enhanced when the field strength becomes stronger, which forms a thinning and long swarm pattern. After multiple oscillation cycles, the magnetic and hydrodynamic interactions between the nanoparticle chains balance each other, eventually forming a dynamic-equilibrium ribbon-like microswarm (Fig. 1.2d). For this swarming system, the magnetic interaction between the building blocks is the main factor in the swarm formation and is also dominant in maintaining the swarm pattern stability. Therefore, the ribbon-like swarm is a typical magnetic field-induced swarm, which we will describe in detail in Chap. 3. Another swarming system with ribbon-like patterns consists of peanut-shaped hematite particles (Xie et al. 2019). In addition to the peculiar peanut-like shape, the hematite particles are ferromagnetic and have permanent magnetic moments along their short axis, which is different from Fe_3O_4 nanoparticles. Therefore, the hematite particles randomly aggregate into clusters through dipole–dipole interactions in the absence of an external magnetic field. When a conical rotating magnetic field with the central axis in the horizontal plane is applied, each particle rolls in the plane along its long axis. The particles are arranged in a ribbon-like array due to magnetic attraction, and the moving direction of the entire array is perpendicular to its long axis. Ferrofluid droplets can form circular swarm pattern under the rotating magnetic field (Sun et al. 2021). When the rotating frequency of the magnetic field becomes very high, each droplet fluctuates only in small amplitudes. In this case, the swarm turns into a solid-like state, in which the droplets are tightly bound through magnetic force, and the hydrodynamic effect is almost negligible.

1.2.2 Medium-Induced Microswarm

In the medium-induced microswarms, the hydrodynamic interaction between magnetic agents is dominant. As shown in Fig. 1.3a, the paramagnetic Fe_3O_4 nanoparticles form a vortex-like microswarm under the actuation of in-plane rotating magnetic fields (Yu et al. 2018b). The nanoparticles also assemble into chains first, and each rotating chain generates a local vortex around it. The hydrodynamic interaction between vortices is attractive over long distances, which drives them to approach each other and merge into a single vortex after reaching a certain distance. The vortices generated by multiple chains merge with each other, forming a dynamically balanced circular vortex-like swarm pattern. In the vortex-like swarm, each particle chain not only rotates around its own center but also orbits around the swarm center. The main vortex exerts an inward force on all nanoparticle chains in the swarm. Thus, even though the hydrodynamic interaction between chains is repulsive in the short range, the inward force plays the role of confining nanoparticle particle chains within the swarm. When adjusting the magnetic field parameters, such as field strength and rotating frequency, the rotating state of the chains changes correspondingly, resulting in changes in the swarm pattern (such as area and shape). The vortex-like nanoparticle swarm is a representative medium-induced microswarm whose swarm pattern

1.2 Magnetic Field-Driven Microswarms

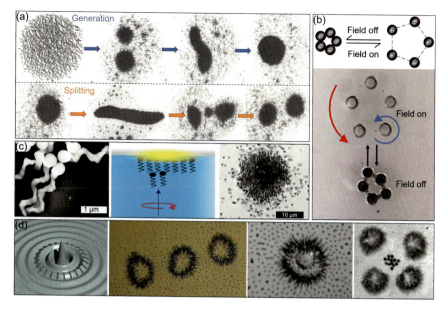

Fig. 1.3 Medium-induced microswarms. **a** The active generation and splitting of the vortex-like paramagnetic Fe$_3$O$_4$ nanoparticle swarm under rotating magnetic fields. Reproduced with permission from reference (Yu et al. 2018b). **b** Ordered swarm pattern formed by magnetic droplets under a conical rotating field. Reproduced with permission from reference (Wang et al. 2019). **c** A circular swarm pattern formed by helical micropropellers at the air–liquid interface. Reproduced with permission from reference (Vach et al. 2017). **d** The aster-like microswarm formed by ferromagnetic nickel microparticles at the liquid–liquid interface. Reproduced with permission from reference (Snezhko and Aranson 2011)

is mainly maintained by the hydrodynamic force, which will be described in detail in Chap. 2.

Short-range hydrodynamic repulsion can be used to induce dynamically balanced swarming states of magnetic agents. As shown in Fig. 1.3b, droplets containing ferromagnetic particles are placed below the liquid–air interface (Wang et al. 2019). When no magnetic field is applied, the droplets gather together and contact each other due to the attraction between the ferromagnetic particles and the capillary interaction between the droplets. After applying a conical rotating magnetic field (the central axis is perpendicular to the horizontal plane), the particles rotate with the magnetic field and drive the droplet to rotate. The fluid vortices induced by the rotating droplets exert repulsive hydrodynamic forces on adjacent droplets, resulting in an ordered swarm pattern in which the originally aggregated droplets are separated from each other. Hydrodynamic interactions between droplets can be tuned by adjusting the magnetic field, and the swarm pattern changes accordingly, such as expanding and contracting. Actuated by the rotating magnetic field, the helical micropropellers can swim upward and reach the top air–liquid interface (Vach et al. 2017). Under the suitable field strength and rotating frequency, the hydrodynamic interaction among these identical helical swimmers enables them to form a corotating microswarm (Fig. 1.3c), which is similar to the vortex-like microswarm. Here, the hydrodynamic interaction mainly depends on the rotation-translation coupling effects (Jager et al. 2013), showing similarity to the dynamic self-assembly of sphere particles (Yan et al. 2015b).

The hydrodynamic interaction plays a more important role when microswarms are located at the interface of the two liquids. As shown in Fig. 1.3d, ferromagnetic microspheres are placed at the interface of two immiscible liquids (Snezhko and Aranson 2011). Similar to paramagnetic nanoparticles, the ferromagnetic nickel spherical microparticles form chain-like structures through magnetic dipole–dipole interactions. Under the actuation of a vertical alternating magnetic field, the chains perform periodic oscillations at the liquid–liquid interface, resulting in the deformation of the interface. Then, a resonant excitation of interfacial waves is caused by the deformation, and quasi-static hydrodynamic streaming flows are excited by periodic oscillations of the interface. Subsequently, the circular waves are further excited, causing particle chains to assemble into a dynamic swarm at the interface. The chains are distributed on different slopes of the same wave, which form an aster-like pattern (Fig. 1.3d). Compared with microswarms on solid substrates or liquid–air interfaces, the easily deformable liquid–liquid interface leads to very different internal hydrodynamic interactions, where unique microswarms are generated. Furthermore, many other studies show that the collective behavior can be adjusted to form various microswarms by changing the hydrodynamic interaction among building blocks. For example, ferromagnetic particles excited by an external rotating magnetic field can generate strong fluid flow, and repulsive hydrodynamic interactions among multiple rotating particles lead to the formation of dynamic lattices rather than aggregates due to magnetic attraction (Han et al. 2020). Ferromagnetic particles actuated by a uniaxial alternating magnetic field can also generate fluid vortices, and the size of the vortices and the particle density can be tuned by changing the magnetic field parameters (Kokot et al. 2019).

1.2.3 Weakly Interacted Microswarm

In a magnetic microswarm, if the interactions between the building blocks are too weak, the behavior of each individual is hardly affected by other individuals around them. In this case, building blocks in the microswarm move independently under the actuation of external fields, and we categorize this type of magnetic microswarm as the weakly interacted microswarm. A swarm of micropropellers capable of penetrating the vitreous humor and reaching the retina has been reported (Wu et al. 2018). The micropropeller has a spherical head and a helical tail, and the diameter of its head (~ 500 nm) is comparable to the mesh size of the biopolymer network of the vitreous humor. The perfluorocarbon coating on the surface of the micropropellers greatly reduces the interaction of the propellers with the biopolymer network, allowing a swarm of micropropellers to perform efficient propulsion in the vitreous humor. In the micropropeller swarm, the interaction between the individual micropropellers is negligible. They are driven in the same direction by an external magnetic field, but have no influence on each other. Since there is no significant difference in their moving velocity, they appear in the form of weakly interacted microswarm. Similarly, helical FePt nanopropellers demonstrate active cell targeting and magnetic transfection of lung carcinoma cells (Fig. 1.4a) (Kadiri et al. 2020). The surface wettability of artificial bacterial flagella (ABF) significantly affects their swimming behavior (Wang et al. 2018). Different chemistries are used to change wettability properties of the ABF surface to obtain ABFs with different step-out frequencies and moving velocities. Since the movements of the ABFs within the swarm do not interfere with each other, selective control of the ABFs can be achieved by adjusting the frequency of the magnetic field (Fig. 1.4b). In a magnetotactic bacteria swarm, each individual moves in the same direction guided by the gradient of a static field without interacting with each other (Fig. 1.4c) (Taherkhani et al. 2014). A swarm of nickel nanorods actuated by a rotating magnetic field moves toward the blood clot in tPA solution (Cheng et al. 2014). Although there is no obvious interaction between the nanorods, the swarming effect enables the individual functions to be enhanced. Specifically, the convection induced by the swarming nanorods enhances the mass transport of the tPA molecules to the clot, increasing the efficiency of thrombolysis.

In the rotating field-actuated paramagnetic nanoparticle swarm, when the rotating plane of the magnetic field is perpendicular to the horizontal plane, the nanoparticle chains will rotate with the magnetic field in the vertical plane. The end of the chain near the bottom experiences more resistance than the end at the top due to the presence of the substrate, causing the chain to perform a tumbling motion along its long axis (Fig. 1.4d) (Yu et al. 2017). Due to the large separation distance between the chains, both the magnetic and hydrodynamic interactions are small. Especially after being disassembled by the dynamic magnetic field, the length of the particle chain becomes shorter, and the difference becomes smaller. All particle chains in the swarm move parallel to the rotating plane of the magnetic field and do not interfere with each other. It is worth noting that if the separation distance between the building blocks of weakly interacted microswarms is reduced, then the magnetic and hydrodynamic interactions will affect the behavior of the swarm. When the motions of the individual agents are not independent, the weakly interacted microswarm will transform into magnetic field-induced or medium-induced microswarms. Similarly, when the driving magnetic field changes, the microswarm can also transition between magnetic field induced and medium induced (Jin et al. 2021; Yu et al. 2021).

1.3 Light-Driven Microswarms

Light is capable of providing adjustable radiant energy in a wireless and remote propagation manner and is widely used to actuate micro/nanorobot swarms (Xu et al. 2017a). The building blocks of light-driven microswarms are generally photosensitive materials, such as AgCl and TiO_2 microparticles, and TiO_2/Pt, TiO_2/Au Janus

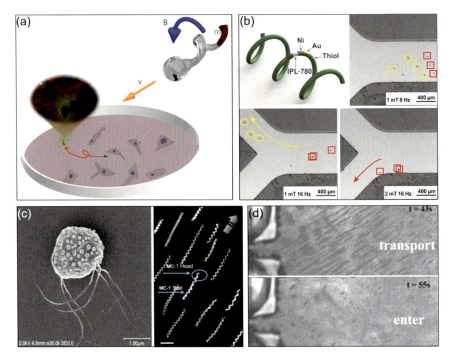

Fig. 1.4 Weakly interacted microswarms. **a** Helical FePt nanopropellers navigate through cell media. Reproduced with permission from reference (Kadiri et al. 2020). **b** Selective control of the artificial bacterial flagella in a swarm. Reproduced with permission from reference (Wang et al. 2018). **c** The swarm of magnetotactic bacteria moves in the direction of magnetic field. Reproduced with permission from reference (Taherkhani et al. 2014). **d** The swarm of Fe$_3$O$_4$ nanoparticle chains performs a tumbling motion. Reproduced with permission from reference (Yu et al. 2017)

particles. Light can achieve population generation, dispersion, and directional motion by inducing solute concentration gradients around these photosensitive materials. Light can induce solute concentration gradients around these photoactive materials to achieve reversible generation and directional locomotion of microswarms.

The Janus particles have surfaces with two or more distinct physical properties, which endows them with asymmetric structures or functions and allows them to exhibit collective behaviors under illumination. Graphite-SiO$_2$ Janus particles exhibit propulsion in a water-lutidine mixture under the irradiation of a laser beam, and the moving velocity can be adjusted by changing the illumination power (Buttinoni et al. 2013). The propulsion principle is that the carbon-coated hemisphere of the Janus particle absorbs the laser and generates thermal which heats up the water-lutidine mixture above a critical temperature. The demixing then provides a phoretic force that actuates the particles. These Janus particles form swarm pattern at low particle densities. When the particle densities are high, the whole system transforms into several big swarm patterns surrounded by a dilute gas phase (Fig. 1.5a). The swarm patterns disappear after the light is turned off. The light-driven swarm

1.3 Light-Driven Microswarms

is formed by self-trapping effect of the Janus particles. Particles are blocked with relatively strong collision due to the persistence of their orientations. And higher densities and velocities of the particle increase collision probability, leading to larger swarm patterns. Under UV irradiation, SiO_2–TiO_2 Janus particles exhibit reversible expansion–contraction behavior (Fig. 1.5b) (Hong et al. 2010). The SiO_2–TiO_2 Janus particles spontaneously aggregate together in a close-packed manner in the absence of UV light. Once exposed to UV light, the tightly packed particles quickly move away from each other, forming an expanded swarm pattern. After the UV light is turned off, the particles return to the aggregated pattern. This expansion–contraction swarming behavior is attributed to the synergy of multiple mechanisms, including photocatalytic-induced diffusiophoresis of TiO_2, osmotic propulsion due to unequal concentrations of solute molecules on different sides of the particles, and UV-induced surface charge interaction.

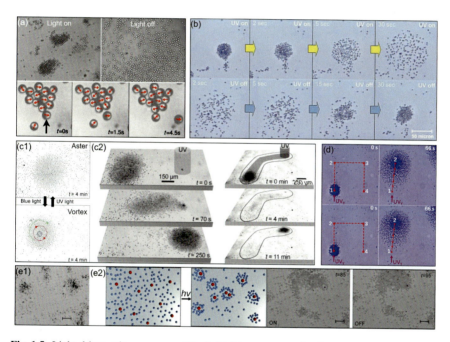

Fig. 1.5 Light-driven microswarms. **a** The light-driven swarm of carbon-coated Janus particles. Reproduced with permission from reference (Buttinoni et al. 2013). **b** The reversible expansion–contraction behavior of a swarm of SiO_2–TiO_2 Janus particles under UV light. Reproduced with permission from reference (Hong et al. 2010). **c** Polystyrene microparticle swarm. **c1** Colloidal aster and vortex induced by light and an AC electric field. **c2** Swarm navigation by changing the location of the UV spot. Reproduced with permission from reference (Hernàndez-Navarro et al. 2014). **d** The TiO_2 particle swarm navigated by pulsed UV irradiation along a predesigned path. Reproduced with permission from reference (Mou et al. 2019b). **e** The light-powered AgCl particle swarm in water. **e1** The swarm of AgCl particles under UV illumination. **e2** The predator–prey-like behavior of AgCl particles and silica spheres. Reproduced with permission from reference (Ibele et al. 2009)

The moving direction of self-diffusiophoresis particles is hard to control, which makes it difficult to navigate light-driven microswarms. The patterned substrate and moving light spot are beneficial to realize directional locomotion of light-driven microswarms. The formation and navigation can be achieved using a photosensitive azosilane monolayer (Hernàndez-Navarro et al. 2014). Under the irradiation of UV, the functionalized substrate adopts a *cis* configuration (planar) and reverts to a *trans* configuration (homeotropic) under blue light. When applying alternating electric field, polystyrene microparticles form star-like (aster) or vortex-like microswarm patterns (Fig. 1.5c1). The transformation between the two types of swarm patterns can be realized by changing the radiation type (UV or blue light). The microswarms are able to be navigated through changing the light spot location or moving the irradiated area along a predesigned path (Fig. 1.5c2). During the navigation, particles in the swarm move in a leader–follower manner, rather than as a strongly interacting interacted ensemble. Motion control of the light-driven microswarms can also be realized by programming the light radiation. TiO_2 particles spontaneously gather into swarms in water through electrolyte diffusiophoresis. The TiO_2 particle swarm can move in a spreading state actuated by a sidewise-applied UV light (Mou et al. 2019b). The locomotion stops when turning off the UV light, and the particles perform random Brownian motion. By irradiating pulsed UV light with on–off repeating cycles, the swarm is able to move phototactically along a predesigned path (Fig. 1.5d). The swarm exhibits dilatation-gathering behaviors during locomotion, with its center moving toward different directions by applying UV light from different sides. In addition to UV light, near-infrared light can also induce pattern formation and navigation of TiO_2/Pt microparticle swarms (Deng et al. 2018). The swarm locomotion under near-infrared irradiation is mainly attributed to the light-induced natural and Marangoni convection. The irradiation of the near-infrared light leads to a rise in temperature, which generates a local convection flow across exposed and unexposed areas. Then, the microparticles move with the flow toward the center of the exposed area.

The heterogeneous light-driven microswarms consist of multiple types of building blocks, which endows them with the ability of multiresponsiveness and the potential for building intelligent micromachines. The UV light can induce asymmetric photolysis of AgCl particles (~ 1 μm in diameter) in deionized water, which may be caused by the inhomogeneity of the particle surface or the uneven exposure of the particles (Ibele et al. 2009). The overall reaction is expressed as

$$4AgCl + 2H_2O \xrightarrow{h\nu, Ag^+} 4Ag + 4H^+ + 4Cl^- + O_2. \qquad (1.6)$$

The asymmetric photodissociation generates a local ion (H^+ and Cl^-) gradient around the AgCl particles, resulting in self-diffusiophoresis movement. AgCl particles move at speeds up to 100 μm/s and gradually aggregate into swarm patterns (Fig. 1.5e1). When the silica spheres are mixed with AgCl particles and then irradiated with UV light, the silica spheres will be attracted by the AgCl particles and surround them. However, due to electrostatic repulsion, the silica spheres have no physical contact with the AgCl particles, and there is an exclusion zone around the AgCl particles (Fig. 1.5e2). When the light is turned off, the exclusion zone will disappear. The collective behavior of silica spheres and AgCl particles under UV light is similar to the predator–prey-like behavior in nature.

1.4 Acoustic Field-Driven Microswarms

Acoustic waves have been used in the field of micro/nanorobots to achieve actuation and micromanipulation of tiny agents in a wireless and non-contact manner (Rao et al. 2015; Xu et al. 2017b). The principle of acoustic actuation is based on the fact that micrometer-scale particles are subjected to acoustic radiation forces in the acoustic field (Friend and Yeo 2011). The force exerted on tiny objects in standing acoustic waves is much greater than that in traveling acoustic waves (Ding et al. 2012). Therefore, current related research mainly focuses on actuation and micromanipulation based on standing waves. The acoustic radiation force on an object in the acoustic field can be written as (Bruus 2012)

$$F_a = -\left(\frac{\pi p_0^2 V \beta_w}{2\lambda}\right) \Phi(\beta, \rho) \sin(2kd), \tag{1.7}$$

where p_0 and V are the pressure amplitude and volume of the object, respectively. λ is the wavelength, k is the wavenumber, and d is the distance between the particle and the node or antinode. $\Phi(\beta, \rho) = (5\rho_c - 2\rho_w)/(2\rho_c + 2\rho_w) - \beta_c/\beta_w$, where ρ_c and ρ_w are the density of the object and medium and β_c and β_w are the compressibility of the object and medium, respectively. Therefore, $\Phi(\beta, \rho)$ indicates the density and compressibility relationship between the object and the medium. The above expression shows that the magnitude of the acoustic radiation force of an object is proportional to its size, and objects with a size smaller than the wavelength are difficult to be driven by the acoustic field.

The Au-Ru microrods are levitated rapidly to the middle plane by acoustic waves and settle down slowly when the acoustic field is turned off (Wang et al. 2015). The microrods can gather into a ring-like swarm pattern at the center node plane, where the ultrasound forms a standing wave (Fig. 1.6a). The ring-like morphology depends on acoustic nodal structure and reflection of sound waves. In addition, the microrods are also subject to a chemical propulsion force due to the catalytic reaction in the H_2O_2 solution. Therefore, microrods in the swarm rotate around the chain axis at kilohertz frequency and simultaneously exhibit motion along the chain. After the acoustic field is turned off, the swarm pattern disappears, and the microrods perform random propulsion. Reversible swarm formation and separation of Pt-Au nanowires can also be triggered by the acoustic field (Xu et al. 2015). The catalytic nanowires display autonomous motion in the H_2O_2 solution due to the asymmetric decomposition of H_2O_2 and exhibit a spreading state. When the acoustic field is applied, a standing wave is generated, which leads to the migration of nanowires toward the pressure nodes. Then, a high-density circular swarm pattern is formed, as shown in Fig. 1.6b1. The acoustic force is divided into primary radiation force and secondary radiation force. The first one is the main force generated by the acoustic field, which is responsible for driving the nanowires toward the nearest pressure nodes and antinodes and causing their aggregation within the nodal plane. The latter is generated by sound waves scattered by objects in the acoustic field, promoting further gathering of the nanowires through agent-agent acoustic interactions. Figure 1.6b1 shows the continuous transformation of the swarm between dispersed and aggregated states. When the acoustic field is applied, the nanowires gather at the nodes and maintain the swarm pattern. After turning off the acoustic field, the swarm pattern disappears rapidly due to electrophoretic propulsion of the nanowires. Besides, the locomotion of the nanowire swarm can be achieved by changing frequency of the applied acoustic field. The microswarm is able to move at a speed of around 45 μm/s. During the locomotion, the pattern remains stable due to the secondary radiation force. However, it is hard to navigate the microswarm along a predesigned path due to difficulties in precisely adjusting the position of the pressure node.

The liquid metal agents are also able to serve as building blocks of acoustic field-driven microswarms (Li et al. 2020). Eutectic gallium-indium alloy liquid metal

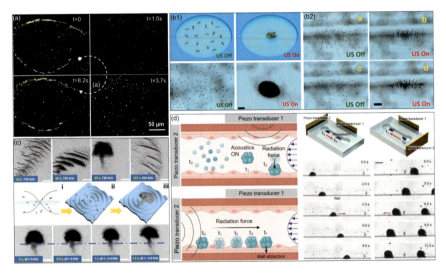

Fig. 1.6 Acoustic field-driven microswarms. **a** The Au-Ru microrods exhibit reversible aggregation-dispersion behavior and form a ring-like pattern under an acoustic field. Reproduced with permission from reference (Wang et al. 2015). **b** The acoustic field-driven self-propelled Pt-Au nanowire swarm. **b1** Formation of a circular pattern under the acoustic field. **b2** Reversible swarming and dispersion of the nanowires when turning the acoustic field on and off. Reproduced with permission from reference (Xu et al. 2015). **c** The dandelion flower-like microswarm formed by liquid metal colloidal under acoustic fields. Reproduced with permission from reference (Li et al. 2020). **d** The microbubble swarm capable of performing upstream navigation in physiological conditions actuated by acoustic fields. Reproduced with permission from reference (Fonseca et al. 2022)

nanorods with a core–shell structure can form striped patterns under an acoustic field with a frequency of 730 kHz (Fig. 1.6c). These stripes gradually merged into larger ones by decreasing the frequency to 728 kHz. When the frequency is further modulated to 722 kHz, a dandelion flower-like swarm pattern is formed. The swarming behaviors of nanorods are ascribed to changes in acoustic energy density and node distribution when tuning the frequency of the stand-wave acoustic field. Furthermore, the translational motion of the swarm pattern can be achieved by fine-tuning the frequency. Finally, the dandelion flower-like pattern is dispersed when the frequency is decreased to 680 kHz. Clinically approved microbubbles are capable of circulating through the vasculature due to acoustic forces-induced self-assembly (Fonseca et al. 2022). Multiple piezoelectric transducers are used to generate the acoustic wave. The microbubbles self-assemble into a microswarm in the acoustic wavefield and then move toward the side wall due to the primary and secondary acoustic radiation forces (Fig. 1.6d). Another piezoelectric transducer is turned on when the microbubbles reach the side wall, and the newly generated acoustic radiation force allows the microbubble swarm to move along the side wall. It is noted that the microswarm in motion is always close to the side wall, where the flow velocity approaches zero in laminar flow environments. By using this strategy, upstream navigation of the microswarm in physiological conditions can be realized.

1.5 Electric Field-Driven Microswarms

The electric field-driven microswarms depend on electrostatic force-based interactions and field-induced electrohydrodynamic (EHD) interactions between the building blocks. The attractive or repulsive interaction forces are mainly determined by the charges of agents and the distance between them, in which the latter one is able to be generated by the induced EHD flows. Electrostatic imbalance provides fundamental mechanisms for inducing various collective behaviors, which is essential for forming an electrostatic force-governed microswarm.

Metal-dielectric Janus colloidal particles (silica spheres with one hemisphere coated with metal) in vertical AC electric fields exhibit diverse swarming states which can be tuned by changing the frequency of the input electric field (Fig. 1.7a) (Yan et al. 2016). The particles perform self-propulsion due to induced charge electrophoresis (Gangwal et al. 2008), i.e., the local field-induced ionic flows near the two hemispheres of the Janus particle are different. First, particles that approximate isotropic spheres display random configurations like gas. The gas state emerges when the frequency of the field is low (~ kHz) due to the negligible electrostatics and strong ionic screening effects (Nishiguchi and Sano 2015). When increasing the frequency (20–50 kHz), ionic screening is significantly reduced, and the stronger dipolar interactions affect the collective behavior. The particles keep the metal side facing forward in motion, and the repulsion between the metal hemispheres leads to the formation of a coherent swarm state after binary collisions. When the frequency is further increased to the megahertz range, the opposite dipoles of the metallic and dielectric hemispheres yield attractive interactions between particles, resulting in the formation of a chain-like state. Besides adjusting frequency of the electric field, the cluster state can be induced by adding salt to the medium to tune the dipolar interactions. The polarization of particles with geometric asymmetry provides various swarming states under the electric field. Chiral and achiral structures are formed by colloidal dimers with different geometric ratios. This process is governed by the in-plane dipolar repulsion between petals as well as the out-of-plane attraction between the central dimer and surrounding petals (Ma et al. 2015). The assembled swarming structures are influenced by the number of dimers and the ratio between lobe radii. Achiral structures are more likely to form when dimers are more than five, while chiral structures have four or fewer petals. If the ratio of radius approaches zero or one, chirality disappears. The rotation propulsion of the assembled structures is based on the unbalanced hydrodynamic flow arising from the chirality of the clusters.

The induced directional EHD flows around particles lead to self-assembly under the actuation of an AC electric field. Colloidal particles can assemble into ordered aggregates under the actuation of EHD (Yeh et al. 1997). For a dielectric particle with radius R and polarization coefficient K, the velocity of the EHD flow around it is expressed as (Ristenpart et al. 2007)

$$U_{\text{EHD}} \sim \frac{C}{\mu} \left(\frac{K' + \tilde{\sigma} K''}{1 + \overline{\omega}^2} \right) \left(\frac{3(r/R)}{?[1 + (r/R)^2]^{5/2}} \right), \tag{1.8}$$

where μ is the viscosity of the medium, $\overline{\omega}$ is the frequency, and r is the distance between particle and the affected region. $C = \varepsilon\varepsilon_0(E_p/2H)2\kappa H$, where H is the separation distance between electrodes, E_p is the amplitude, and $\varepsilon\varepsilon_0$ is the permittivity of the medium. The above expression indicates that U_{EHD} is strongly dependent on the particle type, which means that the combination of different types of particles may lead to unbalanced EHD flows that can be used for object propulsion. Microparticles with different sizes and dielectric properties can assemble into microswarms due to the EHD interaction (Liang et al. 2020). The generated electric field-driven microswarm has a hierarchical leader–follower-like structure (Fig. 1.7b). The swarm performs locomotion due to the imbalance of surrounding EHD flows, with the "leader particle" leading and the "follower particles" following. The moving velocity can be adjusted by changing the AC voltage, frequency, and number of followers. The particles with the same size and dielectric properties also tend to assemble into microswarms. The formed microswarm displays no net propulsion due to the symmetric EHD flow. A strategy to break the symmetry of EHD flow is adding new particles with different sizes or dielectric properties, and the newly formed microswarm is able to move. The microswarm absorbs new particles by EHD attraction and demonstrates coordinated motion. Besides, electric fields can also introduce agent-agent hydrodynamic interactions in the microswarm. The charge distribution on the surface of insulating particles inside the conducting fluid is disrupted when applying an electric field (Bricard et al. 2013). The particles are actuated by electrostatic torque and rotate in random directions, leading to swarming states (Fig. 1.7c).

1.5 Electric Field-Driven Microswarms

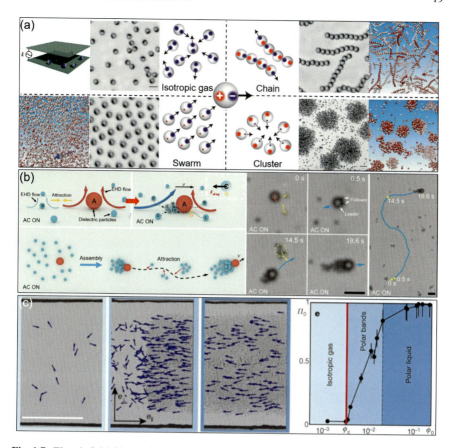

Fig. 1.7 Electric field-driven microswarms. **a** Different swarming states of Janus colloidal particles induced by electrostatic imbalance. Reproduced with permission from reference (Yan et al. 2016). **b** Formation and locomotion of the hierarchical leader–follower-like microswarm by AC electric field-induced EHD interactions. Reproduced with permission from reference (Liang et al. 2020). **c** Rolling particles show different swarming states due to electric field-induced hydrodynamic interactions. Reproduced with permission from reference (Bricard et al. 2013)

When the particle density is low, the particles roll in random directions at the same speed, and an isotropic gaseous phase state is formed. By increasing the particle fraction above a critical value, a macroscopic band forms and propagates at a constant velocity through an isotropic phase. The behavior is particle velocity independent and field amplitude independent. When further increasing the particle fraction, a homogeneous polar-liquid phase is formed, in which the head of the propagating bands catches up with the tail of themselves. This research shows that electric field-induced hydrodynamic interactions can essentially affect individual particles in performing a directed swarming motion.

1.6 Hybrid Field-Driven Microswarms

The building blocks of a microswarm may be responsive to more than one external field. We classify micro/nanorobot swarms controlled by two or more external power sources as hybrid field-driven microswarms. Hybrid field-driven microswarms combine the advantages of various driving strategies, enabling them to adapt to more complex environments, achieve more functions, and have better controllability. To date, hybrid field-driven microswarms are generally actuated by two external fields. A common strategy is that one field induces the formation of the swarm pattern, and the other field actuates the locomotion or pattern transformation of clusters based on task requirements and application sites.

The magnetic field has good compatibility and can be combined with many other driving methods. For example, hybrid nanomotors consisting of a magnetic helical body and a concave nanorod end demonstrate three different swarming states, i.e., rotating pattern under the magnetic-acoustic hybrid field (magnetic and acoustic fields), translational motion under the magnetic field, and loosely gathered state under acoustic field (Fig. 1.8a) (Li et al. 2015). The nanomotors aggregate toward pressure nodes in the acoustic field, while the rotating magnetic field induces attractive interactions between nanomotors and the driven torque for pattern rotation. Therefore, the swarming behavior of the nanomotors depends on the relationship between the magnetic and acoustic forces. The two forces are equal at the boundary of an equilibrium microswarm, leading to a circular pattern. The secondary radiation force causes the enhanced gathering effect. The reversible swarming behaviors indicate that hybrid actuation approaches provide new microswarm patterns and navigation methods that benefit targeted delivery in complex environments. Superparamagnetic microparticles demonstrate neutrophil-inspired swarming behavior based on the actuation of magnetic and acoustic fields (Fig. 1.8b) (Ahmed et al. 2017). First, the circular swarm pattern is formed by microparticles under the rotating magnetic field. Then, the swarm is propelled to the side boundary by the acoustic radiation force. Since the swarm rotates as a whole, it rolls along the boundary after approaching it, which is similar to the motion of neutrophils rolling on endovascular walls. A higher moving velocity is obtained when enhancing the contact between the microswarm and the boundary or increasing the input rotating frequency. This strategy is also applied to realize upstream motion of microswarms (Ahmed et al. 2021). In a capillary with flowing fluid, the microswarm can roll along the wall in the opposite direction of the flow in a combination of acoustic and magnetic fields (Fig. 1.8c). Such magnetic-acoustic hybrid actuation methods show potential for swarm control in the vascular system, where the delivery tasks should be conducted in a predicted manner. The Au-Ru-Ni nanorods in water self-assemble into different structures through magnetic interactions in the absence of an applied magnetic field (Fig. 1.8d) (Ahmed et al. 2014). The assembly mode is alerted and behaves like a structure with flexible hinges under the actuation of acoustic and magnetic fields. The changes in swarming structure affect the direction of locomotion and lead to a loop-like motion trajectory.

1.6 Hybrid Field-Driven Microswarms

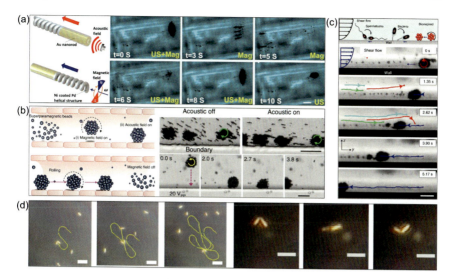

Fig. 1.8 Magnetic-acoustic fields-driven microswarms. **a** Different swarming states of hybrid micromotors in the magnetic and acoustic fields. Reproduced with permission from reference (Li et al. 2015). **b** The microswarm actuated by magnetic and acoustic fields performs neutrophil-like motion. Reproduced with permission from reference (Ahmed et al. 2017). **c** The acousto-magnetic microswarm with upstream motility. Reproduced with permission from reference (Ahmed et al. 2021). **d** Metal nanorods tipped with ferromagnetic segments exhibit assembly and propulsion under the acoustic field and with the guidance of magnetic field. Reproduced with permission from reference (Ahmed et al. 2014)

Magnetic fields can also be combined with light to actuate microswarms. Fe_3O_4@SiO_2 nanoparticles form a tornado-like microswarm with a 3D structure under the actuation of magnetic field and light (Fig. 1.9a) (Ji et al. 2020). The nanoparticles gather into a circular swarm pattern in the rotating magnetic field due to agent-agent magnetic and hydrodynamic interactions. When the light is applied, the particles in the swarm are affected by light-induced convection and gather toward the center of the light and rise vertically. The vertical displacement of the particles leads to the tornado-like three-dimensional structure of the microswarm. This light-magnetic-driven microswarm is capable of adjusting the rate of chemical reactions, which will be introduced in detail in Chap. 5. The hematite cube encapsulated inside the polymer microrobot with part exposed to the hydrogen peroxide shows collective behaviors under light and magnetic field (Palacci et al. 2013). Chemical concentration gradients with osmotic and phoretic effects are generated due to the decomposition of hydrogen dioxide under blue light illumination, and these microrobots gradually gather into crystallite structures. These structures disassemble due to thermal diffusion when turning off the light illumination. The rotational diffusion is suppressed by adding a static magnetic field because the alignment of the hematite cube tilts the orientation of the microrobots. The crystallite structure remains stable and performs locomotion along the field direction. The pattern disassembles under the influence of the magnetic field and reforms when the light is turned on. However, if the initial separation distance of the microrobots exceeds the critical value, they will move independently along the field direction instead of forming a pattern. Besides, the inherent magnetic interactions among building blocks can play an important role in the light-driven microswarm. Self-propelled peanut-shaped hematite particles assemble into ribbon-like patterns perpendicular to their long axis in hydrogen dioxide solution under the illumination of blue light (Fig. 1.9b) (Lin et al. 2017). The ribbon-like patterns are formed by frequent collisions between the particles and are aligned side by side with each other. The swarming particles exhibit a phototactic orientation, that is, movement toward the center of the light spot. The magnetic dipole–dipole interactions between hematite particles act to maintain the stability of the swarm pattern.

The hierarchical leader–follower-like microswarm actuated by the AC electric field we introduced above (as shown in Fig. 1.7b) can also be driven by light. When the UV light is irradiated vertically, the photoresponsive follower particles tend to surround the leader particle, which causes the swarm pattern to change from asymmetric to symmetric and the movement to stop (Fig. 1.10a) (Liang et al. 2020). When UV light is illuminated from the side, the more phototactic particles in the swarm move to the side away from the light source, and the microswarm aligns parallel or antiparallel to the light direction and demonstrates collective positive or negative phototaxis. This light-electric field-driven strategy can also be applied to

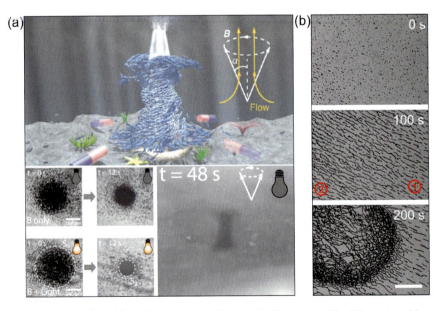

Fig. 1.9 Magneto-light-driven microswarms. **a** The tornado-like swarm with a 3D structure driven by magnetic field and light. Reproduced with permission from reference (Ji et al. 2020). **b** Ribbon-like patterns formed by peanut-shaped hematite particles due to diffusio-osmotic propulsion, phoretic attraction, and magnetic dipole–dipole interaction. Reproduced with permission from reference (Lin et al. 2017)

form different leader–follower-like microswarms by actuating a variety of dissimilar microparticles, such as L-TiO$_2$/S-TiO$_2$, L-TiO$_2$/S-SiO$_2$, and L-SiO$_2$/S-TiO$_2$. The light illumination, together with the acoustic field, is able to form swarms of Janus microrobots (Tang et al. 2019). The bowl-like TiO$_2$/Au microrobot performs locomotion toward its exterior side under an acoustic field. This directional motion is independent of the materials since it is caused by edges-generated second-order acoustic streaming flow (Kaynak et al. 2017). The UV light-actuated self-prophetic motion of the microrobot is material dependent, which is different from the acoustic-driven motion. In the presence of H$_2$O$_2$ and UV light, the microrobot exhibits motion toward the TiO$_2$ side. Therefore, the moving directions of acoustic driving and optical driving can be made the same or opposite by adjusting the positions of Au and TiO$_2$. Two groups of the Janus microrobots gather to the low-pressure nodes, with the exterior side facing the pattern's center under the acoustic field (Fig. 1.10b). After turning on the UV light, the two patterns consisting of Janus microrobots with a reversed position of TiO$_2$ and Au layers exhibit different behaviors: pattern expansion and pattern contraction with the exterior layers of Au and TiO$_2$, respectively. This light-modulated gathering effect is able to be controlled by turning on/off the UV light.

24 1 Introduction to Micro/Nanorobot Swarms

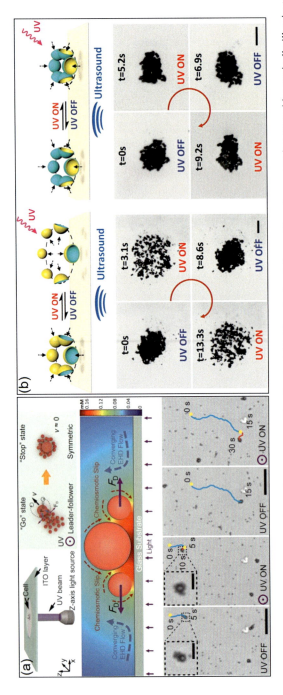

Fig. 1.10 Light-electric/acoustic field-driven microswarms. **a** The electric field-driven leader–follower-like microswarm is actuated by vertically illuminated ultraviolet light. Reproduced with permission from reference (Liang et al. 2020). **b** Light-modulated swarming behaviors of bowl-like TiO$_2$/Au (left) and Au/TiO$_2$ (right) Janus microrobots under an acoustic field. Reproduced with permission from reference (Lin et al. 2017)

References

Abbott JJ, Diller E, Petruska AJ (2020) Magnetic methods in robotics. In: Leonard NE (ed) Annual review of control, robotics, and autonomous systems, vol 3, pp 57–90

Ahmed S, Gentekos DT, Fink CA, Mallouk TE (2014) Self-assembly of nanorod motors into geometrically regular multimers and their propulsion by ultrasound. ACS Nano 8(11):11053–11060

Ahmed D, Baasch T, Blondel N, Laubli N, Dual J, Nelson BJ (2017) Neutrophil-inspired propulsion in a combined acoustic and magnetic field. Nat Commun 8:770

Ahmed D, Sukhov A, Hauri D, Rodrigue D, Maranta G, Harting J, Nelson BJ (2021) Bioinspired acousto-magnetic microswarm robots with upstream motility. Nat Mach Intelligence 3(2):116–124

Alapan Y, Yasa O, Schauer O, Giltinan J, Tabak AF, Sourjik V, Sitti M (2018) Soft erythrocyte-based bacterial microswimmers for cargo delivery. Sci Robot 3(17):eaar4423

Anderson C, Theraulaz G, Deneubourg J-L (2002) Self-Assemblages in Insect Societies. Insectes Soc 49(2):99–110

Aranson IS (2013) Collective behavior in out-of-equilibrium colloidal suspensions. C R Phys 14(6):518–527

Barbot A, Tan HJ, Power M, Seichepine F, Yang GZ (2019) Floating magnetic microrobots for fiber functionalization. Sci Robot 4(34):eaax8336

Bente K, Codutti A, Bachmann F, Faivre D (2018) Biohybrid and bioinspired magnetic microswimmers. Small 14(29):1704374

Bergeles C, Yang GZ (2014) From passive tool holders to microsurgeons: safer, smaller, smarter surgical robots. IEEE Trans Biomed Eng 61(5):1565–1576

Bowick MJ, Fakhri N, Marchetti MC, Ramaswamy S (2022) Symmetry, thermodynamics, and topology in active matter. Phys Rev X 12(1):010501

Bricard A, Caussin J-B, Desreumaux N, Dauchot O, Bartolo D (2013) Emergence of macroscopic directed motion in populations of motile colloids. Nature 503(7474):95–98

Brierley AS, Cox MJ (2010) Shapes of krill swarms and fish schools emerge as aggregation members avoid predators and access oxygen. Curr Biol 20(19):1758–1762

Bruus H (2012) Acoustofluidics 7: the acoustic radiation force on small particles. Lab Chip 12(6):1014–1021

Buttinoni I, Bialké J, Kümmel F, Löwen H, Bechinger C, Speck T (2013) Dynamical clustering and phase separation in suspensions of self-propelled colloidal particles. Phys Rev Lett 110(23):238301

Chen XZ, Jang BM, Ahmed D, Hu CZ, De Marco C, Hoop M, Mushtaq F, Nelson BJ, Pane S (2018a) Small-scale machines driven by external power sources. Adv Mater 30(15):1705061

Chen CR, Karshalev E, Guan JG, Wang J (2018b) Magnesium-based micromotors: water-powered propulsion, multifunctionality, and biomedical and environmental applications. Small 14(23):1704252

Cheng R, Huang W, Huang L, Yang B, Mao L, Jin K, ZhuGe Q, Zhao Y (2014) Acceleration of tissue plasminogen activator-mediated thrombolysis by magnetically powered nanomotors. ACS Nano 8(8):7746–7754

de Avila BEF, Lopez-Ramirez MA, Mundaca-Uribe R, Wei XL, Ramirez-Herrera DE, Karshalev E, Nguyen B, Fang RH, Zhang LF, Wang J (2020) Multicompartment tubular micromotors toward enhanced localized active delivery. Adv Mater 32(25):2000091

Deng Z, Mou F, Tang S, Xu L, Luo M, Guan J (2018) Swarming and collective migration of micromotors under near infrared light. Appl Mater Today 13:45–53

Diller E, Sitti M (2013) Micro-scale mobile robotics. Found Trends® Robot 2(3):143–259

Ding XY, Lin SCS, Kiraly B, Yue HJ, Li SX, Chiang IK, Shi JJ, Benkovic SJ, Huang TJ (2012) On-chip manipulation of single microparticles, cells, and organisms using surface acoustic waves. Proc Natl Acad Sci USA 109(28):11105–11109

Dobnikar J, Snezhko A, Yethiraj A (2013) Emergent colloidal dynamics in electromagnetic fields. Soft Matter 9(14):3693–3704

Dong X, Sitti M (2020) Controlling two-dimensional collective formation and cooperative behavior of magnetic microrobot swarms. Int J Robot Res 39(5):617–638

Esteban-Fernández de Ávila B, Lopez-Ramirez MA, Báez DF, Jodra A, Singh VV, Kaufmann K, Wang J (2016) Aptamer-modified graphene-based catalytic micromotors: off-on fluorescent detection of ricin. ACS Sensors 1(3):217–221

Field RD, Anandakumaran PN, Sia SK (2019) Soft medical microrobots: design components and system integration. Appl Phys Rev 6(4):041305

Fish have it easy in schools. Nature 506(134). https://doi.org/10.1038/506134a

Fonseca ADC, Kohler T, Ahmed D (2022) Ultrasound-controlled swarmbots under physiological flow conditions. Adv Mater Interfaces 9(26):2200877

Friend J, Yeo LY (2011) Microscale acoustofluidics: microfluidics driven via acoustics and ultrasonics. Rev Mod Phys 83(2):647–704

Fu Y, Yu H, Zhang X, Malgaretti P, Kishore V, Wang W (2022) Microscopic swarms: from active matter physics to biomedical and environmental applications. Micromachines 13(2):295

Gangwal S, Cayre OJ, Bazant MZ, Velev OD (2008) Induced-charge electrophoresis of metallodielectric particles. Phys Rev Lett 100(5):058302

Garattoni L, Birattari M (2018) Autonomous task sequencing in a robot swarm. Sci Robot 3(20):eaat0430

Giovanazzi S, Gorlitz A, Pfau T (2002) Tuning the dipolar interaction in quantum gases. Phys Rev Lett 89(13):130401

Han K, Kokot G, Das S, Winkler RG, Gompper G, Snezhko A (2020) Reconfigurable structure and tunable transport in synchronized active spinner materials. Sci Adv 6(12):eaaz8535

Hemelrijk C, Reid D, Hildenbrandt H, Padding J (2015) The increased efficiency of fish swimming in a school. Fish Fish 16(3):511–521

Hernàndez-Navarro S, Tierno P, Farrera JA, Ignés-Mullol J, Sagués F (2014) Reconfigurable swarms of nematic colloids controlled by photoactivated surface patterns. Angew Chem Int Ed 53(40):10696–10700

Hong Y, Diaz M, Córdova-Figueroa UM, Sen A (2010) Light-driven titanium-dioxide-based reversible microfireworks and micromotor/micropump systems. Adv Func Mater 20(10):1568–1576

Ibele M, Mallouk TE, Sen A (2009) Schooling behavior of light-powered autonomous micromotors in water. Angew Chem 121(18):3358–3362

Jager S, Stark H, Klapp SHL (2013) Dynamics of cluster formation in driven magnetic colloids dispersed on a monolayer. J Phys-Condensed Matter 25(19):195104

Ji F, Jin D, Wang B, Zhang L (2020) Light-driven hovering of a magnetic microswarm in fluid. ACS Nano 14(6):6990–6998

Jin D, Zhang L (2022) Collective behaviors of magnetic active matter: recent progress toward reconfigurable, adaptive, and multifunctional swarming micro/nanorobots. Acc Chem Res 55:98–109

Jin D, Yuan K, Du X, Wang Q, Wang S, Zhang L (2021) Domino reaction encoded heterogeneous colloidal microswarm with on-demand morphological adaptability. Adv Mater 33(37):2100070

Jing WM, Chowdhury S, Guix M, Wang JX, An Z, Johnson BV, Cappelleri DJ (2019) A microforce-sensing mobile microrobot for automated micromanipulation tasks. IEEE Trans Autom Sci Eng 16(2):518–530

Kadiri VM, Bussi C, Holle AW, Son K, Kwon H, Schütz G, Gutierrez MG, Fischer P (2020) Biocompatible magnetic micro-and nanodevices: fabrication of fept nanopropellers and cell transfection. Adv Mater 32(25):2001114

Karshalev E, de Avila BEF, Beltran-Gastelum M, Angsantikul P, Tang SS, Mundaca-Uribe R, Zhang FY, Zhao J, Zhang LF, Wang J (2018) Micromotor pills as a dynamic oral delivery platform. ACS Nano 12(8):8397–8405

References

Kaynak M, Ozcelik A, Nourhani A, Lammert PE, Crespi VH, Huang TJ (2017) Acoustic actuation of bioinspired microswimmers. Lab Chip 17(3):395–400

Kei Cheang U, Lee K, Julius AA, Kim MJ (2014) Multiple-robot drug delivery strategy through coordinated teams of microswimmers. Appl Phys Lett 105(8):083705

Kim Y, Yuk H, Zhao R, Chester SA, Zhao X (2018) Printing ferromagnetic domains for untethered fast-transforming soft materials. Nature 558(7709):274–279

Kim M, Julius A (2012) Microbiorobotics: biologically inspired microscale robotic systems. William Andrew

Klapp SH (2016) Collective dynamics of dipolar and multipolar colloids: from passive to active systems. Curr Opin Colloid Interface Sci 21:76–85

Kokot GP, Sokolov A, Snezhko A (2019) Guided self-assembly and control of vortices in ensembles of active magnetic rollers. Langmuir 36(25):6957–6962

Kong L, Mayorga-Martinez CC, Guan JG, Pumera M (2020) Photocatalytic micromotors activated by uv to visible light for environmental remediation, micropumps, reversible assembly, transportation, and biomimicry. Small 16(27):1903179

Kummer MP, Abbott JJ, Kratochvil BE, Borer R, Sengul A, Nelson BJ (2010) Octomag: an electromagnetic system for 5-dof wireless micromanipulation. IEEE Trans Rob 26(6):1006–1017

Li J, Li T, Xu T, Kiristi M, Liu W, Wu Z, Wang J (2015) Magneto–acoustic hybrid nanomotor. Nano Lett 15(7):4814–4821

Li S, Batra R, Brown D, Chang H-D, Ranganathan N, Hoberman C, Rus D, Lipson H (2019) Particle robotics based on statistical mechanics of loosely coupled components. Nature 567(7748):361–365

Li Z, Zhang H, Wang D, Gao C, Sun M, Wu Z, He Q (2020) Reconfigurable assembly of active liquid metal colloidal cluster. Angew Chem Int Ed 59(45):19884–19888

Li JX, de Avila BEF, Gao W, Zhang LF, Wang J (2017) Micro/nanorobots for biomedicine: delivery, surgery, sensing, and detoxification. Sci Robot 2(4):eaam6431

Li JY, Li XJ, Luo T, Wang R, Liu CC, Chen SX, Li DF, Yue JB, Cheng SH, Sun D (2018) Development of a magnetic microrobot for carrying and delivering targeted cells. Sci Robot 3(19):eaat8829

Liang X, Mou F, Huang Z, Zhang J, You M, Xu L, Luo M, Guan J (2020) Hierarchical microswarms with leader–follower-like structures: electrohydrodynamic self-organization and multimode collective photoresponses. Adv Func Mater 30(16):1908602

Liljestrom V, Chen C, Dommersnes P, Fossum JO, Groschel AH (2019) Active structuring of colloids through field-driven self-assembly. Curr Opin Colloid Interface Sci 40:25–41

Lin XK, Wu ZG, Wu YJ, Xuan MJ, He Q (2016) Self-propelled micro-/nanomotors based on controlled assembled architectures. Adv Mater 28(6):1060–1072

Lin Z, Si T, Wu Z, Gao C, Lin X, He Q (2017) Light-activated active colloid ribbons. Angew Chem 129(43):13702–13705

Ma F, Wang S, Wu DT, Wu N (2015) Electric-field-induced assembly and propulsion of chiral colloidal clusters. Proc Natl Acad Sci 112(20):6307–6312

Mahoney AW, Abbott JJ (2016) Five-degree-of-freedom manipulation of an untethered magnetic device in fluid using a single permanent magnet with application in stomach capsule endoscopy. Int J Robot Res 35(1–3):129–147

Marchetti MC, Fily Y, Henkes S, Patch A, Yllanes D (2016) Minimal model of active colloids highlights the role of mechanical interactions in controlling the emergent behavior of active matter. Curr Opin Colloid Interface Sci 21:34–43

Massana-Cid H, Meng F, Matsunaga D, Golestanian R, Tierno P (2019) Tunable self-healing of magnetically propelling colloidal carpets. Nat Commun 10:2444

Medina-Sanchez M, Schwarz L, Meyer AK, Hebenstreit F, Schmidt OG (2016) Cellular cargo delivery: toward assisted fertilization by sperm-carrying micromotors. Nano Lett 16(1):555–561

Mirzaeinia A, Heppner F, Hassanalian M (2020) An analytical study on leader and follower switching in v-shaped canada goose flocks for energy management purposes. Swarm Intell 14(2):117–141

Mlot NJ, Tovey CA, Hu DL (2011) Fire ants self-assemble into waterproof rafts to survive floods. Proc Natl Acad Sci 108(19):7669–7673

Molinero-Fernández Á, Jodra A, Moreno-Guzmán M, López MÁ, Escarpa A (2018) Magnetic reduced graphene oxide/nickel/platinum nanoparticles micromotors for mycotoxin analysis. Chem Euro J 24(28):7172–7176

Mou F, Li X, Xie Q, Zhang J, Xiong K, Xu L, Guan J (2019a) Active micromotor systems built from passive particles with biomimetic predator-prey interactions. ACS Nano 14(1):406–414

Mou F, Zhang J, Wu Z, Du S, Zhang Z, Xu L, Guan J (2019b) Phototactic flocking of photochemical micromotors. Iscience 19:415–424

Nagy M, Akos Z, Biro D, Vicsek T (2010) Hierarchical group dynamics in pigeon flocks. Nature 464(7290):890–893

Nelson BJ, Kaliakatsos IK, Abbott JJ (2010) Microrobots for minimally invasive medicine. Annu Rev Biomed Eng 12:55–85

Nishiguchi D, Sano M (2015) Mesoscopic turbulence and local order in janus particles self-propelling under an ac electric field. Phys Rev E 92(5):052309

Ou JF, Liu K, Jiang JM, Wilson DA, Liu L, Wang F, Wang SH, Tu YF, Peng F (2020) Micro-/nanomotors toward biomedical applications: the recent progress in biocompatibility. Small 16(27):1906184

Pacchierotti C, Ongaro F, van den Brink F, Yoon C, Prattichizzo D, Gracias DH, Misra S (2018) Steering and control of miniaturized untethered soft magnetic grippers with haptic assistance. IEEE Trans Autom Sci Eng 15(1):290–306

Palacci J, Sacanna S, Steinberg AP, Pine DJ, Chaikin PM (2013) Living crystals of light-activated colloidal surfers. Science 339(6122):936–940

Parrish JK, Edelstein-Keshet L (1999) Complexity, pattern, and evolutionary trade-offs in animal aggregation. Science 284(5411):99–101

Petersen KH, Napp N, Stuart-Smith R, Rus D, Kovac M (2019) A review of collective robotic construction. Sci Robot 4(28):eaau8479

Rao KJ, Li F, Meng L, Zheng HR, Cai FY, Wang W (2015) A force to be reckoned with: a review of synthetic microswimmers powered by ultrasound. Small 11(24):2836–2846

Ricotti L, Trimmer B, Feinberg AW, Raman R, Parker KK, Bashir R, Sitti M, Martel S, Dario P, Menciassi A (2017) Biohybrid actuators for robotics: a review of devices actuated by living cells. Sci Robot 2(12):eaaq0495

Ristenpart W, Aksay IA, Saville D (2007) Electrohydrodynamic flow around a colloidal particle near an electrode with an oscillating potential. J Fluid Mech 575:83–109

Rubenstein M, Cornejo A, Nagpal R (2014a) Programmable self-assembly in a thousand-robot swarm. Science 345(6198):795–799

Rubenstein M, Ahler C, Hoff N, Cabrera A, Nagpal R (2014b) Kilobot: a low cost robot with scalable operations designed for collective behaviors. Robot Auton Syst 62(7):966–975

Rubenstein M, Ahler C, Nagpal R (2012) Kilobot: a low cost scalable robot system for collective behaviors. In: 2012 IEEE international conference on robotics and automation. IEEE, 3293–3298

Shapiro B, Kulkarni S, Nacev A, Sarwar A, Preciado D, Depireux DA (2014) Shaping magnetic fields to direct therapy to ears and eyes. Annu Rev Biomed Eng 16(1):455–481

Shields CW, Velev OD (2017) The evolution of active particles: toward externally powered self-propelling and self-reconfiguring particle systems. Chem 3(4):539–559

Sitti M (2017) Mobile microrobotics. MIT Press

Sitti M (2018) Miniature soft robots—road to the clinic. Nat Rev Mater 3(6):74–75

Sitti M, Wiersma DS (2020) Pros and cons: magnetic versus optical microrobots. Adv Mater 32(20):1906766

Sitti M, Ceylan H, Hu W, Giltinan J, Turan M, Yim S, Diller E (2015) Biomedical applications of untethered mobile milli/microrobots. Proc IEEE 103(2):205–224

Snezhko A, Aranson IS (2011) Magnetic manipulation of self-assembled colloidal asters. Nat Mater 10(9):698–703

References

Steager EB, Sakar MS, Magee C, Kennedy M, Cowley A, Kumar V (2013) Automated biomanipulation of single cells using magnetic microrobots. Int J Robot Res 32(3):346–359

Sumpter DJ (2006) The principles of collective animal behaviour. Philos Trans R Soc B Biol Sci 361(1465):5–22

Sumpter DJ (2010) Collective animal behavior. Princeton University Press

Sun M, Fan X, Tian C, Yang M, Sun L, Xie H (2021) Swarming microdroplets to a dexterous micromanipulator. Adv Func Mater 31(19):2011193

Szuba T (2017) Ant-inspired, invisible-hand-controlled robotic system to support rescue works after earthquake. In: International conference on computational collective intelligence. Springer, 507–517

Taherkhani S, Mohammadi M, Daoud J, Martel S, Tabrizian M (2014) Covalent binding of nanoliposomes to the surface of magnetotactic bacteria for the synthesis of self-propelled therapeutic agents. ACS Nano 8(5):5049–5060

Tang S, Zhang F, Zhao J, Talaat W, Soto F, Karshalev E, Chen C, Hu Z, Lu X, Li J (2019) Structure-dependent optical modulation of propulsion and collective behavior of acoustic/light-driven hybrid microbowls. Adv Func Mater 29(23):1809003

Tierno P, Schreiber S, Zimmermann W, Fischer TM (2009) Shape discrimination with hexapole-dipole interactions in magic angle spinning colloidal magnetic resonance. J Am Chem Soc 131(15):5366–5367

Usherwood JR, Stavrou M, Lowe JC, Roskilly K, Wilson AM (2011) Flying in a flock comes at a cost in pigeons. Nature 474(7352):494–497

Vach PJ, Walker D, Fischer P, Fratzl P, Faivre D (2017) Pattern formation and collective effects in populations of magnetic microswimmers. J Phys D Appl Phys 50(11):11LT03

Vicsek T, Zafeiris A (2012) Collective motion. Phys Rep 517(3–4):71–140

Wang H, Pumera M (2020) Coordinated behaviors of artificial micro/nanomachines: from mutual interactions to interactions with the environment. Chem Soc Rev 49(10):3211–3230

Wang Q, Zhang L (2018) External power-driven microrobotic swarm: from fundamental understanding to imaging-guided delivery. ACS Nano 15(1):149–174

Wang W, Duan W, Zhang Z, Sun M, Sen A, Mallouk TE (2015) A tale of two forces: simultaneous chemical and acoustic propulsion of bimetallic micromotors. Chem Commun 51(6):1020–1023

Wang XP, Hu CZ, Schurz L, De Marco C, Chen XZ, Pane S, Nelson BJ (2018) Surface-chemistry-mediated control of individual magnetic helical microswimmers in a swarm. ACS Nano 12(6):6210–6217

Wang Q, Yang L, Wang B, Yu E, Yu J, Zhang L (2019) Collective behavior of reconfigurable magnetic droplets via dynamic self-assembly. ACS Appl Mater Interfaces 11(1):1630–1637

Ward A, Webster M (2016) Sociality: the behaviour of group-living animals

Weimerskirch H, Martin J, Clerquin Y, Alexandre P, Jiraskova S (2001) Energy saving in flight formation. Nature 413(6857):697–698

Wu Z, Troll J, Jeong H-H, Wei Q, Stang M, Ziemssen F, Wang Z, Dong M, Schnichels S, Qiu T, Fischer P (2018) A swarm of slippery micropropellers penetrates the vitreous body of the eye. Sci Adv 4(11):eaat4388

Xie H, Sun M, Fan X, Lin Z, Chen W, Wang L, Dong L, He Q (2019) Reconfigurable magnetic microrobot swarm: multimode transformation, locomotion, and manipulation. Sci Robot 4(28):eaav8006

Xu T, Soto F, Gao W, Dong R, Garcia-Gradilla V, Magana E, Zhang X, Wang J (2015) Reversible swarming and separation of self-propelled chemically powered nanomotors under acoustic fields. J Am Chem Soc 137(6):2163–2166

Xu L, Mou F, Gong H, Luo M, Guan J (2017a) Light-driven micro/nanomotors: from fundamentals to applications. Chem Soc Rev 46(22):6905–6926

Xu TL, Xu LP, Zhang XJ (2017b) Ultrasound propulsion of micro-/nanomotor. Appl Mater Today 9:493–503

Xu D, Wang Y, Liang C, You Y, Sanchez S, Ma X (2020) Self-propelled micro/nanomotors for on-demand biomedical cargo transportation. Small 16(27):1902464

Yan J, Bae SC, Granick S (2015a) Colloidal superstructures programmed into magnetic janus particles. Adv Mater 27(5):874–879

Yan J, Bae SC, Granick S (2015b) Rotating crystals of magnetic janus colloids. Soft Matter 11(1):147–153

Yan J, Han M, Zhang J, Xu C, Luijten E, Granick S (2016) Reconfiguring active particles by electrostatic imbalance. Nat Mater 15(10):1095–1099

Yang LD, Wang QQ, Zhang L (2018) Model-free trajectory tracking control of two-particle magnetic microrobot. IEEE Trans Nanotechnol 17(4):697–700

Yang LD, Zhang YB, Wang QQ, Chan KF, Zhang L (2020) Automated control of magnetic spore-based microrobot using fluorescence imaging for targeted delivery with cellular resolution. IEEE Trans Autom Sci Eng 17(1):490–501

Yang L, Zhang Y, Wang Q, Zhang L (2019) An automated microrobotic platform for rapid detection of C. Diff toxins. IEEE Trans Biomed Eng 67(5):1517–1527

Yeh S-R, Seul M, Shraiman BI (1997) Assembly of ordered colloidal aggregates by electric-field-induced fluid flow. Nature 386(6620):57–59

Yigit B, Alapan Y, Sitti M (2020) Cohesive self-organization of mobile microrobotic swarms. Soft Matter 16(8):1996–2004

Yu J, Xu T, Lu Z, Vong CI, Zhang L (2017) On-demand disassembly of paramagnetic nanoparticle chains for microrobotic cargo delivery. IEEE Trans Rob 33(5):1213–1225

Yu J, Wang B, Du X, Wang Q, Zhang L (2018a) Ultra-extensible ribbon-like magnetic microswarm. Nat Commun 9:3260

Yu J, Yang L, Zhang L (2018b) Pattern generation and motion control of a vortex-like paramagnetic nanoparticle swarm. Int J Robot Res 37(8):912–930

Yu J, Jin D, Chan KF, Wang Q, Yuan K, Zhang L (2019) Active generation and magnetic actuation of microrobotic swarms in bio-fluids. Nat Commun 10:5631

Yu J, Yang L, Du X, Chen H, Xu T, Zhang L (2021) Adaptive pattern and motion control of magnetic microrobotic swarms. IEEE Trans Rob 38(3):1552–1570

Zhang YB, Yuan K, Zhang L (2019) Micro/nanomachines: from functionalization to sensing and removal. Adv Mater Technol 4(4):1800636

Chapter 2
Pattern Formation and Control of a Vortex-Like Paramagnetic Nanoparticle Swarm

Abstract This chapter describes a vortex-like paramagnetic nanoparticle swarm driven by rotating magnetic fields. The formation principle and control mechanism of this microswarm are clarified in detail through theoretical analysis and experimental results. The vortex-like microswarm is capable of performing reversible pattern expansion, contraction, merging, and splitting by adjusting the input field. Controlled locomotion of the microswarm in a curved and branched channel with high access rates is presented. Besides, this chapter further describes programming the rotating magnetic field to obtain an elliptical swarm pattern with an adjustable aspect ratio.

Keywords Pattern formation · Vortex · Paramagnetic nanoparticles · Rotating field · Swarm control

2.1 Introduction

Micro/nanorobots have great potential in biomedical applications, such as targeted drug delivery, biosensing, and cell manipulation, and thus have drawn extensive attention in recent years (Nelson et al. 2010; Sitti et al. 2015; Chen et al. 2021; Wu et al. 2020). At the micro/nanoscale, it is difficult to integrate onboard processors, actuators, and sensors on tiny agents. Therefore, micro/nanorobots (microrobots) are generally untethered and actuated by external fields or chemical gradients. To date, scientists have designed different types of microrobots that are actuated in a variety of methods to perform various tasks, such as microparticles (Folio and Ferreira 2017; Sadelli et al. 2016), helical-shaped microswimmers (Zhang et al. 2009; Peyer et al. 2010), spore-based microrobots (Zhang et al. 2019), a single bacterium (Khalil et al. 2013), and nanowires (Petit et al. 2012). However, individual micro/nanorobots are small in size and thus have limited functionality. For example, when applied to targeted drug delivery in the human body, the drug-loading capacity of a single micro/nanorobot is very limited. Furthermore, tracking and navigating individual micro/nanorobots in complex in vivo environments using existing imaging methods is also a great challenge. Compared with a single micro/nanorobot, the micro/nanorobot swarm is capable of loading much more cargo (such as drugs), which can significantly improve the efficiency of targeted delivery (Lanauze et al. 2014; Snezhko and Aranson 2011). On the other hand, the increase in the number of micro/nanorobots can enhance the contrast of medical imaging, making localization easier. Medical imaging methods like magnetic resonance imaging (MRI) and ultrasound imaging can be used to achieve accurate and efficient tracking of micro/nanorobot swarms in the human body (Martel et al. 2007; Wang and Zhang 2018). And higher-precision wireless navigation can be realized based on real-time feedback from medical images (Martel 2013; Martel et al. 2009).

It is worth noting that the micro/nanorobot swarm here does not refer to a simple aggregation of numerous individual agents but an emerging swarming structure, which is usually divided into two types: equilibrium and active (out of thermodynamic equilibrium). The equilibrium swarm generally has a relatively fixed structure after formation and is usually static and lacks the ability to perform the controlled motion, which makes it difficult to serve as a mobile robotic end effector (Chen et al. 2011; Mao et al. 2013; Crassous et al. 2014; Ma et al. 2015; Palacci et al. 2013; Yan et al. 2012). The active swarm needs to be continuously supplied with energy by the external field, so that the swarm has a relatively stable and dynamically balanced structure, which is more suitable for applications in biomedicine and other fields (Snezhko and Aranson 2011; Belkin et al. 2007; Yu et al. 2018; Du et al. 2021). In our opinion, the ideal micro/nanorobot swarm should have properties such as being able to perform directional motion as a whole, reversibly change its morphology, and disassemble into individual micro/nanorobots. Enabling all micro/nanorobots inside the swarm to move synchronously and reach the targeted area through global control is the key point for achieving on-demand delivery with high access rates, which is crucial for applications in complex environments. Besides, the method of generating and controlling micro/nanorobot swarms should be simple and efficient. Therefore, investigations on the agent-environment interaction and the interaction between building blocks of the swarm are required to develop dexterous micro/nanorobot swarm platforms.

This chapter will introduce a micro/nanorobot swarm system driven by the simple rotating magnetic field, which consists of paramagnetic Fe_3O_4 nanoparticles with a diameter of 500 nm. The swarm has a vortex-like pattern, and the particles in it are all confined in the core region, which can perform synchronized motion under the actuation of the magnetic field. By adjusting the frequency of the magnetic field, the vortex-like swarm can perform reversible pattern expansion, contraction, merging, and splitting. When a small pitch angle is added to the rotating magnetic field, the vortex-like swarm can move as an entity with high controllability. We demonstrate the controlled navigation of the swarm in a curved and branched channel, with more than 90% of particles in the swarm successfully reaching the targeted area. This indicates that this microswarm has the potential for adaptive pattern control and navigation in complex environments and is suitable for targeted delivery with high access rates. In addition, by programming the magnetic field, that is, changing the circular rotating field to elliptical, the vortex-like swarm will exhibit an elliptical

2.2 Swarm Formation

dynamic-equilibrium pattern. Such pattern transformation ability greatly increases the environmental adaptability of the swarm, making it possible to apply in complex environments. In general, the vortex-like swarm introduced in this chapter preliminarily tackles above-mentioned challenges of the micro/nanorobot swarm, including swarm formation, pattern transformation, and motion control, which shows great potential to serve as a future multifunctional biomedical application platform.

2.2 Swarm Formation

Paramagnetic nanoparticles in the magnetic field form a chain-like structure through the interaction between magnetic dipoles. In our analysis, nanoparticle chains are considered as the basic unit for building the microswarm. As shown in Fig. 2.1a, nanoparticle chains in the fluidic environment rotate with the dynamic magnetic field, thereby creating vortices around them. The flow velocity decreases gradually from the ends of the particle chain to the center, and the velocity in the central region of the chain is almost zero. The velocity distribution of the flow field is $\vec{u}$, and its vorticity $\vec{\xi}$ is expressed as

$$\vec{\xi} = \nabla \times \vec{u} = \left(\frac{\partial u_z}{\partial y} - \frac{\partial u_y}{\partial z}, \frac{\partial u_x}{\partial z} - \frac{\partial u_z}{\partial x}, \frac{\partial u_y}{\partial x} - \frac{\partial u_x}{\partial y} \right), \quad (2.1)$$

where u_x, u_y, and u_z are the orthogonal components of $\vec{u}$ along the three axes of the Cartesian coordinate system. We only consider two-dimensional vortices (in the x–y plane), so only the component of vorticity in the z-axis direction is not equal to zero, and Eq. (2.1) can be simplified to

$$\vec{\xi} = \left(\frac{\partial v_y}{\partial x} - \frac{\partial v_x}{\partial y} \right) \vec{z}. \quad (2.2)$$

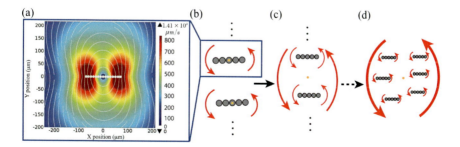

Fig. 2.1 Schematic illustration of vortices generated by the rotation of nanoparticle chains. **a** Fluid flow generated by a single particle chain with a rotating frequency of 5 Hz. The color profile indicates the magnitude of the flow velocity field. **b–d** Corotation of multiple particle chains. Gray circles represent nanoparticles, and red arrows represent vortex flows created by nanoparticle chains

According to the definition, the circulation is the line integral of the velocity field over a closed curve, and it can be expressed as follows in rigid body rotation (Saffman 1995):

$$\Gamma = \oint_C \vec{u} dl = \int_A \xi \cdot \vec{n} ds, \tag{2.3}$$

where A represents a reducible closed surface with the contour C. The vorticity is twice the angular velocity of the vortex ω ($\xi = 2\omega$). We consider the swarm to be the core part of the entire vortex and rotating like a rigid body, so that the vorticity of the swarm can be estimated based on these assumptions.

The vortex-like swarm is generated by the merging of vortices induced by the rotation of nanoparticle chains, as shown in Fig. 2.1b–d. Nanoparticles in the fluid are in a dispersed state in the absence of magnetic fields. When a rotating magnetic field is applied, the particle chain first rotates around its own center, inducing the vortex correspondingly. The vortices exert a long-range attraction on the other chains, bringing the chains closer to each other, and thus the distance between the chains gradually decreases. When the distance between the two chains is smaller than a critical value, the chains will rotate coaxially, indicating that the merging of the two vortices has been completed, as shown in (Fig. 2.1b, c). In the merged vortex, each chain rotates around its own center, and both rotate around their common center, resulting in a larger flow velocity. With the merging of vortices continuously proceeding, a main vortex containing multiple particle chains is finally generated (Fig. 2.1d). The force analysis of particle chains inside a stable vortex-like swarm is shown in Fig. 2.2. Take the ith particle chain as an example, it is subjected to the repulsive hydrodynamic force $F^v_{i+1,i}$ generated by the short-range interaction between the two vortices and the magnetic force $F^r_{i+1,i}$ exerted by the adjacent $(i + 1)$th chain. In addition, the main vortex exerts an inward force F^a_i on the ith chain. The dynamic balance of these three forces causes the ith chain to rotate around the swarm center.

The merging of two vortices depends on the distance between them and the size of each vortex. We denote the distance between the two vortices as d, the radius of the vortex core as a, and the ratio of a to d (a/d) can represent the relative position of the

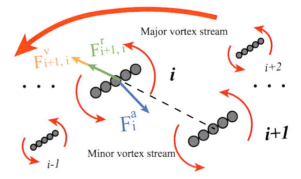

Fig. 2.2 Schematic diagram of forces exerted on the particle chain in the swarm. The blue arrow indicates the inward force exerted by the main vortex, the yellow arrow indicates the repulsive hydrodynamic force, and the green arrow indicates the magnetic force

2.2 Swarm Formation

two vortices. When a/d is small, the interaction between the vortices is weak, and the vortices rotate independently. Merging of vortices occurs when a/d exceeds a critical value $(a/d)_c$. It has been reported that the value of $(a/d)_c$ is about 0.3 (Melander et al. 1988; Saffman and Szeto 1980). Accordingly, we simulated the merging process of two identical vortices when a/d exceeds the critical value $(a/d)_c$, and the results are shown in Fig. 2.3. In the initial state, the two vortices are separated from each other (Fig. 2.3a). Due to the long-range hydrodynamic attraction between the vortices, the vortices approach each other and deform rapidly upon contact (Fig. 2.3b). The vortices then merged into an elliptical shape, with two strong filaments of vorticity ejected from both sides of the ellipse. The filaments gradually roll up and dissipate near the vortex core, and an axisymmetric circular vortex is finally formed (Fig. 2.3c, d). The simulation results in Fig. 2.3 theoretically support the formation mechanism of the vortex-like swarm, that is, the swarm can be formed through the merging of vortices induced by rotating nanoparticle chains.

According to the above analysis, the formation of the vortex-like swarm requires that the distance between the nanoparticle chains is close enough, i.e., the concentration of paramagnetic nanoparticles is large enough. The minimum concentration of nanoparticles required to form the vortex-like swarm can be estimated from the critical value $(a/d)_c$. First, the total mass M ($M = C_0 V_0$) and the total volume V_p (V

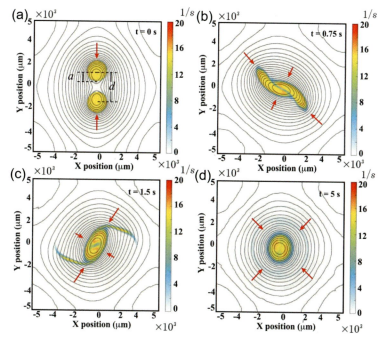

Fig. 2.3 Simulation of the merging process of two identical vortexes. The colored legend represents the vorticity field, and the initial vorticity distribution is uniform. (a/d) is set to 0.3, and the original radius of the vortex is set to 800 μm. The red arrows indicate the directions of fluid pressure.

Table 2.1 Parameters used to calculate the minimum nanoparticle concentration required to generate the vortex-like swarm

Parameter	Value
C_0	4.5 μg/μL
V_0	12 μL
ρ	5180 kg/m^3
r_c	5 μm
h	50 μm
$(a/d)_c$	0.3

$= M/\rho$) of the nanoparticles can be calculated. C_0 is the concentration of nanoparticles in the initial colloidal suspension, V_0 is the volume of suspension used in the experiment, and ρ is the density of the particles. Considering the nanoparticle chain as a cylinder with a cross-sectional radius r_c and a length h, its volume V_c can be calculated as $V_c = \pi r^2 h$. h is estimated from a previously reported Mason number model (rotating frequency: 10 Hz) (Reenen et al. 2014), and equal nanoparticle chain lengths under the same magnetic field are assumed. Based on this, we can obtain the total number of nanoparticle chains $N_c = V_0/V_c$. Assuming that the nanoparticle chains are uniformly distributed on the substrate, the distance between two adjacent vortex centers is set as $(a/d)_c$. Then, the maximum distribution area of the nanoparticle chains is calculated. From the parameters listed in Table 2.1, we estimated the minimum particle concentration required to form the vortex-like swarm to be 3.2 μg/mm^2 (note: particle concentration here refers to the ratio of particle mass to the distribution area of particles on the substrate).

2.3 Characterization of the Vortex-Like Swarm

In order to enable the microswarm to perform various tasks as a microrobot, effective control of its motion and shape is required. So, we characterize its properties, such as equations of motion and the interaction of fluids inside and outside the vortex core. First, the momentum equation in the inertial reference frame can be expressed as (Grabowski and Berger 1976):

$$\frac{D\vec{u}}{Dt} = \frac{\partial \vec{u}}{\partial t} + (\vec{u} \cdot \nabla)\vec{u} = \frac{-\nabla P}{\rho} + \frac{\nabla \cdot \tau}{\rho} + \vec{g}. \tag{2.4}$$

P is the local pressure, ρ is the fluid density, τ is the stress tensor, and $\vec{g}$ is the body force per unit mass. According to $(\vec{u} \cdot \nabla)\vec{u} = \nabla(0.5\vec{u} \cdot \vec{u}) - \vec{u} \times \vec{\xi}$, Eq. (2.4) can be expressed as

$$\frac{\partial \vec{u}}{\partial t} + \nabla\left(\frac{\vec{u} \cdot \vec{u}}{2}\right) - \vec{u} \times \vec{\xi} = \frac{-\nabla P}{\rho} + \frac{\nabla \cdot \tau}{\rho} + \vec{g}. \tag{2.5}$$

2.3 Characterization of the Vortex-Like Swarm

Then, we take the curl of Eq. (2.5)

$$\frac{D\vec{\xi}}{Dt} = -\vec{\xi}(\nabla \cdot \vec{u}) + (\vec{\xi} \cdot \nabla) + \frac{1}{\rho^2}\nabla\rho \times \nabla P \\ - \frac{1}{\rho^2}\nabla\rho \times (\nabla \cdot \tau) + \frac{1}{\rho}\nabla \times (\nabla \cdot \tau) + \nabla \times \vec{g} \qquad (2.6)$$

Since Newtonian fluids have constant density and viscosity, the second, fourth, fifth, and seventh terms above can be eliminated. According to $\nabla\tau = \mu\nabla^2\vec{u}$, Eq. (2.6) can be further simplified as

$$\frac{D\vec{\xi}}{Dt} = (\vec{\xi} \cdot \nabla)\vec{u} + v\nabla^2\vec{\xi}, \qquad (2.7)$$

where v is the dynamic viscosity of the fluid. Equation (2.7) clearly shows that the vortex diffuses due to viscosity, which is very similar to the diffusion phenomenon of Lamb–Oseen vortex (Green 1995). For a two-dimensional axisymmetric vortex in an inviscid infinite fluid, considering its viscosity from the onset of diffusion, Eq. (2.7) can be written as

$$\frac{\partial\vec{\xi}}{\partial t} + (\vec{u} \cdot \nabla)\vec{\xi} = (\vec{\xi} \cdot \nabla)\vec{u} + v\nabla^2\vec{\xi}. \qquad (2.8)$$

Because the Lamb–Oseen vortex is two dimensional, $\nabla\vec{u}$ is in the plane with $\vec{\xi}$ perpendicular to it, and the third term of Eq. (2.8) is eliminated. Besides, $\Delta\vec{\xi}$ is radially outward, $\vec{u}$ is tangential, so the second term can also be eliminated. Therefore, Eq. (2.8) can be further simplified as

$$\partial\xi_z(r \cdot t)/\partial t = v\nabla^2\vec{\xi}(r, t). \qquad (2.9)$$

The Poisson equation can be solved with a similar variable solution $\xi_z = f(\eta)/t$, where $\eta = r/\sqrt{vt}$, r is the distance to the center of the vortex. The expression for the approximate solution is

$$\xi_z(r, t) = \frac{\Gamma_0}{4\pi vt}\exp\left(\frac{-r^2}{4vt}\right) \quad (r > R) \qquad (2.10)$$

$$u_\theta = \left(\frac{\Gamma_0}{2\pi r}\right)\left[1 - \exp\left(-\frac{r^2}{4vt}\right)\right] \quad (r > R), \qquad (2.11)$$

where u_θ indicates the tangential velocity, R is the radius of the vortex core, and Γ_0 is the initial circulation of the vortex. Due to the effect of fluid viscosity, vorticity decreases with increasing distance from the vortex core. Since the inertial force is negligible at low Reynolds numbers, the presented laminar Navier–Stokes equation is applicable in this model. In addition, the rotating frequency in the model is fixed, so no axial and radial velocities are generated. Therefore, the model we built can describe the fluid flow around the swarm well.

Notably, because the vortex core is driven by the external magnetic field, its vorticity should not decrease. Therefore, we further assume that the core part of the vortex performs rigid rotation, the vorticity remains unchanged at any position in the core region, and the velocity is linearly related to the distance to the center of the vortex. Then the vorticity and velocity distribution at the center of the vortex can be expressed as

$$\xi_z = \frac{\Gamma_0}{\pi R^2} \quad (r < R) \tag{2.12}$$

$$u_\theta = \left(\frac{\Gamma_0 r}{2\pi R^2}\right) \quad (r < R). \tag{2.13}$$

The vortex can be described by an analytical model of tangential velocity and vorticity, as shown in Fig. 2.4a, b, which presents a vortex-like swarm with a radius of 8×10^{-4} m. In the core region of the swarm ($r < 8 \times 10^{-4}$ m), the tangential flow velocity increases linearly, and the vorticity remains constant. Outside the core region ($r > 8 \times 10^{-4}$ m), the tangential velocity gradually decreases, and the vorticity decreases sharply (becomes about 1/s at 1.5 mm from the center). Figure 2.4c shows the velocity distribution of the entire swarming system. The size of the arrow indicates the magnitude of the flow velocity, the blue arrow indicates the velocity distribution in the vortex core, and the red arrow indicates the velocity distribution around the core. Figure 2.4a, c shows that at the contour of the swarm, the tangential velocity reaches the maximum value.

2.4 Pattern Transformation of the Vortex-Like Swarm

2.4.1 Changes in Swarm Size

In some application scenarios, such as navigating through microchannels with varying diameters or enhancing imaging contrast, it is desirable for microswarms to be able to change their size. Derivation of the circulation is expressed as

$$\frac{d\Gamma}{dt} = \frac{d}{dt}\oint_C \vec{u} dl. \tag{2.14}$$

According to Kelvin's theorem in rotational frame

$$\Gamma_a = \Gamma + f_c A_n, \tag{2.15}$$

2.4 Pattern Transformation of the Vortex-Like Swarm

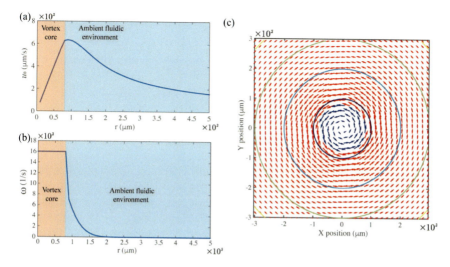

Fig. 2.4 Analytical model of the vortex-like swarm. **a** Tangential velocity distribution. **b** Vorticity distribution. **c** Schematic diagram of vortex tangential velocity. The size of the arrow indicates the magnitude of the flow velocity, the blue arrow indicates the velocity distribution in the vortex core, the red arrow indicates the velocity distribution around the core, and the colored circles represent gradients of vortex vorticity

where f_c is the Coriolis parameter and A_n is the vortex area. Taking the derivation of Eq. (2.15):

$$d\Gamma/dt = -f_c(dA_n/dt). \tag{2.16}$$

Then, we get

$$\frac{d}{dt}2\pi r u = -f_c \frac{d}{dt}\pi r^2. \tag{2.17}$$

Due to circulation conservation, the circulation of the swarm remains unchanged when its size varies

$$2\pi r u + \pi f_c r^2 = 2\pi r_0 u_0 + \pi f_c r_0^2, \tag{2.18}$$

where u_0 is the initial tangential flow velocity and r_0 is the initial radius of the vortex core. Based on this, we can get the relationship between the rotating frequency of the swarm and its size (coverage area), as shown in Eq. (2.19). The comparison of experimental results with this model is presented in Sect. 2.5.

$$\omega = \frac{f_c(r_0^2 - r^2) + 2v_0 r_0}{2r^2} \tag{2.19}$$

2.4.2 Spread Swarm Pattern

By adjusting the inner interactions, the morphology of the microswarm can be changed, such as pattern expansion. As shown in Fig. 8.1, particle chains in a stable vortex-like swarm are subjected to magnetic force and hydrodynamic force between chains, the inward force exerted by the main vortex, and these forces balance each other. The inward force is the key to maintaining the swarm pattern, which can be adjusted to change the swarm morphology. For example, the vorticity can be reduced by decreasing the rotating frequency, which reduces the inward force. In this case, the short-range repulsive force between nanoparticle chains becomes dominant, leading to significant swelling of the swarm pattern.

It is noted that simply reducing the input rotating frequency f_f of the magnetic field results in longer particle chains due to the reduced fluid drag. The longer length will increase the repulsion between chains, and the vortex-like swarm cannot be effectively generated. Another way to adjust the inward force is to bring the chain into a step-out state, which can be realized by increasing f_f. The actual rotating frequency of the chain (f_{cr}) in the step-out state is lower, thereby reducing the inward force. Therefore, the relationship between f_f and f_{cr} can affect the expansion area of the swarm. We characterize the step-out behavior of the nanoparticle chain in detail, and Fig. 2.5 is the schematic diagram of the step-out rotation of the chain. Initially, the long axis of the chain is parallel to the direction of the magnetic field (Fig. 2.5a). The magnetic field rotates at an angular velocity ω_f, while the chain rotates at a smaller angular velocity ω_c due to fluid drag. When the magnetic field rotates from position P1 to P2, the angle between the magnetic field and the chain (i.e., the phase lag) reaches 90° (θ_1). When the phase lag exceeds 90°, the chain rotates in the opposite (counterclockwise) direction, as shown in Fig. 2.5b. Subsequently, the direction of the chain's long axis is parallel to the magnetic field again, and the angle of reverse rotation is θ_b, as shown in Fig. 2.5c. We define several key parameters here: t_1 denotes the time for the phase lag to change from 0° to 90° (P1–P2), t_2 denotes the time for the phase lag to change from 90° to 0° (P2–P4), T_0 is the time of a step-out cycle, and θ_p is the angle that the chain rotates in a step-out cycle. The expressions for these parameters are given

$$t_1 = \frac{\pi/2}{2\pi(f_f - f_{cr})} \tag{2.20}$$

$$t_2 = \frac{\pi/2}{2\pi(f_f + f_{cr})} \tag{2.21}$$

$$\theta_p = \theta_c - \theta_b$$
$$= \frac{\pi f_{cr}^2}{f_f^2 - f_{cr}^2} \left(\theta_c = \omega_c t_1 = \frac{\pi f_{cr}}{2(f_f - f_{cr})}, \theta_b = \omega_c t_2 = \frac{\pi f_{cr}}{2(f_f + f_{cr})} \right) \tag{2.22}$$

2.4 Pattern Transformation of the Vortex-Like Swarm

$$T_0 = t_1 - t_2 = \frac{f_f}{2\left(f_f^2 - f_{cr}^2\right)} \quad (f_{cr} = \omega_c/2\pi, \ f_f = \omega_f/2\pi). \tag{2.23}$$

One rotation of the chain takes $N = 2\pi/\theta_p$ cycles, so the average rotating frequency of the nanoparticle chain in the step-out state can be expressed as

$$f_r = \frac{1}{NT_0} = \frac{f_{cr}^2}{f_f}. \tag{2.24}$$

We assume that the inward force exerted by the vortex on the nanoparticle chain is proportional to the rotating frequency of the vortex f_v (proved in Sect. 2.4.3), and the input rotating frequency is linearly related to the rotating frequency of the vortex (verified experimentally). Based on these assumptions, the relationship between the vortex inward force and the input rotating frequency can be obtained.

We perform numerical simulation using COMSOL, in which the short chains are treated as individual particles that repel each other due to magnetic forces. The key parameters used in the simulation are listed in Table 2.2, including magnetic susceptibility χ_p, permeability μ_0, particle volume V_p, and magnetic field strength B. The swarm pattern expands significantly at high input rotating frequencies (e.g., 70 Hz), which is caused by the reduced inward force of the vortex. According to the above analysis and assumptions, the change in the vortex inward force can be estimated. The inward force reaches a maximum value when the chain just reaches its step-out frequency (about 20 Hz). In the simulation, when the input frequency is 20 Hz, each particle is subjected to a centripetal force in equilibrium with the magnetic repulsion force, based on which the maximum value of the vortex inward force can be estimated. Since the inward force is proportional to the rotating frequency of the swarm f_v, a further estimate can be made using the change rate of the input rotating frequency f_f. Applying different vortex inward forces in the simulation can get the change of pattern area with f_f. Actually, the input rotating frequency f_f and the swarm rotating frequency f_v do not coincide, and in most cases $f_f > f_v$. Applying f_f directly to the simulation results in an overestimated inward force, so we calibrate these two frequencies based on experimental results (see Sect. 2.5). As shown in Fig. 2.6, the curves of both calibrated simulation and experimental result increase with the input rotating frequency. The swarm area remains stable when the input

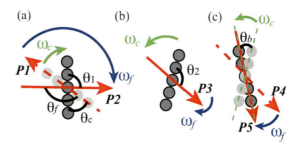

Fig. 2.5 Schematic diagram showing the step-out rotation process of the paramagnetic nanoparticle chain

Table 2.2 Parameters used to estimate the inward force and pattern area

Parameter	Value
χ_p	0.8
Original area of the swarm	3.6×10^{-7} mm^2
μ_0	1.257×10^{-6} V s/(A m)
B	10^{-3} T
V_p	2.12×10^{-14} mm^3

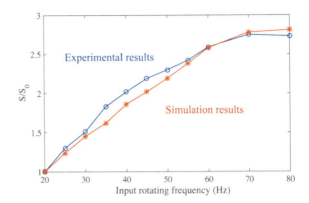

Fig. 2.6 Change of pattern area of the vortex-like swarm in the step-out state. The blue line indicates the experimental result, and the red line indicates the simulation result. S_0 indicates the swarm area when the inward force is maximum (i.e., the input rotating frequency is 20 Hz)

rotating frequency reaches 70 Hz. The experimental results are in good agreement with the simulation results.

2.4.3 Inward Force

The inward force provided by the vortex is crucial to maintain the pattern stability of the vortex-like microswarm. We characterize the inward force for better pattern control. The general form of the motion equation of a rigid sphere in a fluid vortex has been reported by previous studies (Lecuona et al. 2002). The velocity of particles in the core region of the microswarm can be expressed as

$$V = u + St\left(\frac{3}{2}\varepsilon - 1\right)(u \cdot \nabla u - I) + O\left(St^{3/2}\right) \quad (2.25)$$

$$V_\theta = u_\theta(r) - I_\theta St\left(\frac{3}{2}\varepsilon - 1\right) + O\left(St^{3/2}\right) \quad (2.26)$$

$$V_r = St\left(\frac{3}{2}\varepsilon - 1\right)\left(-\frac{u_\theta^2(r)}{r} - I_r\right) + O\left(St^{3/2}\right), \quad (2.27)$$

2.5 Experimental Results

Table 2.3 Parameters used to estimate the influences of the vortex flow and particle inertia

Parameter	Value
ω_f	10 Hz
U_0	8×10^{-3} m/S
D	2×10^{-4} m
v_c	10^{-6} kg/(s m)
ρ_f	10^3 kg/m^3
ρ_P	1.05×10^3 kg/m^3
R_0	8×10^{-4} m

where V represents the velocity of the particle and its tangential and radial components are denoted by V_θ and V_r, respectively. The radial and tangential components of inertia are represented by I_r and I_θ, respectively. I_r and I_θ are equal to each other for spherical and uniform beads. O stands for the order of magnitude. The drag force on a sphere is proportional to its velocity, which is balanced by the inward force of the vortex (based on Stokes' law). Therefore, the inward force of a vortex is proportional to the velocity of the particles inside it. According to Eq. (2.25), the velocity of nanoparticles has a linear relationship with the vortex flow. Therefore, the inward force exerted by the vortex on the nanoparticles is proportional to the rotating frequency of the vortex. From Eq. (2.26), it can be obtained that the tangential velocity of non-magnetic particles is affected by the local tangential flow velocity $u_\theta(r)$ and inertia effect $\mathrm{St}I_\theta(1.5\varepsilon - 1)$. However, the two parts are in different orders of magnitude, i.e., $u_\theta(r) \sim O(10^{-4})$ $(r = R_0)$ and $\mathrm{St}I_\theta(1.5\varepsilon - 1) \sim O(10^{-4})$ $(r = R_0)$. Therefore, the influence of inertia is negligible, and the tangential velocity of the particles is mainly determined by the local flow velocity. For non-magnetic particles, according to Eq. (2.27), the radial velocity depends on the local tangential flow velocity $4r\mathrm{St}$ and inertial effects $\mathrm{St}I_r$. The two parts are of the same order of magnitude: $\mathrm{St}u_\theta^2/r \sim O(10^{-4})$ $(r = R_0)$ and $\mathrm{St}I_r \sim O(10^{-4})$ $(r = R_0)$. These data are calculated using the parameters in Table 2.3. The above analysis shows that the critical method to control the force of the vortex exerted on the non-magnetic particle is to adjust the input rotating frequency. In addition, it can be seen from Eq. (2.27) that the vortex force exerted on particles with larger inertia is smaller. The controllable inward force endows the vortex-like swarm with better environmental adaptability.

2.5 Experimental Results

2.5.1 Formation of the Vortex-Like Microswarm

In this section, we experimentally demonstrate the formation process of the vortex-like microswarm, as shown in Fig. 2.7. When no magnetic field is applied, the magnetic nanoparticles on the substrate are dispersed in water ($t = 0$ s). When a

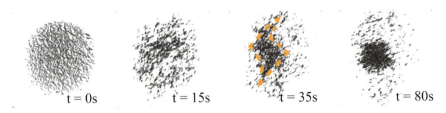

Fig. 2.7 Formation process of the vortex-like microswarm

rotating magnetic field is applied, the nanoparticles form particle chains that rotate with the magnetic field ($t = 15$ s). After a period of vortex merging, the nanoparticle chains gather together to initially form a swarm pattern ($t = 35$ s). The rotating pattern continuously attracts surrounding particle chains, eventually forming a dynamic-equilibrium vortex-like swarm ($t = 80$ s). Finally, most of the particles on the substrate are gathered into the swarm, while part of the particle chain is outside the swarm due to insufficient inward forces. These particle chains rotate with the magnetic field and simultaneously rotate around the swarm under the effect of the vortex. The outer contour of the swarm is clearly visible, and the particle chains inside it are relatively densely arranged but do not form a solid entity. Smaller interspaces can still be seen in the swarm, which are spaces between particle chains. The particles around the swarm will be continuously sucked into the pattern due to the attraction of the vortex, and the particles inside the swarm will be continuously thrown out due to the centrifugal force. This suggests that the vortex-like swarm is a dynamic-equilibrium system, which is consistent with the proposed theoretical model (Sect. 2.2).

According to the theoretical analysis, the formation of the vortex-like microswarm requires the balance of various inner interactions, which means the parameters of the rotating magnetic field need to meet certain conditions. We experimentally investigated the effect of magnetic field strength and rotating frequency on the formation of vortex-like swarms, as shown in Fig. 2.8. The phase diagram shows five regions representing the five different states of particle chains under the rotating magnetic field. Region I indicates that when the rotating frequency is low (such as 1–2 Hz), the nanoparticles can only form chain-like structures and cannot generate the vortex-like swarm. This is because the effect of vortex generated by the rotation of the nanoparticle chains is weak, and the vortices cannot merge, so the coverage area does not shrink, as shown in the inset (6 mT, 2 Hz). When the frequency increases to about 3 Hz (region II), the inward force of the vortex gradually increases, and some vortices merge and gather into smaller regions. However, there are many particle chains (about 40%) that are not attracted by the main vortex. The particle chains inside the swarm are also only loosely gathered together, resulting in an unstable swarm pattern, as shown in the inset (10 mT, 3 Hz). Region III represents the magnetic field parameters that can form a stable vortex-like swarm; i.e., most nanoparticles can be successfully aggregated into the swarm, as shown in the inset (10 mT, 3 Hz). When the magnetic field strength is low (such as 4 mT), the corresponding feasible rotating frequency range is small (3.5–5 Hz). When the magnetic field strength is 10 mT, the feasible

2.5 Experimental Results

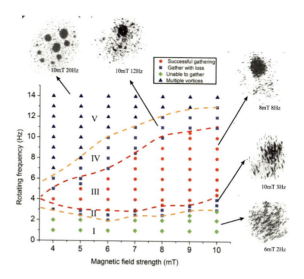

Fig. 2.8 Phase diagram shows the formation of the vortex-like swarm under different parameters of the rotating magnetic field. Different symbols are used to represent four different states: successful gathering, gather with loss, unable to gather, and generating multiple vortices. The inset shows the experimental image

frequency range increases to 4 to 10 Hz. Region IV indicates that the swarm formation is accompanied by more particle loss, as shown in the inset (10 mT, 12 Hz). This is due to higher fluid drag induced by the increased rotating frequency, and the nanoparticle chains are disassembled into shorter chains, resulting in insufficient vortex inward force. As the rotating frequency continues to increase (region V), the chain length decreases, and the dispersed nanoparticles form multiple vortex-like swarms, as shown in the inset (10 mT, 20 Hz).

2.5.2 Characterization of the Vortex-Like Swarm

We further characterize the vortex-like swarm under the magnetic field parameters in the successful region (region III of Fig. 2.8). The swarm is generated using different initial nanoparticle concentrations and rotating frequencies of the magnetic field. Theoretically, even with different initial concentrations, the area of generated swarm under the same initial conditions (i.e., rotating frequency, magnetic field strength, and the total number of nanoparticles used) should be identical if all particles are aggregated into the vortex. The experimental results are shown in Fig. 2.9. As the initial concentration of particles increases, the final area of the swarm remains almost unchanged. This result validates our inference that the final area of the swarm is not related to the initial particle concentration. Meanwhile, this also indicates that the formation of a vortex-like swarm can efficiently aggregate dispersed paramagnetic nanoparticles. The curve in Fig. 2.9 rises slightly (less than 10%) with particle concentration because some particles cannot be successfully aggregated at low concentrations. With the same total number of particles, a lower initial particle concentration means a larger initial coverage area. Insufficient fluid force prevents particles in the peripheral region from being successfully attracted. The rotating frequency has an effect on the final area of the swarm. Comparing the experimental results at different rotating frequencies in Fig. 2.9, it can be found that the swarm area is the smallest when the frequency is 8 Hz and the largest when the frequency is 6 Hz. This experimental result is consistent with the circulation conservation of Eq. (2.15).

The time required to generate vortex-like swarms at different initial particle concentrations (4–11 $\mu g/mm^2$) is investigated, and the results are shown in Fig. 2.10. The time from applying the magnetic field to the formation of a stable swarm decreases with increasing initial particle concentration. A high initial particle concentration means that the chain-to-chain distance is shorter, and the attraction between two vortices is stronger. Thus, the time it takes for the two vortices to approach each other is shorter. The curve in Fig. 2.10 gradually flattens as the initial particle concentration increases to a larger value. Due to particle loss and measurement inaccuracy during the formation process, the results have some errors. Therefore, we perform linear fitting based on a large number of experimental results to statistically demonstrate the time spent on swarm formation.

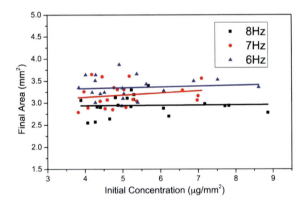

Fig. 2.9 Relationship between the initial particle concentration and the area of the generated vortex-like microswarm. The magnetic field strength is 8 mT. The curves are fitted from the data points

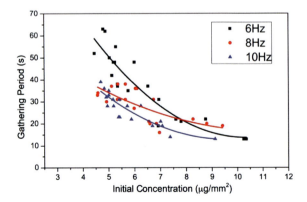

Fig. 2.10 Formation time of the vertex-like microswarm at different initial particle concentrations and rotating frequencies. The magnetic field strength is 8 mT

2.5 Experimental Results

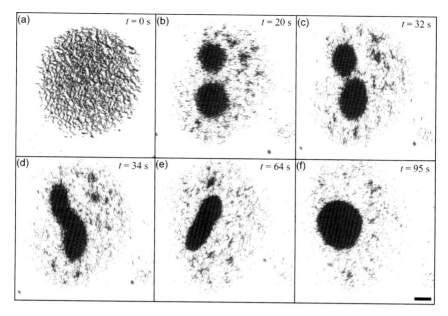

Fig. 2.11 Merging process of two separate vortex-like microswarm. The magnetic field strength is 7.5 mT, and the rotating frequency is 8 Hz during merging. The scale bar is 500 μm

The merging experiment of two independent vortex-like swarms is performed, and the results are shown in Fig. 2.11. The magnetic field parameters are in region V of Fig. 2.8 (i.e., 12 Hz and 7.5 mT). A low initial nanoparticle concentration (i.e., 5 μg/mm^2) is used to increase the probability of generating multiple independent swarms (Fig. 2.11a). As shown in Fig. 2.11b, the relative distance between two independent swarms generated in the experiment satisfies $(a/d)_c$. The two swarms then attract each other, and the distance between them gradually decreases. Their swarm patterns deform due to the attraction force before contact, as shown in Fig. 2.11c. After the two swarms touch and merge, they first have an elongated pattern, as shown in Fig. 2.11d, e. Finally, the elongated pattern gradually transforms into an axisymmetric circular pattern (Fig. 2.11f). The experimental results agree well with the simulation results in Fig. 2.3.

2.5.3 Pattern Transformation

The relationship between swarm area and input rotating frequency is given in Sect. 2.4.1, as shown in Eq. (2.19). The calculation result according to Eq. (2.19) is shown in Fig. 2.12 (the blue curve). The results show that the area of the swarm decreases as the input rotating frequency increases. For example, increasing the rotating frequency from 5 to 10 Hz reduces the swarm area by 50% of the initial area. This model is experimentally verified, and the experimental results are shown in Fig. 2.12 (red circle). When the rotating frequency is low (below 6 Hz), the experimental and modeling results are in good agreement, with a difference of only about 5%. The difference reaches more than 20% when the input frequency is larger than 7 Hz, which is mainly caused by the difference between the vortex rotating frequency f_v and the input rotating frequency f_f. The mathematical model actually (the blue curve) describes the relationship between f_v and S/S_0. The input frequency f_f in the experiment is higher than the actual vortex rotating frequency f_v. Therefore, increasing f_f in the experiment can make the swarm area correspond to the area calculated by the model. The error of the original mathematical model mainly comes from the difference between f_v and f_f, so the experimental data are used to calibrate it (Fig. 2.13). We obtained a calibrated model by linear fitting, and the calculation results according to the calibrated model are shown in Fig. 2.12 (yellow curve). The modified model can well estimate the variation of swarm area with frequency over all input frequency ranges.

Theoretical analysis shows that the area of the vortex-like swarm is closely related to the frequency of the rotating magnetic field. We experimentally investigate the relationship between the swarm area and the input rotating frequency (from 0 to

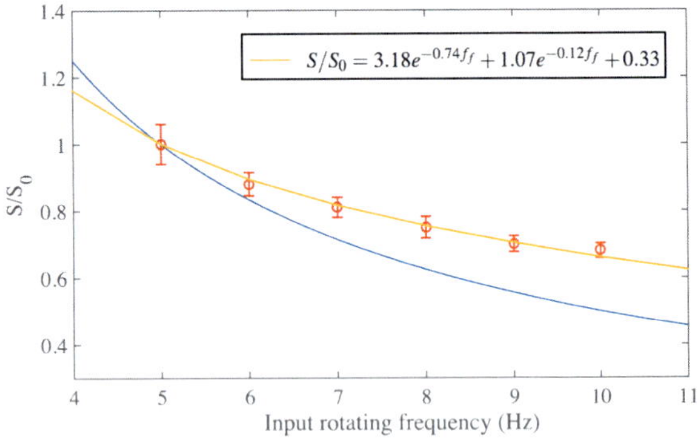

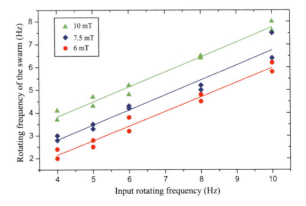

Fig. 2.12 Relationship between the swarm area and the input rotating frequency of the magnetic field. The blue curve represents the original model, the yellow curve represents the modified model, and the red circle represents the experimental data. S_0 represents the initial area, i.e., the coverage area of the swarm when the input rotating frequency is 5 Hz

Fig. 2.13 Calibration results showing the relationship between the input rotating frequency and rotating frequency of the vortex-like swarm under different field strengths

2.5 Experimental Results

90 Hz), as shown in Fig. 2.14. S_0 in the figure represents the initial area of the swarm, i.e., the swarm area when the rotating frequency is 5 Hz. According to Eq. (2.19), increasing the rotating frequency reduces the swarm area. Experimental data show that the swarm area decreases to around 0.65 S_0 when the rotating frequency increases to 10 Hz. When increasing the frequency from 10 Hz to about 20 Hz (red area in Fig. 2.14), the area of the swarm remains almost the same. As shown in the inset in Fig. 2.14, swarms in this rotating frequency range have a dynamic-equilibrium circular pattern with high particle concentrations. An input rotating frequency above 20 Hz makes the swarm exhibit a spread state. As shown in the inset in Fig. 2.14, the swarm pattern changes from circular to spindle shaped with an unstable contour, leading to a significant increase in the swarm area. The swarm area at 70 Hz is about three times as large as that at 20 Hz (i.e., 1.85 S_0). The pattern transformation and area increase of swarms in the spread state are caused by the step-out behavior of the nanoparticle chains. The inward force of the vortex is greatly reduced, making the swarm unable to maintain a stable pattern like that in the red region of Fig. 2.14. Therefore, only part of the particle chains in the swarm rotates around the center of the vortex, and massive loss of particles in the swarm is easy to occur during locomotion.

It is worth noting that by reducing the input rotating frequency below 20 Hz, the swarm in the spread state can transform back to the normal dynamic-equilibrium state. Figure 2.15a1–a6 shows the reversible transformation between the dynamic-equilibrium state and spread state of a vortex-like swarm by adjusting the input rotating frequency. First, a circular pattern is formed by nanoparticles when the rotating frequency is 8 Hz (Fig. 2.15a1, $t = 0$ s). Increasing the rotating frequency to 25 Hz brings the swarm into a spread state (Fig. 2.15a2, $t = 31$ s). Immediately after increasing the input frequency further to 40 Hz, the swarm area becomes larger, showing an unstable gear-like contour (Fig. 2.15a3, a4). The swarm tends to return to a circular shape when the input frequency is reduced to 8 Hz (Fig. 2.15a5, $t = 45$ s). After about 10 s, the swarm changes from the spread state back to the dynamic-equilibrium state with a circular pattern (Fig. 2.15a6, $t = 56$ s).

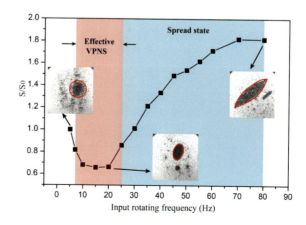

Fig. 2.14 Relationship between swarm area and input rotating frequency. The red area indicates that the swarm is in a dynamic-equilibrium state, and the blue area indicates that the swarm is in a spread state. Insets show swarm patterns at different input rotating frequencies. The magnetic field strength is 7.5 mT

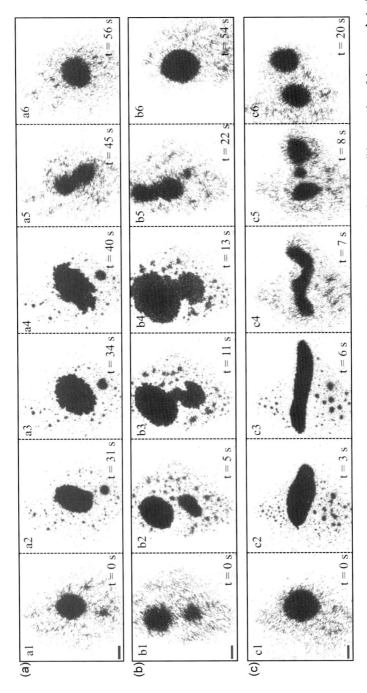

Fig. 2.15 Pattern transformation of the vortex-like swarm by using the spread state. **a** Experimental results of the reversible expansion of the swarm. **b** Active merging of two swarms. **c** One swarm actively splits into two smaller swarms. The magnetic field strength is 7.5 mT, and the scale bar is 600 μm

2.5 Experimental Results

For better control and navigation of the swarm, a single swarm pattern is required rather than multiple independent patterns. The rotating magnetic field parameters in region V of Fig. 2.8 can make the nanoparticles generate multiple vortex-like swarms, which increases the difficulty of swarm motion control and path planning. Therefore, the method for combining multiple swarms into one main swarm is needed. We experimentally demonstrate the active merging process of two independent vortex-like swarms, as shown in Fig. 2.15b1–b6. First, two independently rotating swarms are generated under a magnetic field with a rotating frequency of 12 Hz. Since the value of (a/d) of these two swarms does not reach the critical value $(a/d)_c$, they cannot merge spontaneously for a period of time (more than 60 s). We utilize the spread state of the swarm to realize active merging. The swarms exhibit the spread state when increasing the input frequency to 40 Hz, and the swarm area begins to increase (Fig. 2.15b2, $t = 5$ s). As the two swarms expand, the swarm edges begin to touch (Fig. 2.15b3, $t = 11$ s). Their connection area increases rapidly, indicating that the two swarms have merged into one (Fig. 2.15b4, $t = 13$ s). Subsequently, the input frequency is reduced to 6 Hz to make the swarm exit spread state (Fig. 2.15b5, $t = 22$ s). After about 30 s, the swarm transforms from an irregular pattern to a circular pattern, which indicates that the merging process is complete (Fig. 2.15b6, $t = 54$ s).

Active splitting of the vortex-like swarm can also be achieved using the spread state, as shown in Fig. 2.15c1–c6. At the initial moment, the swarm has a circular pattern (Fig. 2.15c1, $t = 0$ s). An elliptical rotating magnetic field with a frequency of 80 Hz is used to bring the swarm into the spread state (Fig. 2.15c2, $t = 3$ s). The elliptical field is used here in order to generate large deformation (elongation) of the swarm. The vortex-like swarm under an elliptical field will be introduced in detail in Sect. 2.6. When an elongated pattern is formed (Fig. 2.15c3, $t = 6$ s), the magnetic field is immediately changed back to a uniform circular rotating field with a frequency of 12 Hz. Then, the swarm tends to turn back into a circular pattern, and thus the elongated pattern is distorted (Fig. 2.15c4, $t = 7$ s). As the swarm elongates, the particle chains at both ends become farther apart, resulting in insufficient long-range hydrodynamic attraction between them. By appropriately reducing the recovery frequency (the frequency at which the swarm exits the spread state), the nanoparticle chains become longer, enhancing the hydrodynamic force of the vortex and reversibly changing the elongated pattern back to a circular pattern. Therefore, at an appropriate recovery frequency, the two vortices appear at opposite ends of the swarm and do not merge spontaneously. Finally, one vortex-like swarm splits into two smaller vortex-like swarms, indicating that the splitting process is complete (Fig. 2.15c6, $t = 20$ s).

2.5.4 Swarm Locomotion

In the low Reynolds number environment, the locomotion of tiny objects is generally achieved by breaking the symmetry of motion. Besides, objects experience increased resistance as they approach the boundary. According to this, the locomotion of vortex-like microswarms on a two-dimensional plane can be realized by adding a pitch angle to the rotating magnetic field. The mechanism is that the rotating nanoparticle chain tilts out of the plane with the magnetic field, resulting in an imbalance of drag forces on different parts of it. Specifically, the end of the nanoparticle chain close to the substrate moves slowly, while the end away from the substrate moves faster. Thus, a net displacement can be generated on the plane. The velocity of the nanoparticle chain depends on its length and the value of the pitch angle. In fact, nanoparticle chains inside the swarm are not uniform in length and thus have different moving velocities. Due to the influence of the vortex inward force, nanoparticle chains with different velocities can be confined in the swarm, so that the swarm can maintain a stable pattern during locomotion. The increase of the pitch angle makes the difference in moving speed between nanoparticle chains larger. If the vortex inward force is insufficient to trap these particle chains, particle loss or even swarm pattern collapse can occur. Therefore, the stability of the vortex-like swarm during locomotion is influenced by the value of the pitch angle.

Figure 2.16 shows the swarm pattern when moving at different pitch angles. When the pitch angle is 90°, the nanoparticle chain performs a tumbling motion. And the chains attract each other to form longer chains due to the magnetic force during the tumbling process, resulting in a larger change in length. As a result, the moving velocities of the chains have a significant difference, and the vortex-like swarm cannot be generated, as shown in Fig. 2.16a1, a2. When the pitch angle is reduced to 14°, the swarm continuously loses nanoparticles during the locomotion and cannot maintain a stable pattern as well (Fig. 2.16b1, b2). The experimental data show that when the pitch angle is lower than 10°, the vortex-like swarm can maintain a dynamic-equilibrium state during the movement. Figure 2.16c1, c2 shows the locomotion of a vortex-like swarm at a pitch angle of 2°. Due to the slower moving velocity, most of the particle chains are trapped by the vortex-like swarm and perform synchronized directional motion at the same speed.

We quantitatively characterize the area change during swarm locomotion at the three pitch angles of Fig. 2.16 (i.e., 90°, 14°, and 2°) (Fig. 2.17). When the pitch angle is 90°, the swarm area increases dramatically (almost doubles in 1 s). When the pitch angle is 14°, the swarm area increases slowly in the first 0.5 s. Subsequently, the swarm pattern collapses, and the area increases rapidly. The rapid increase of the swarm area indicates that the structure of the pattern is unstable, and the inner interactions change significantly. Strictly speaking, the swarming structure no longer exists in this case. When the pitch angle is 2°, the swarm area remains stable (varies less than 5%). Figure 2.18 shows the relationship between the moving velocity of the swarm and the pitch angle at different rotating frequencies. It can be seen that the moving velocity increases with the pitch angle but has no obvious relationship with the rotating frequency. An increase in the rotating frequency enables the nanoparticle chain to complete more rotations in the same period of time, but the shorter chain length results in a smaller net displacement per rotation. Therefore, the moving velocity of the nanoparticle chains remains almost constant when the rotating frequency is changed.

2.5 Experimental Results

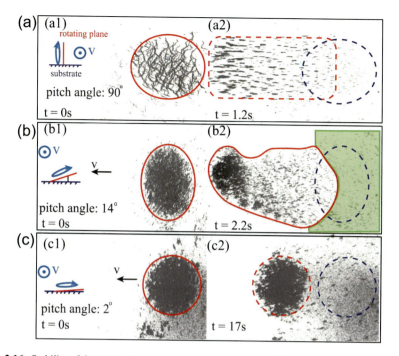

Fig. 2.16 Stability of the vortex-like swarm during locomotion at different pitch angles. **a1, b1, c1** indicate the initial state of the swarm. **a2, b2, c2** indicate the swarm after a period of locomotion. The insets show schematic diagrams of magnetic fields corresponding to each pitch angle

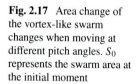

Fig. 2.17 Area change of the vortex-like swarm changes when moving at different pitch angles. S_0 represents the swarm area at the initial moment

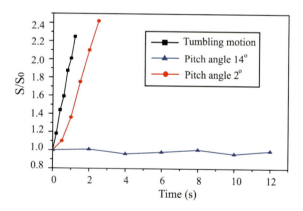

In practical applications, nanoparticles need to be delivered precisely to the targeted area with high access rates, so swarms that can effectively deliver nanoparticles are required. In addition, increasing the pitch angle can improve the moving velocity of the swarm, but the swarm pattern can be stretched, and particles are easily lost, which also requires the swarm to have a stronger trapping force. Theoretical

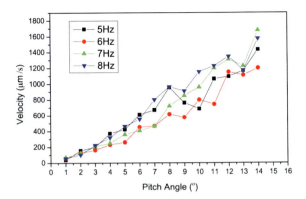

Fig. 2.18 Relationship between the moving velocity of the vortex-like swarm and the pitch angle at different rotating frequencies

analysis shows that the inward force of the vortex is tunable (Sect. 2.4.3) and can be used to modulate the trapping force of the vortex-like swarm to improve the ability of targeted delivery. The influence of the magnetic field parameters on the swarm trapping force is experimentally investigated, as shown in Fig. 2.19. Non-magnetic polystyrene microparticles were used in the experiments to visually demonstrate the trapping force of the vortex-like swarm. The polystyrene microparticles are randomly distributed on the substrate at the beginning (Fig. 2.19a1, b1, c1, d1). Four groups of swarms are then actuated by the same magnetic field (magnetic field strength: 6 mT, rotating frequency: 8 Hz, and pitch angle: 5°) to get close to the microparticles. All microparticles are drawn into the swarm due to the fluid attraction induced by the vortex (Fig. 2.19a2, b2, c2, d2). The microparticles trapped by the vortex move with the swarm when keeping the magnetic field parameters constant (Fig. 2.19a3). Decreasing the input rotating frequency to 4 Hz reduces the inward force of the vortex, leading to the loss of two microparticles during locomotion (Fig. 2.19b3). This shows that increasing the rotating frequency of the swarm can enhance the trapping force. Increasing the pitch angle to 8° can increase the moving velocity of the swarm, but it also leads to microparticle loss during locomotion (Fig. 2.19c3). Smaller diameters (100 μm in diameter) are used in Fig. 2.19d1–d3. After the microparticles are trapped, the parameters of the actuating magnetic field are the same as that in Fig. 2.19b3. Due to the smaller diameter, the microparticles in the swarm are difficult to observe, so a vertical magnetic field is applied after actuation to observe their position in the swarm. Figure 2.19d3 shows that all the microparticles move with the swarm without loss, indicating that microparticles with smaller inertia are more easily trapped by the swarm. According to the above results, the stability of the vortex-like swarm during locomotion can be enhanced by using a higher input rotating frequency (e.g., 8 Hz), a smaller pitch angle, and smaller (nanoscale) building blocks.

The motion stability and controllability of the vortex-like swarm enable applications in complex environments. Figure 2.20 shows the navigation of a vortex-like swarm in a branched channel. The channel has two branches, and the swarm is initially generated at the left entrance of the semicircular channel (Fig. 2.20a1). Then, by adjusting the direction angle of the magnetic field, the swarm passes through

2.5 Experimental Results

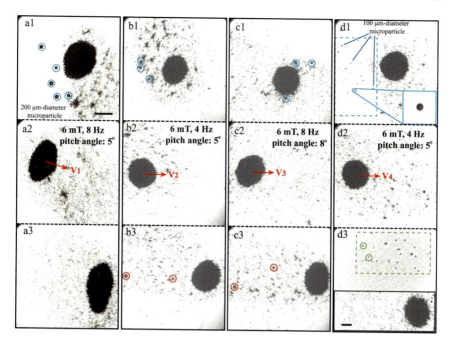

Fig. 2.19 Tunable trapping force of the vortex-like swarm is experimentally validated by investigating the behavior of non-magnetic polystyrene microparticles trapped by the swarm. The blue circles label microparticles at the initial moment, the green circles label microparticles trapped by the swarm, and the red circles label microparticles escaping from the swarm. The scale bar is 800 μm, and the scale bar within the inset is 500 μm

the channel without contacting the side wall, as shown in Fig. 2.20a2–a5. The green spots in Fig. 2.20b1–b5 represent the real-time positions of the swarm in the channel. Finally, the swarm reaches the right side through the channel. The swarm maintains a stable pattern when passing through the channel, and most of the particles successfully reach the target area. The process of the swarm passing through an angled channel is shown in Fig. 2.20c1–c5. Figure 2.20d1–d5 shows the real-time position of the swarm. Similarly, the vortex-like swarm carries most of the particles through the channel. The access rate of particles can be obtained by calculating and comparing the swarm area before and after entering the channel. The results show that the access rates in the semicircular channel and the angled channel are 91.6% and 96.8%, respectively, indicating that the vortex-like swarm can effectively perform targeted delivery. We use the tumbling motion of particle chains (i.e., the pitch angle is 90°) as a control experiment. The tumbling motion causes the swarm area to expand significantly. Since the channels are not straight and have branches or corners, many particle chains are stuck by the side wall, resulting in low access rates. The experimental results show that the access rate of particles passing through the semicircular channel is less than 10% and that through the angled channel is about 50%. This indicates that simple tumbling motion cannot effectively achieve

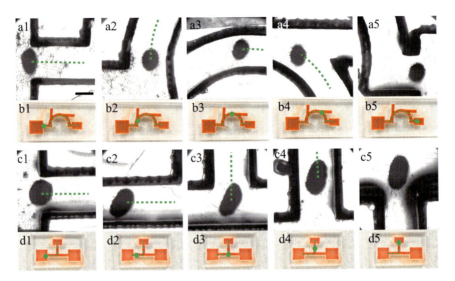

Fig. 2.20 Navigation of the vortex-like swarm in two different channels. The green dotted lines represent the predetermined path, and the green spots represent the real-time position of the swarm in the channel. The scale bar is 800 μm

large-scale targeted delivery of particles in complex environments. The dynamic-equilibrium vortex-like swarm enables the targeted delivery of particles with high accuracy and high access rates.

2.6 Elliptical Vortex-Like Swarm

2.6.1 Elliptical Rotating Magnetic Field

Section 2.5.3 introduces the pattern transformation of the vortex-like microswarm, including reversible expansion and contraction, as well as splitting and merging, by increasing the rotating frequency to bring the swarm into a spread state. We use an elliptical rotating magnetic field to generate larger deformation, i.e., pattern elongation. In fact, low-frequency elliptical rotating magnetic fields (as shown in Fig. 2.21) can also enable the vortex-like swarm to perform similar pattern transformation (Yu et al. 2021). The elliptical magnetic field can be expressed as

$$\mathbf{B}(\theta)_{x-y} = \begin{bmatrix} \cos\beta & -\sin\beta & 0 \\ \sin\beta & \cos\beta & 0 \\ 0 & 0 & 1 \end{bmatrix} \begin{bmatrix} a\cos\theta \cdot \mathbf{x} \\ b\sin\theta \cdot \mathbf{y} \\ 0 \end{bmatrix}_{x'-y'} v, \quad (2.28)$$

2.6 Elliptical Vortex-Like Swarm

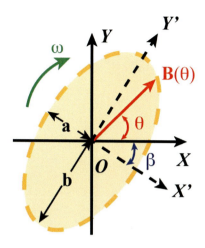

Fig. 2.21 Schematic illustration of the elliptical rotating magnetic field. The red arrow indicates the magnetic field direction at a certain moment, and the yellow dotted line indicates the magnetic field trajectory

where β is the angle between the $x-y$ coordinate and the $x'-y'$ coordinate, i.e., the angle between the minor axis of the elliptical field and the x-axis. θ here is the angle between the direction of the magnetic field and the x-axis. a and b denote the magnitude of the magnetic field when it is in the direction of the minor axis and the major axis of its trajectory, respectively. Here, we define a field ratio $\gamma = a/b$, which can be tuned to change the shape of the magnetic field. When $\gamma = 1$ (i.e., $a = b$), the field is a circular rotating field that we previously used.

Paramagnetic nanoparticle chains in the magnetic field encounter the drag torque and the magnetic torque. Thus, for a nanoparticle chain consisting of $(2N + 1)$ nanoparticles, the Mason number can be expressed as (Yu et al. 2017)

$$R_T = 64 \frac{\mu_0 \eta \omega}{\chi_p^2 B^2} \frac{N^2}{\left(\ln(N) + \frac{1.2}{N}\right)}, \quad (2.29)$$

where B indicates the magnitude of the field strength, η is the viscosity of surrounding fluids, and ω is the angular velocity of the nanoparticle chains. For a stable particle chain, the drag torque and the magnetic torque should balance each other, and thus the Mason number R_T is equal to 1. According to Eq. (2.29), a larger B leads to an increase in N, which means that the chain length is proportional to the magnetic field strength. For the elliptical field, the magnetic field strength varies with time, i.e., B is a function of θ, which can be expressed as

$$B = \sqrt{a^2 \cos^2 \theta + b^2 \sin^2 \theta} \quad (2.30)$$

Therefore, nanoparticle chains in the elliptical field have a time-varying length. The change in chain length in one actuation cycle (the change in θ is 2π) is shown in Fig. 2.22. When the magnetic field rotates to the major axis of the ellipse ($\theta = \pi/2$ or $3\pi/2$), the field strength reaches the maximum value, and the chain length is the

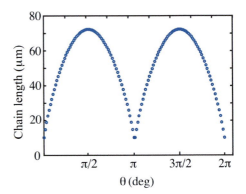

Fig. 2.22 Variation of the chain length with θ in an elliptical rotating field. $\gamma = 0.5$ (a is 5 mT and b is 10 mT)

longest at this time. When the magnetic field rotates to the minor axis of the ellipse ($\theta = \pi$ or 2π), the chain length becomes the shortest. The variation of the particle chain length with the magnetic field is the basis for generating an elliptical pattern.

2.6.2 Formation of the Elliptical Vortex-Like Microswarm

Figure 2.23 shows the mechanism of the formation of the elliptical vortex-like swarm. We first analyze the fluid interactions caused by changes in the chain length. When the magnetic field is along the major axis of the ellipse (Y' direction in the figure), the field strength reaches the maximum value, enhancing the chain-chain interaction. At this point, the nanoparticles form longer chains (Fig. 2.23, Part 1). The flow velocity on the contour of the swarm is expressed as

$$\mathbf{U}(\theta) \propto \pi f L(\theta), \tag{2.31}$$

where $L(\theta)$ indicates the chain length calculated at different θ (Fig. 2.22). Since the angular velocity of rotating chains is constant, longer chains can induce higher flow velocity. Previous analyses have shown that faster flow results in larger vortex inward force (Sect. 2.4.3). Therefore, the high-flow velocity induced by the particle chains rotated to the major axis of the elliptical field can cause a larger inward force of the vortex, making the pattern contract in the direction perpendicular to the chain. When the magnetic field rotates to the minor axis direction of the ellipse, the field strength becomes lower, and the length of the particle chain also becomes shorter, as shown in Fig. 2.23, Part 2. At this time, the flow velocity generated by the chain is low, leading to a smaller vortex inward force along the minor axis of the elliptical field. The constraining of particle chains by the swarm in this direction is also weakened. The difference in trapping forces of the swarm in the two directions results in the elongation of the circular pattern into an elliptical pattern. The major axis direction

2.6 Elliptical Vortex-Like Swarm

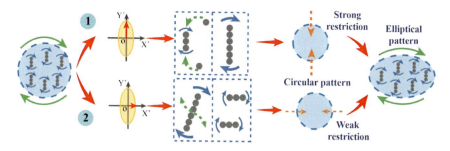

Fig. 2.23 Schematic diagram of the fluid interaction during the formation of the elliptical swarm. Gray circles represent nanoparticles. The blue areas indicate the swarm area. Green arrows indicate the rotation of the swarm. The blue arrows indicate the flow induced by the chain's rotation, and the thickness of the arrows represents the magnitude of the flow velocity. The green dashed lines indicate the assembly and disassembly of nanoparticle chains. Orange dashed lines indicate the swarm trapping force

of the elliptical swarm is perpendicular to the major axis of the elliptical magnetic field.

The swarm-induced flow field is further characterized. Similar to the analysis in Sect. 2.2, the elliptical swarm is considered to perform rigid body rotation. The coordinate of the swarm center is set to (0, 0). The dynamic of a tiny unit (fluid particle) in the swarm is given by

$$\dot{x} = \frac{\partial \Psi}{\partial y}, \quad \dot{y} = -\frac{\partial \Psi}{\partial x}, \tag{2.32}$$

where (x, y) indicates the location of the fluid particle. Ψ denotes the stream function. The characteristics of fluid influenced by the swarm can be represented by the fluid particle. For an elliptical vortex-like swarm, the stream function is expressed as

$$\psi_v = -\frac{1}{4}e^{-2k}\cos 2v - \frac{1}{2}\kappa, \tag{2.33}$$

where (κ, v) is the elliptic coordinate relative to the vortex and can be transferred to Cartesian coordinates

$$\begin{cases} X' = p \sinh \kappa \sin v \\ Y' = p \cosh \kappa \cos v \end{cases}, \tag{2.34}$$

where $p^2 = (1 - \xi^2)/\xi$ (ξ is the vorticity). Here, $X' - Y'$ indicates the coordinate system rotated with the elliptical vortex (Fig. 2.21). X' is along the minor axis of the ellipse, and Y' is along the major axis. If there is no external flow, the governing motion equation of the fluid particle is expressed as

$$\begin{cases} \dot{\kappa} = \frac{h^2}{2}\left(-\Omega + p^{-2}e^{-2t}\right)\sin 2v \\ \dot{v} = \frac{h^2}{2}\left[-\Omega \sinh 2\kappa - p^{-2}\left(e^{-2\kappa \cos 2v} - 1\right)\right] \end{cases}, \tag{2.35}$$

where $\Omega = r/(r+1)^2$, $h^2 = \left(\cosh^2 \kappa - \cos^2 v\right)^{-1}$.

According to the assumption of rigid body rotation, the vorticity remains unchanged inside the swarm. To characterize the flow velocity inside the swarm, two key factors need to be considered. First, according to Eq. (2.31), the flow velocity induced by the chain's rotation is proportional to the velocity at the end of the chain by a proportionality factor K_c, i.e., $U_m(\theta) = K_c \cdot \pi f L(\theta)$. The value of K_c can be obtained from experimental data. The flow rate induced by the swarm can be monitored experimentally using non-magnetic particles (such as polystyrene microparticles used in Sect. 2.5.2), based on which the relationship between $U_m(\theta)$, the frequency f, and chain length $L(\theta)$ can be obtained, which can be used to calibrate K_c. To simplify the analysis, K_c is set to 1 here. On the other hand, the rotation of the swarm as an entity induces a large overall flow field around the pattern. The nanoparticle chains inside the swarm rotate around their own centers as well as the swarm center, proving the existence of the overall background flow. Therefore, the vorticity ξ_z and the flow velocity u_θ can be expressed as

$$\xi_z = \frac{\Gamma_0}{\pi ab}(r(\theta) < R(\theta)) \tag{2.36}$$

$$u_\theta = \pi K_c f L(\theta) + \frac{\Gamma_0}{C} \cdot \frac{r(\theta)}{R(\theta)}(r(\theta) < R(\theta)) \tag{2.37}$$

$$C = \pi[3(a+b) - \sqrt{(3a+b)(a+3b)}], \tag{2.38}$$

where Γ_0 is the swarm circulation, $r(\theta)$ indicates the position of a point in the elliptical coordinate system at (r, θ), and $R(\theta)$ represents the distance from the center of the swarm to the point (R, θ) on the contour of the swarm. The vorticity of the surrounding flow generated by the swarm rotation can be denoted by Eq. (2.7), and the solutions are shown in Eqs. (2.10) and (2.11). According to these equations, the distribution of vorticity and flow velocity induced by the elliptical vortex-like swarm can be obtained, as shown in Fig. 2.24a, b, respectively.

In the calculation, the maximum value of the chain length is set to 200 μm, the rotation frequency is set to 5 Hz, and the aspect ratio of the swarm is set to 2. Figure 2.24a shows that the vorticity inside the swarm ($r(\theta) < R(\theta)$) is constant and uniformly distributed. Outside the swarm ($r(\theta) > R(\theta)$), the vorticity decays with increasing distance from the swarm center. Figure 2.25 shows the distribution of flow velocity inside (blue arrows) and outside (red arrows) the swarm. The thickness of the arrow represents the magnitude of the flow velocity. It can be seen from the figure that the value of the flow velocity along the short axis of the swarm is larger than that along the long axis, which is due to the longer chain length when the chains are along the short axis of the swarm.

2.6 Elliptical Vortex-Like Swarm

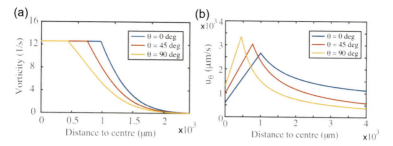

Fig. 2.24 Vorticity and tangential velocity distribution. **a** Vorticity distribution of the elliptical vortex. **b** Tangential velocity distribution of the flow induced by the elliptical vortex-like swarm

Fig. 2.25 Distribution of flow velocity inside and outside an elliptical vortex-like swarm. The blue arrows represent the flow velocity inside the swarm, the red arrows indicate the flow velocity outside the swarm, and the thickness of the arrow indicates the magnitude of the flow velocity

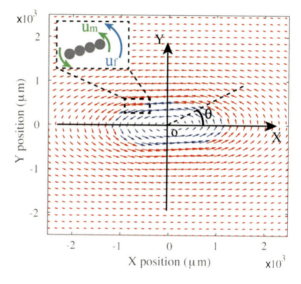

After the initial transformation from the circular pattern to the elliptical pattern, the swarm will further elongate along its long axis, especially when the field ratio γ is small. The decrease of γ makes chains along the minor axis of the elliptical field (i.e., the long-axis direction of the swarm) become shorter. In this case, the chain-chain hydrodynamic force becomes weaker, and the magnetic force becomes dominant. According to the previous analysis, longer chains are generated when the magnetic field is along the major axis of the elliptical field (see Fig. 2.26a, Part 3). As the magnetic field continues to rotate to the minor axis of the elliptical field, the field strength will decrease significantly, resulting in insufficient magnetic torque to drive the long chain to rotate in synchrony with the magnetic field. At this moment, the long chain will go into the spread state and oscillate slightly about the major axis of the elliptical field. In the center of the swarm, nanoparticle chains are more compactly arranged and thus more easily to be step-out. As the field strength becomes larger, the repulsion force between parallel chains is enhanced, and the chain-chain

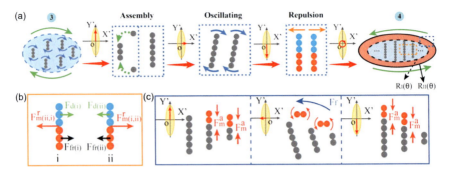

Fig. 2.26 a Schematic showing the mechanism of pattern elongation along the long axis of the elliptical swarm. The blue arrows indicate the rotation of particle chains, the green arrows indicate the rotation of the swarm, the orange arrows indicate the repulsion between nanoparticle chains, and the green dashed arrows indicate the assembly of nanoparticles. **b** Schematic illustration of the force between two adjacent nanoparticle chains. The red arrows indicate repulsive magnetic force F_m^r, the green arrows indicate the fluid drag F_d, and the black arrows indicate the friction force F_{fr}. **c** Schematic illustration of the behavior of nanoparticles located in the ring region of the swarm. Gray circles represent particles in the oscillating region, and red circles represent particles in the ring region. The red arrows indicate the interaction between the magnetic dipoles

distance increases (Fig. 2.26a, Part 4). Macroscopically, it appears as the elongation of the swarm pattern along its long axis (i.e., the minor axis of the elliptical field). After a number of actuation cycles, the elliptical swarm with a larger aspect ratio can be obtained.

The elliptical vortex-like swarm system can be divided into three regions. The first is the oscillating region ($r < R_I$) inside the swarm core (the blue area in Part 4 of Fig. 2.26a). Particle chains in this region oscillate (become step-out) when the field strength decreases and repel each other when the field strength increases. The force diagram of the particle chain is shown in Fig. 2.26b, including the fluid drag F_d, the friction force between the particle chain and the substrate F_{fr}, and the repulsive magnetic force between the particle chains F_m^r. Besides, the particle chains are also subjected to the vortex trapping force. The second is the ring region ($R_I < r < R_{II}$), as shown in Part 4 of Fig. 2.26a (the red area). This part is the peripheral region of the swarm pattern, in which the distribution of particle chains is relatively loosely arranged compared to the swarm center. Particle chains in this region rotate around the swarm center with the applied magnetic field. Figure 2.26c shows the motion of two particle chains in the ring region. The red circles represent particles in the ring region, while the gray circles represent particles in the oscillating region. When the magnetic field strength is maximum (pointing to the Y' axis), the gray and red particles assemble into longer chains. The field strength gradually decreases, and the long chains oscillate and then disassemble. After disassembly, the red part of the particle chain orbits around the swarm. The last one is the ambient region ($r > R_{II}$), i.e., the surrounding fluid affected by the rotation of the swarm.

There are differences in the flow fields induced in the above three regions of the swarm. The nanoparticle chains in the oscillating region are in the step-out state and

2.6 Elliptical Vortex-Like Swarm

cannot form a stable flow. The flow in the ring region is mainly caused by the rotation of the chain, so the flow velocity can be considered to be proportional to the chain length. The induced flow field in the ambient region follows Eq. (2.7). Therefore, the flow velocity distribution of the whole swarming system can be expressed as

$$\begin{cases} \text{No stable flow} & (r(\theta) < R_I(\theta)) \\ u_\theta = u_{II} = \pi K_c f L(\theta) & (R_I(\theta) < r(\theta) < R_{II}(\theta)) \\ u_\theta = u_{II}\left[1 - \exp\left(-\frac{r(\theta)^2}{4vt}\right)\right] & (r(\theta) < R_{II}(\theta)) \end{cases} \qquad (2.39)$$

The flow distribution is calculated based on Eq. (2.39), as shown in Fig. 2.27a. Curves of different colors correspond to different θ values. In the oscillating region, there is still induced flow due to the motion of chains. Therefore, the colored areas in Fig. 2.27a indicate that the induced flow is not computable. The distribution of flow velocity at the boundaries between different regions is shown in Fig. 2.27b, corresponding to the area highlighted by the black dashed rectangle in Fig. 2.27a. The changes in flow velocity in different regions of the swarm can be clearly seen in Fig. 2.27b. The overall schematic of the flow field is shown in Fig. 2.27c. When the field ratio γ is low (e.g., $\gamma = 0.3$), chain-like structures consisting of nanoparticles can be clearly observed, as shown in the inset in Fig. 2.27c, which strongly supports our analysis. This section clarifies the mechanism of elliptical vortex-like swarm formation, and the related experimental results are presented in the next section.

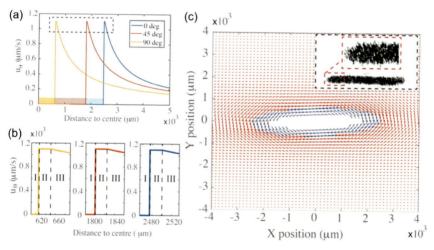

Fig. 2.27 a Distribution of u_θ in three different regions of the elliptical swarm. b Changes in flow velocity at the boundaries of different swarm regions, i.e., the enlarged view of the black dashed rectangular area in (a). c Distribution of u_θ in a plane. The blue arrows represent the flow velocity inside the swarm, the red arrows indicate the flow velocity outside the swarm, and the thickness of the arrow indicates the magnitude of the flow velocity

2.6.3 Experimental Results

The experimental result of the swarm elongation from a circular pattern to an elliptical pattern is shown in Fig. 2.28a–c. Here, we use α to denote the aspect ratio of the swarm. At the beginning ($t = 0$ s), the swarm has a circular pattern ($\alpha = 1$) under a circular rotating field ($\gamma = 1$). Then, the field is turned to be elliptical by decreasing γ to 0.75, and the swarm starts to elongate. After 10 s, an elliptical pattern with an aspect ratio of around 3 is obtained. The aspect ratio of the elliptical swarm is mainly related to the field ratio of the rotating field. When further decreasing γ to 0.4, the swarm continues to elongate, and finally an elliptical pattern with an aspect ratio of around 7 is obtained ($t = 60$ s). The elliptical swarm pattern can reversibly shrink back to a circular pattern, as shown in Fig. 2.28d–f. The vortex inward force is enhanced by increasing γ, and then the elliptical swarm tends to contract. Compared with the pattern elongation process, the shrinkage process takes a longer time. After 100 s, the elliptical swarm successfully transforms back to circular. During the shrinkage process, the orientation of the swarm changes significantly due to the reconfiguration of particle chains.

The stable motion of the elliptical swarm can also be achieved by adding a small pitch angle (1°–6°) to the elliptical magnetic field. In order to realize better motion control of the elliptical swarm, an actuation algorithm is proposed to decouple the locomotion direction and the long-axis direction of the elliptical swarm. The elliptical

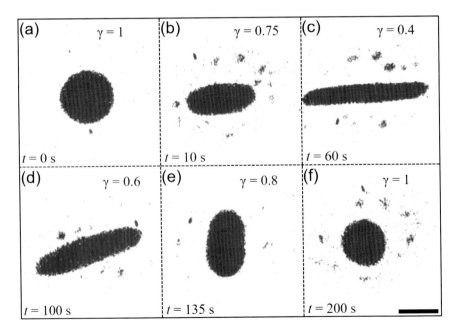

Fig. 2.28 Reversible elongation and shrinkage of the elliptical vortex-like swarm. The field strength and rotating frequency are 8 mT and 7 Hz, respectively. The scale bar is 500 μm

2.6 Elliptical Vortex-Like Swarm

field used to actuate the elliptical swarm can be expressed as

$$\boldsymbol{B}_E = \begin{bmatrix} a\cos(2\pi ft) & b\sin(2\pi ft) & 0 \end{bmatrix} \begin{bmatrix} x \\ y \\ 0 \end{bmatrix} \tag{2.40}$$

Then, the equation of the magnetic field that enables omnidirectional locomotion of the swarm is given

$$\boldsymbol{B}_{\text{Omni}} = \boldsymbol{R}_z(\kappa) \cdot \boldsymbol{R}_y(\varphi) \cdot \boldsymbol{R}_z(\tau) \cdot \boldsymbol{B}_E, \tag{2.41}$$

where $\boldsymbol{R}_z(\kappa)$ describes the long-axis direction of the swarm pattern (β, where $\beta = \tau + \kappa$), $\boldsymbol{R}_y(\varphi)$ indicates the pitch angle, and $\boldsymbol{R}_z(\tau)$ denotes the motion direction of the swarm. Their expressions are

$$\boldsymbol{R}_z(\kappa) = \begin{bmatrix} \cos\kappa & -\sin\kappa & 0 \\ \sin\kappa & \cos\kappa & 0 \\ 0 & 0 & 1 \end{bmatrix} \tag{2.42}$$

$$\boldsymbol{R}_y(\varphi) = \begin{bmatrix} \cos\varphi & 0 & \sin\varphi \\ 0 & 1 & 0 \\ -\sin\varphi & 0 & \cos\varphi \end{bmatrix} \tag{2.43}$$

$$\boldsymbol{R}_z(\tau) = \begin{bmatrix} \cos\tau & -\sin\tau & 0 \\ \sin\tau & \cos\tau & 0 \\ 0 & 0 & 1 \end{bmatrix} \tag{2.44}$$

According to this, the swarm can move omnidirectionally with its long axis in any direction. For other types of field-driven microswarms, this method is still applicable. The velocity distribution of the swarm driven by the magnetic field $\boldsymbol{B}_{\text{Omni}}$ is shown in Fig. 2.29. The blue and red curves represent the cases of $\gamma = 0.7$ and $\gamma = 0.45$, respectively. The angle between the locomotion direction and the long axis of the swarm is written as $|\beta - \tau|$. The moving velocity of the swarm is negatively related to the value of $|\beta - \tau|$. Figure 2.29a, b shows the locomotion of the swarm when $|\beta - \tau| = 0°$ and $|\beta - \tau| = 50°$, respectively. The swarm has a lower maximum velocity under a smaller γ. The possible reason is that the particle chains are more easily to be step-out due to the decrease of the field strength with γ. Besides, the hydrodynamic interaction is also reduced by the shorter chain length.

The circular and elliptical swarms have different characteristics. The elliptical swarm is slender in shape, which enables locomotion in constrained environments. The circular swarm has a faster moving velocity and is more flexible in changing the moving direction. Figure 2.30 shows the navigation of a vortex-like swarm in a complex channel using the transformation between circular and elliptical patterns. Figure 2.30a1–a4 shows the swarm with a circular pattern successfully passing

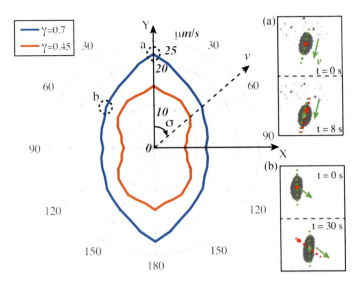

Fig. 2.29 Moving velocity of the elliptical swarm in all directions. **a, b** Experimental results when the angle between the moving direction and the long axis of the swarm is 0° or 50°, respectively

through a sharp corner in the channel (the real-time positions of the swarm are shown in Fig. 2.30b1–b4). If the swarm is driven by an elliptical field at the corner, the elongated elliptical pattern will not be suitable for turning direction. Figure 2.30a5 shows that the elliptical swarm collapses due to contact with the sidewall when turning direction. Subsequently, the swarm needs to pass through a narrow area in the channel (Fig. 2.30c1–c4). When the channel diameter becomes smaller, the swarm is elongated to an elliptical pattern by adjusting the field ratio to enter the narrow area. After passing through the narrow channel, the swarm shrinks back to a circular pattern. If the swarm remains in its circular pattern when entering the narrow area, the confinement of the boundaries on both sides of the channel can cause the swarm to collapse, as shown in Fig. 2.30c5. The reversible transformation between the elliptical pattern and the circular pattern greatly improves the environmental adaptability of the vortex-like swarm for applications in complex environments.

2.6 Elliptical Vortex-Like Swarm

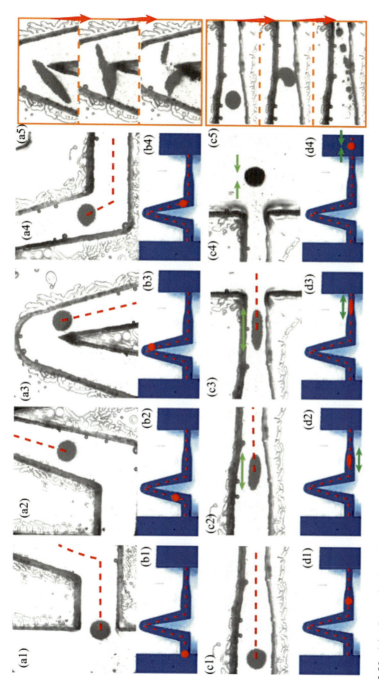

Fig. 2.30 Adaptive navigation of the vortex-like swarm in a complex channel. **a1–a4** and **c1–c4** show the experimental images. **b1–b4** and **d1–d4** show the real-time position of the swarm in the channel. **a5**, **c5** show cases in which the swarm fails to pass through the channel

2.7 Conclusion

This chapter introduces a vortex-like paramagnetic nanoparticle swarm actuated by a rotating field. The nanoparticle chains inside the vortex-like swarm perform synchronized motion under a simple rotating field, resulting in a dynamic-equilibrium state of the swarm. Research outputs on the formation and control of the vortex-like swarm are presented in detail. The mathematical model we introduced here provides a theoretical basis for characterizing various properties of the swarm, and the theoretical analysis agrees well with the experimental results. By adjusting the parameters of the magnetic field, the swarm can perform various types of pattern transformations and locomotion. For example, increasing the input rotating frequency can bring the swarm into a spread state, which enables the swarm to achieve reversible expansion and contraction, as well as splitting and merging. Actuated by an elliptical rotating magnetic field, the swarm can be reversibly transformed from a circular pattern to a stable elliptical pattern. The flexible multimode pattern transformation ability endows micro/nanorobot swarms with better environmental adaptability. By adding a pitch angle to the magnetic field, the swarm can be actuated to perform directional locomotion. At small pitch angles, the swarm moves as an entity that remains stable in motion. Taking advantage of these characteristics, the vortex-like swarm is capable of navigating in diverse microchannels and reaching the targeted area with a high access rate, demonstrating great potential for applications in complex environments.

Although we have demonstrated the potential of the vortex-like swarm as a future clinical targeted delivery tool, there are still many challenges in realizing real medical applications of micro/nanorobot swarms. Most experiments on the micro/nanorobot swarms are conducted in relatively ideal environments (such as an open container), and the swarm is observed by using optical imaging methods. However, in potential application environments of micro/nanorobot swarms, such as the human body, other suitable imaging methods are required. For example, magnetic resonance imaging (MRI) systems commonly used in clinical medicine have high resolution without ionizing radiation and can be used to track micro/nanorobot swarms in vivo. But the strong magnetic field generated by MRI can interfere with magnetic building blocks in the swarm, impeding the clusters from performing normal pattern formation, transformation, and locomotion. Using fluoroscopy imaging to track micro/nanorobot swarms in real time also has some limitations, such as magnetic field interference that may degrade imaging quality (Choi et al. 2010). In addition, prolonged ionizing radiation can cause damage to the human body, and thus the operation time of the swarm is limited. Ultrasound imaging is considered a promising method for tracking micro/nanorobots inside the human body. For instance, the needle path planning in a three-dimensional non-static environment is conducted by using ultrasound imaging feedback, which may endow the micro/nanorobots with obstacle avoidance ability (Vrooijink et al. 2014). Compared with MRI imaging and fluoroscopy imaging, ultrasound imaging is less expensive and more convenient and also has no ionizing radiation. In addition, Doppler ultrasound can also be used to monitor dynamic fluid flow, which is very applicable for imaging micro/nanorobot swarms (Wang et al.

2021). Although the resolution of clinical ultrasound imaging is only about 1–2 mm, the larger overall size of the swarm pattern enables the enhancement of ultrasound imaging contrast. The research progress on imaging of micro/nanorobot swarms will be introduced in detail in the later chapter.

References

Belkin M, Snezhko A, Aranson I, Kwok W-K (2007) Driven magnetic particles on a fluid surface: pattern assisted surface flows. Phys Rev Lett 99(15):158301

Chen Q, Bae SC, Granick S (2011) Directed self-assembly of a colloidal kagome lattice. Nature 469(7330):381–384

Chen H, Zhang H, Xu T, Yu J (2021) An overview of micronanoswarms for biomedical applications. ACS Nano 15(10):15625–15644

Choi J, Jeong S, Cha K, Qin L, Li J, Park J, Park S (2010) Positioning of microrobot in a pulsating flow using EMA system. In: 2010 3rd IEEE RAS & EMBS international conference on biomedical robotics and biomechatronics. IEEE, 588–593

Crassous JJ, Mihut AM, Wernersson E, Pfleiderer P, Vermant J, Linse P, Schurtenberger P (2014) Field-induced assembly of colloidal ellipsoids into well-defined microtubules. Nat Commun 5:5516

De Lanauze D, Felfoul O, Turcot J-P, Mohammadi M, Martel S (2014) Three-dimensional remote aggregation and steering of magnetotactic bacteria microrobots for drug delivery applications. Int J Robot Res 33(3):359–374

Du X, Yu J, Jin D, Chiu PWY, Zhang L (2021) Independent pattern formation of nanorod and nanoparticle swarms under an oscillating field. ACS Nano 15(3):4429–4439

Folio D, Ferreira A (2017) Two-dimensional robust magnetic resonance navigation of a ferromagnetic microrobot using pareto optimality. IEEE Trans Rob 33(3):583–593

Grabowski WJ, Berger S (1976) Solutions of the Navier-Stokes equations for vortex breakdown. J Fluid Mech 75(3):525–544

Green B (1995) Fluid vortices, vol 30. Springer Science & Business Media

Khalil IS, Pichel MP, Abelmann L, Misra S (2013) Closed-loop control of magnetotactic bacteria. Int J Robot Res 32(6):637–649

Lecuona A, Ruiz-Rivas U, Nogueira J (2002) Simulation of particle trajectories in a vortex-induced flow: application to seed-dependent flow measurement techniques. Meas Sci Technol 13(7):1020

Ma F, Wang S, Wu DT, Wu N (2015) Electric-field-induced assembly and propulsion of chiral colloidal clusters. Proc Natl Acad Sci 112(20):6307–6312

Mao X, Chen Q, Granick S (2013) Entropy favours open colloidal lattices. Nat Mater 12(3):217–222

Martel S (2013) Magnetic navigation control of microagents in the vascular network: challenges and strategies for endovascular magnetic navigation control of microscale drug delivery carriers. IEEE Control Syst Mag 33(6):119–134

Martel S, Mathieu J-B, Felfoul O, Chanu A, Aboussouan E, Tamaz S, Pouponneau P, Yahia LH, Beaudoin G, Soulez G (2007) Automatic navigation of an untethered device in the artery of a living animal using a conventional clinical magnetic resonance imaging system. Appl Phys Lett 90(11):114105

Martel S, Felfoul O, Mathieu J-B, Chanu A, Tamaz S, Mohammadi M, Mankiewicz M, Tabatabaei N (2009) MRI-based medical nanorobotic platform for the control of magnetic nanoparticles and flagellated bacteria for target interventions in human capillaries. Int J Robot Res 28(9):1169–1182

Melander MV, Zabusky NJ, McWilliams JC (1988) Symmetric vortex merger in two dimensions: causes and conditions. J Fluid Mech 195:303–340

Nelson BJ, Kaliakatsos IK, Abbott JJ (2010) Microrobots for minimally invasive medicine. Annu Rev Biomed Eng 12(1):55–85

Palacci J, Sacanna S, Steinberg AP, Pine DJ, Chaikin PM (2013) Living crystals of light-activated colloidal surfers. Science 339(6122):936–940

Petit T, Zhang L, Peyer KE, Kratochvil BE, Nelson BJ (2012) Selective trapping and manipulation of microscale objects using mobile microvortices. Nano Lett 12(1):156–160

Peyer KE, Zhang L, Kratochvil BE, Nelson BJ (2010) Non-ideal swimming of artificial bacterial flagella near a surface. In: 2010 IEEE international conference on robotics and automation. IEEE, 96–101

Sadelli L, Fruchard M, Ferreira A (2016) 2D observer-based control of a vascular microrobot. IEEE Trans Autom Control 62(5):2194–2206

Saffman PG (1995) Vortex dynamics. Cambridge University Press

Saffman P, Szeto R (1980) Equilibrium shapes of a pair of equal uniform vortices. Phys Fluids 23(12):2339–2342

Sitti M, Ceylan H, Hu W, Giltinan J, Turan M, Yim S, Diller E (2015) Biomedical applications of untethered mobile milli/microrobots. Proc IEEE 103(2):205–224

Snezhko A, Aranson IS (2011) Magnetic manipulation of self-assembled colloidal asters. Nat Mater 10(9):698–703

Van Reenen A, de Jong AM, den Toonder JM, Prins MW (2014) Integrated lab-on-chip biosensing systems based on magnetic particle actuation–a comprehensive review. Lab Chip 14(12):1966–1986

Vrooijink GJ, Abayazid M, Patil S, Alterovitz R, Misra S (2014) Needle path planning and steering in a three-dimensional non-static environment using two-dimensional ultrasound images. Int J Robot Res 33(10):1361–1374

Wang Q, Chan KF, Schweizer K, Du X, Jin D, Yu SCH, Nelson BJ, Zhang L (2021) Ultrasound Doppler-guided real-time navigation of a magnetic microswarm for active endovascular delivery. Sci Adv 7(9):eabe5914

Wang Q, Zhang L (2018) External power-driven microrobotic swarm: from fundamental understanding to imaging-guided delivery. ACS Nano 15(1):149–174

Wu Z, Chen Y, Mukasa D, Pak OS, Gao W (2020) Medical micro/nanorobots in complex media. Chem Soc Rev 49(22):8088–8112

Yan J, Bloom M, Bae SC, Luijten E, Granick S (2012) Linking synchronization to self-assembly using magnetic Janus colloids. Nature 491(7425):578–581

Yu J, Xu T, Lu Z, Vong CI, Zhang L (2017) On-demand disassembly of paramagnetic nanoparticle chains for microrobotic cargo delivery. IEEE Trans Rob 33(5):1213–1225

Yu J, Wang B, Du X, Wang Q, Zhang L (2018) Ultra-extensible ribbon-like magnetic microswarm. Nat Commun 9:3260

Yu J, Yang L, Du X, Chen H, Xu T, Zhang L (2021) Adaptive pattern and motion control of magnetic microrobotic swarms. IEEE Trans Rob 38(3):1552–1570

Zhang L, Abbott JJ, Dong L, Peyer KE, Kratochvil BE, Zhang H, Bergeles C, Nelson BJ (2009) Characterizing the swimming properties of artificial bacterial flagella. Nano Lett 9(10):3663–3667

Zhang Y, Zhang L, Yang L, Vong CI, Chan KF, Wu WK, Kwong TN, Lo NW, Ip M, Wong SH (2019) Real-time tracking of fluorescent magnetic spore–based microrobots for remote detection of C. diff toxins. Sci Adv 5(1):eaau9650

Chapter 3
Ribbon-Like Magnetic Colloid Microswarm

Abstract This chapter introduces a ribbon-like paramagnetic nanoparticle swarm driven by oscillating magnetic fields. Reversible pattern elongation and shrinkage of the microswarm with high aspect ratio changes are triggered by changing the field ratio of the input fields. The pattern transformation behavior is elucidated by analyzing the interactions among nanoparticle chains and their reconfiguration process. The ultra-extensible property endows the microswarm with controlled splitting and merging abilities. The ribbon-like microswarm has two different locomotion modes and high pattern stability. The microswarm navigating through complex microchannels to multiple target locations with high access rates and manipulating and separating microparticles in a non-contact manner are presented.

Keywords Swarm pattern · Elongation · Shrinkage · Oscillating field · Reconfiguration · Ribbon

3.1 Introduction

Collective behaviors in nature enable living creatures to form various swarming structures, such as bird flocks (Weimerskirch et al. 2001; Nagy et al. 2010), fish schools (Hemelrijk et al. 2010), insect swarms (Anderson et al. 2002; Mlot et al. 2011), and bacteria colonies (Felfoul et al. 2016; Martel et al. 2009), which rely on the local communication between numerous individuals. Animal swarms have abilities that are far beyond the individual and can better adapt to complex and changing environments (Sumpter 2010; Ward and Webster 2016). Emulating such amazing natural phenomenon in robotic systems through algorithm design and wireless actuation enables the development of various types of robot swarms (Groß et al. 2006; Tolley and Lipson 2011; Garattoni and Birattari 2018; Ozkan-Aydin and Goldman 2021). For example, the programmable self-assembly of a thousand-robot swarm at the macroscale has been reported (Rubenstein et al. 2014). A particle robot swarm comprising numerous loosely coupled simple "particles" can perform robust locomotion, object transport, and movement toward a light stimulus (Li et al. 2019).

At the micro/nanoscale, colloidal particles are considered to be ideal materials for constructing micro/nanorobot swarms because they are easy to be functionalized and have fast responses to external fields (Zhang et al. 2017; Gao et al. 2015; Chen et al. 2011; Mao et al. 2013; Ma et al. 2015; Palacci et al. 2013; Yan et al. 2012; Tasci et al. 2016). The vortex-like paramagnetic nanoparticle swarm we introduced in Chap. 2 is such a colloidal swarming system actuated by a simple rotating field (Yu et al. 2018, 2021). However, in order to better understand the underlying principle of natural collective behavior, more kinds of micro/nanorobot swarms need to be constructed. By programming the magnetic field, colloidal particles can exhibit various self-assembly behaviors, based on which the formation of a variety of micro/nanorobot swarms can be realized (Belkin et al. 2007; Snezhko and Aranson 2011; Snezhko et al. 2009). In addition, the morphology of natural swarms is generally transformable, and the degree of their shape deformation can be extremely large (Parrish and Edelstein-Keshet 1999; Lee et al. 2006; Hemelrijk and Hildenbrandt 2011, 2012; Storms et al. 2019; Peleg et al. 2018). For example, the swarm of *Bacillaria paradoxa* (a widely distributed diatom) is capable of performing reversible transformation between a fully stretched pattern and a contracted pattern under light stimulation (Jahn and Schmid 2007; Yamaoka et al. 2016; Kapinga and Gordon 1992; Cai et al. 2013). When the swarm is in a contracted state, the long sides of the slender *B. paradoxa* stick to each other and are tightly combined, forming a pattern that has the length and width within several body lengths of individuals. The *B. paradoxa* is able to slide relative to each other and connect in a nearly end-to-end fashion to form a fully stretched pattern that can be dozens of times longer than the contracted pattern. Imitating such swarm deformation in micro/nanorobot swarm systems can significantly enrich functionality and deepen the understanding of collective behaviors.

This chapter introduces a ribbon-like paramagnetic nanoparticle swarm generated under a programmed oscillating magnetic field. Similar to the vortex-like swarm, the ribbon-like is also in a dynamic-equilibrium state, which is formed based on interactions between magnetic nanoparticle chains and their reconfiguration behavior under the dynamic magnetic field. By adjusting the field ratio of the oscillating field, the swarm can actively perform reversible elongation and shrinkage with extremely high aspect ratio changes of the swarm pattern, which is very similar to the morphology transformation of the *B. paradoxa* swarm. In addition, the ribbon-like swarm can also split into multiple sub-swarms, and the number of sub-swarms can be controlled by adjusting the magnetic field parameters. Thanks to the ultra-extensible property, a number of ribbon-like swarms can be efficiently merged into a single pattern. By adjusting the pitch angle of the magnetic field, the swarm is able to perform controlled locomotion in a two-dimensional plane. There are two locomotion modes, i.e., longitudinal movement along the long axis of the swarm and lateral movement along the short axis of the swarm. Unlike the vortex-like swarm, the ribbon-like swarm has stronger stability and can maintain a stable pattern when colliding with the boundaries of various angles. Finally, this chapter shows the flexibility and versatility of ribbon-like swarm through a series of application demonstrations, such as navigating

3.2 Formation of the Ribbon-Like Swarm

3.2.1 The Oscillating Field and Time-Varying Chain Length

We first introduce the oscillating magnetic field used to generate the ribbon-like swarm, which consists of two orthogonal magnetic fields, i.e., B_A and B_C. B_A is a sinusoidally varying magnetic field, which is expressed as $B_A = A\sin(2\pi ft)$. A is the amplitude of the magnetic field, and f is the oscillation frequency. B_C is a uniform magnetic field perpendicular to B_A and has a constant strength of C. The superposition of the two magnetic fields generates an oscillating magnetic field with time-varying angular velocity and field strength, as shown in Fig. 3.1 (red arrow). $\gamma = A/C$ is defined as the field ratio (amplitude ratio), which is a key parameter for adjusting the aspect ratio of the swarm pattern.

The oscillating angle $\kappa(t)$ indicates the angle between directions of the field and B_C (the y-axis), which is expressed as:

$$\tan\kappa(t) = \gamma \sin(2\pi ft). \tag{3.1}$$

By taking the derivative of $\kappa(t)$ with respect to time, the angular velocity $\omega(t)$ of the oscillating field can be obtained:

$$\omega(t) = \frac{d}{dt}\tau(t) = \frac{2\pi\gamma f\cos(2\pi ft)}{1+r^2\sin^2(2\pi ft)}. \tag{3.2}$$

The changes of the oscillating angle $\kappa(t)$ (blue curve) and angular velocity $\omega(t)$ (red dashed curve) with time are plotted in Fig. 3.2. Both curves are periodically

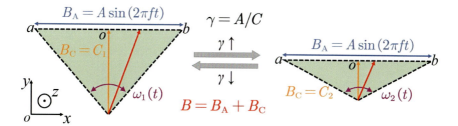

Fig. 3.1 Schematic diagram of the oscillating magnetic field used for generating the ribbon-like swarm. The red arrows indicate the superimposed magnetic fields of B_C and B_A, which vary within the green triangle enclosed by the black dashed lines. The purple arrows indicate the oscillating direction, and $\omega(t)$ is the angular velocity. A is 10 mT in this chapter

Fig. 3.2 Changes of the oscillating angle $\kappa(t)$ (blue curve) and angular velocity $\omega(t)$ (red curve) with time. $\Gamma = 2$ and $f = 5$ Hz

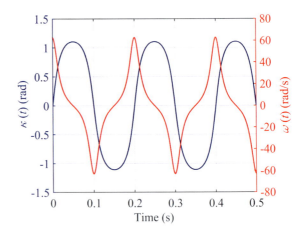

varying. When the magnetic field points to the y-axis (point O in Fig. 3.1), the angular velocity reaches the maximum value, and the field strength is minimum. When the field reaches both sides (points a and b in Fig. 3.1), the angular velocity is minimum, and the strength is maximum. Adjusting the field ratio γ changes the angular velocity and strength of the oscillating field. For example, when γ is increased by keeping A constant and decreasing C (Fig. 3.1, right), the oscillation frequency remains unchanged, but the oscillation angle increases, leading to a higher angular velocity ($\omega_2(t) > \omega_1(t)$). Meanwhile, the overall field strength decreases due to the smaller value of C.

Similar to the analysis of the vortex-like swarm, the chain-like structure formed by paramagnetic nanoparticles under the magnetic field (i.e., the nanoparticle chain) is considered the basic unit of the analysis. In the oscillating magnetic field, particle chains oscillate with the field and are subject to magnetic torque Γ_m and the resistive fluidic drag torque Γ_d, which are indicated by the red and blue arrows in Fig. 3.3, respectively. The angular velocity of the chain also varies with time. Due to the existence of fluid drag, the motion of the particle chain lags behind the magnetic field, and the angle between the chain and the field is defined as the phase lag θ. Here, consider a single uniform chain consisting of $(2N + 1)$ particles, the particle at the center of the chain is located at the origin of the coordinate system, and the rest of the particles are labeled from N to -N. In low Reynolds number environments (the Reynolds number is estimated to be approximately 0.006), the influence of inertia can be negligible. The magnetic torque Γ_m and drag torque Γ_d of the particle chain during the oscillation balance each other, which are expressed as:

3.2 Formation of the Ribbon-Like Swarm

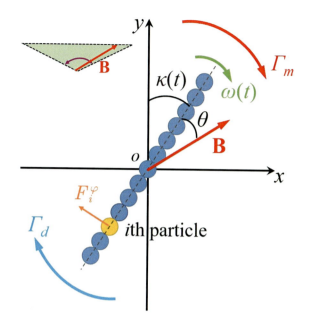

Fig. 3.3 Schematic illustration of the particle chain oscillating with the magnetic field. The phase lag is θ, indicating the angle that the chain lags relative to the magnetic field. The force exerted on the ith particle by other particles is perpendicular to the long axis of the chain. The magnetic torque (red arrow) and drag torque (blue arrow) exert opposite effects on the chain during oscillation

$$\Gamma_m = 2\sum_{i=1}^{N}(F_i^\varphi r_i) = \frac{3\mu_0 m^2}{4\pi}\sin(2\theta)\sum_{i=1}^{N}\left(2r_i \sum_{\substack{j=-N \\ i \neq j}}^{N}\frac{1}{r_{ij}^4}\right) \tag{3.3}$$

$$\Gamma_d = \frac{64\pi a^3}{3}\frac{N^3}{\ln(N)+\frac{1.2}{N}}\eta\omega(t), \tag{3.4}$$

where a is the radius of the single particle. The distances from the ith particle to the chain center and the jth particle are indicated by r_i and r_{ij}, respectively. F_i^φ is the magnetic force exerted on the ith particle by the other $2N$ particles, which is perpendicular to the long axis of the chain.

The length of the particle chain depends on the balance of the drag torque and the magnetic torque, which can be estimated by the model of Mason number (Reenen et al. 2014). Since the strength of the oscillating magnetic field changes with time, the drag torque and the magnetic torque of the particle chain are also time-varying. Therefore, a modified Mason number model is used to denote the relationship between the length of the particle chain (consists of $(2N+1)$ particles) and the oscillating field:

$$64\frac{\mu_0\eta\omega}{\chi_p^2}\frac{N^2}{\left(\ln(N)+\frac{1.2}{N}\right)} - A^2\sin^2(2\pi ft) - C^2 = 0 \tag{3.5}$$

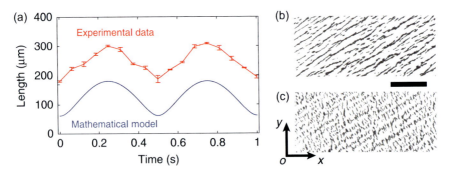

Fig. 3.4 Length of particle chains varies with time. **a** Changes in the chain length with time in the oscillating magnetic field with a frequency of 1 Hz. The blue and red curves indicate the calculation and experimental results, respectively. Each error bar denotes the standard deviation from three trials. **b** Particle chains when the field strength is maximum (the field points to *a* or *b*). **c** Particle chains when the field strength is minimum (the field points to *O*). The scale bar is 400 μm

It is worth noting that this model mainly shows the influence of the magnetic field strength on the chain length, and thus, the angular velocity of the chain oscillation is regarded as a constant to simplify the model. According to this Mason number model, the variation of the chain length with time within a complete oscillation cycle (frequency: 1 Hz) can be calculated, as shown by the blue curve in Fig. 3.4a. The variation trend of the chain length observed in the experiments (red curve in Fig. 3.4a) is consistent with the model predictions. When the magnetic field points to *a* or *b* (Fig. 3.1), the chain length is the longest due to the maximum magnetic field strength (Fig. 3.4b). Similarly, when the magnetic field points to *O*, the field strength becomes smaller, and the particle chains become shorter (Fig. 3.4c). In addition, the experimentally measured chain lengths are longer than the calculated results. This is because the model only considers a single particle chain, while the adjacent chains attract each other to form particle chain bundles in the experiment, causing the chains to become thicker and longer.

3.2.2 Chain-Chain Magnetic Interaction

The magnetic interaction between particle chains in the oscillating field is the crucial factor for the formation of the ribbon-like swarm. When analyzing the chain-chain magnetic interaction, nanoparticle chains are simplified to prolate ellipsoids. As shown in Fig. 3.5, the magnetic interaction between two particle chains is attraction force, denoted as $F_{1,2}$ and $F_{2,1}$. As they approach each other, they encounter fluid drags in the opposite direction of motion, denoted as F_{d1} and F_{d2}. When the external magnetic field is smaller than the saturation magnetization of the particle chain, the magnetization of the chain can be expressed as (Landau et al. 2013):

3.2 Formation of the Ribbon-Like Swarm

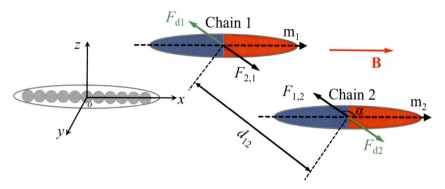

Fig. 3.5 Schematic diagram of the particle chain simplified to an ellipsoid. The distance between the centers of the two chains is d_{12}, and m_1 and m_2 indicate the magnetic dipole moment

$$\bm{m} = \frac{V_c \chi_\alpha}{\mu} \bm{B} = \frac{V_c \chi_\alpha}{\mu_0 (1+\chi)} \bm{B}, \qquad (3.6)$$

where μ is the permeability of the particle chain, χ is the magnetic susceptibility, and χ_α is the apparent susceptibility tensor that takes the shape factor of the particle chain into account. Because the particle chain is simplified to an axisymmetric ellipsoid, χ_α is expressed as (Singh et al. 2005):

$$\chi_\alpha = \begin{pmatrix} \frac{\chi}{(1+n_a \chi)} & 0 & 0 \\ 0 & \frac{\chi}{(1+n_b \chi)} & 0 \\ 0 & 0 & \frac{\chi}{(1+n_b \chi)} \end{pmatrix}, \qquad (3.7)$$

where n_a and n_b are demagnetizing factors along the major axis and minor axis of the ellipsoid, respectively, which are denoted as:

$$\begin{cases} n_a = \frac{1-\varepsilon^2}{2\varepsilon^2}\left(\log\left(\frac{1+\varepsilon}{1-\varepsilon}\right) - 2\varepsilon\right) \\ n_a + 2n_b = 1 \\ \varepsilon = \sqrt{1 - \Lambda^2} \end{cases}, \qquad (3.8)$$

where ε is the eccentricity, Λ is the aspect ratio of the ellipsoid. The magnetic dipole moments of the particle chain can be expressed as:

$$\begin{pmatrix} m_x \\ m_y \\ m_z \end{pmatrix} = \frac{V_m \chi}{\mu_0(1+\chi)} \begin{pmatrix} \frac{B_x}{1+n_a \chi} \\ \frac{B_y}{1+n_b \chi} \\ \frac{B_z}{1+n_b \chi} \end{pmatrix}. \qquad (3.9)$$

For the ith particle chain with dipole moment m_i located at point p_i $(x_i, y_i, z_i)^T$, the introduced magnetic field is expressed as (Schill 2003):

$$B_{pi}(p) \frac{\mu_0}{4\pi} \left(3 \left(m_i \cdot \frac{p - p_i}{||p - p_i||} \right) - m_i \right). \tag{3.10}$$

Therefore, the magnetic force exerted by the ith chain on the $(i + 1)$th chain is expressed as

$$F_{i,i+1} = -\frac{\mu_0}{4\pi} \nabla \left(\left(\nabla \frac{m_i \cdot d_{i,i+1}}{d_{i,i+1}^3} \right) \cdot m_{i+1} \right), \tag{3.11}$$

where $d_{i,i+1} = d_{i,i+1} n_i$ is the distance between the centers of the two chains, as shown in Fig. 3.5. It is clear from the above equation that increasing the magnetic field strength can significantly increase the magnetic interaction force between two particle chains.

3.2.3 Pattern Formation Process

The generating process of the ribbon-like paramagnetic nanoparticle swarm by using an oscillating magnetic field is shown in Fig. 3.6a. Paramagnetic Fe_3O_4 nanoparticles with a diameter of 500 nm are used in the experiments. At the initial moment ($t = 0$ s), nanoparticle chains were randomly distributed in deionized water (DI water) on the substrate. Then, driven by the oscillating magnetic field, the particle chains are rapidly combined into many ribbon-like structures, which can be regarded as individual sub-swarms ($t = 2.0$ s). These sub-swarms eventually merge and form a dynamically stable ribbon-like swarm ($t = 33.0$ s). The detailed self-merging process of sub-swarms is shown in Fig. 3.6b. First, a major sub-swarm (marked by the red rectangle) is generated, which attracts other surrounding smaller sub-swarms (marked by the green rectangle) through hydrodynamic and magnetic interactions. After approaching the main sub-swarm, nanoparticle chains in the smaller sub-swarm are continuously drawn into the main sub-swarm. Finally, a single ribbon-like swarm is generated.

The field ratio γ and the frequency f of the oscillating magnetic field have a crucial influence on the formation of the ribbon-like swarm. Fig. 3.7 is the phase diagram that shows the relationship between pattern formation and field parameters. When γ is small and the frequency is low (region I), the nanoparticles form a large number of unstable chain-like structures, which gradually elongate during the oscillation with the magnetic field. When γ increases (region II), the ribbon-like swarm with a stable pattern can be generated due to stabilized interactions between nanoparticle chains. When the magnetic field parameters are in region III of Fig. 3.7, the nanoparticles form multiple elongated structures, which cannot spontaneously fuse with each other into a single stable pattern. The gray shaded area in the phase diagram indicates that nanoparticles cannot generate regular swarm patterns under field parameters in this region.

3.2 Formation of the Ribbon-Like Swarm

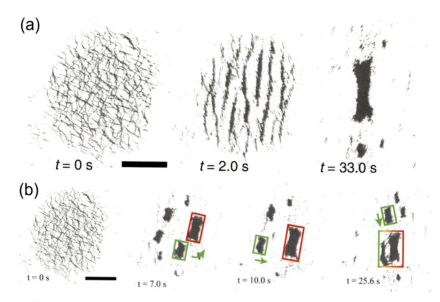

Fig. 3.6 **a** Pattern formation process of a ribbon-like swarm. **b** Self-merging process of the sub-swarms. The scale bars are 800 μm

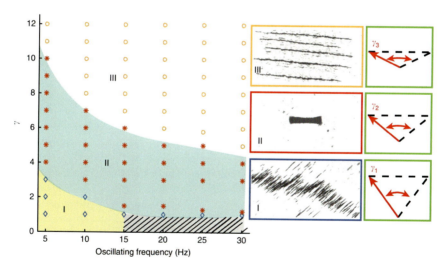

Fig. 3.7 Phase diagram showing the relationship between pattern formation and field parameters. The inset on the right shows the swarm pattern under different magnetic field parameters, and the corresponding magnetic field is schematically displayed in the green rectangle. The gray shaded area indicates that a regular swarm pattern cannot be generated under field parameters in this region

The schematic diagram of the formation process of the ribbon-like swarm is shown in Fig. 3.8. At the initial moment (stage I), nanoparticle chains (indicated by gray bars) are randomly distributed on the substrate. The oscillating magnetic field is then applied (stage II), and the nanoparticle chains start to oscillate with the field. The colorful long rods represent nanoparticle chains in the field, and the red and blue parts represent the different magnetizations of the chains, respectively. The green arrows indicate the oscillation direction, and the yellow dashed lines indicate the distribution area of the chains. In order to clearly explain the formation mechanism of the swarm, only two particle chains are used as an example for analysis. When the chain oscillates to the position of stage III, the long particle chain disassembles into shorter chains due to the decrease of the magnetic field strength and the increase of the angular velocity. In addition, since the field strength is the smallest at this time, it can be assumed that the particle chains are not actuated by the chain-chain interactions, that is, the central position of each particle chain remains unchanged. Then the increasing field strength enhances the magnetic attraction between the chains, resulting in the chains approaching each other and narrowing their distribution area along the x-axis (phase IV and phase V). In subsequent stages (stage VI to stage VIII), nanoparticle chains repeat the reconfiguration behaviors described above. The red dashed rectangle represents the distribution area of the swarm pattern consisting of nanoparticle chains, with obvious shrinkage in the x-axis direction. After going through multiple cycles of circulation, all the long chains are disassembled into short chains with close chain-chain distances. Finally, the ribbon-like swarm is formed. This mechanism is also applicable to the scenario of a larger number of particle chains, which can well explain the principle of the formation of the ribbon-like paramagnetic nanoparticle swarm under the oscillating magnetic field.

3.2.4 Chain-Chain Hydrodynamic Interaction

The chain-chain hydrodynamic interaction is another key factor for the formation of the ribbon-like swarm. For the incompressible fluid system, the governing equations are given by the Navier–Stokes equations:

$$\frac{\partial \boldsymbol{u}}{\partial t} + \boldsymbol{u} \cdot \nabla \boldsymbol{u} = -\frac{1}{\rho}\nabla p + v\nabla^2 \boldsymbol{u},$$
$$\nabla \cdot \boldsymbol{u} = 0$$
(3.12)

where $\boldsymbol{u}$ is the velocity field, p is the pressure, v is the kinematic viscosity of the fluid, and ρ is the fluid density. The solution for the flow velocity field induced by the oscillating chain can be obtained according to the linear transformations of the body frame (Camassa et al. 2008). Figure 3.9 shows a single particle chain (treated as an ellipsoid in the analysis), where x-o-y indicates the world frame, and x*-o-y* indicates the body frame. The chain oscillates with the field in the world frame,

3.2 Formation of the Ribbon-Like Swarm

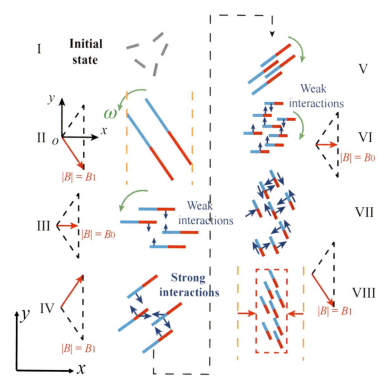

Fig. 3.8 Schematic diagram of the formation process of the ribbon-like swarm. The red and blue parts represent the different magnetizations of the chains. The yellow dashed lines indicate the distribution area of the chains, and the red dashed rectangle indicates the ribbon-like swarm pattern

while the background flow is passive. In the body frame, the chain remains static with respect to the oscillating background flow. The equation of the chain in the body frame is expressed as:

$$\frac{x^{*2}}{a^2} + \frac{y^{*2} + z^{*2}}{b^2} = 1 (a > b). \tag{3.13}$$

$\delta = b/a$ indicates the slenderness, and $e = \sqrt{1-\delta^2}$ is the eccentricity. The oscillation of the x^*-o-y^* plane is denoted by $U(x^*) = -\omega(t)(e_z \times x^*)$, where e_z is along the positive direction of the z-axis. The transformations are given by:

$$R_\kappa = \begin{pmatrix} \sin\kappa(t) & 0 & -\cos\kappa(t) \\ 0 & 1 & 0 \\ \cos\kappa(t) & 0 & \sin\kappa(t) \end{pmatrix}, \tag{3.14}$$

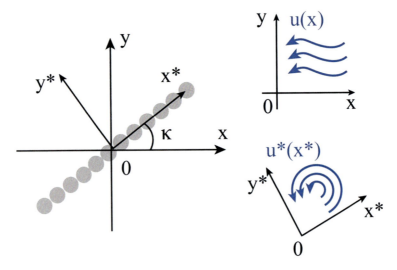

Fig. 3.9 World frame x-o-y and the body frame x*-o-y' of the oscillating chain model

where R_k is a clockwise rotation transformation matrix, and $\kappa(t)$ indicates the oscillating angle. The velocity field $\boldsymbol{u}(x)$ in the body frame is expressed as:

$$u^*(x_0) = 0$$
$$\lim_{x \to \infty} u^*(x^*) = R_k^T U(R_k x^*) = \omega \begin{pmatrix} y \sin \kappa(t) \\ -x \sin \kappa(t) + z \cos \kappa(t) \\ -y \cos \kappa(t) \end{pmatrix}, \quad (3.15)$$

where x_0 locates on the surface of the chain. Therefore, the transformation and the velocity field $\boldsymbol{u}(x)$ in the world frame are expressed as:

$$x = R_\kappa x^*$$
$$\boldsymbol{u}(x) = -U(x) + R_\kappa \boldsymbol{u}^*\left(R_\kappa^T x\right) \qquad (3.16)$$

The construction of the solution that satisfies the boundary conditions in the world frame is given by:

$$\boldsymbol{u}(x_0) = -U(x)$$
$$\lim_{x \to \infty} \boldsymbol{u}(x) = 0 \qquad (3.17)$$

The velocity field induced by the oscillating of the chain at an oscillating angle $\kappa(t)$ can be described by boundary conditions in Eq. (3.15) coupled with the Stokes equations. According to the linearity of the Stokes equations (Zhou et al. 2017), the velocity field can be constructed as the sum of four contributions:

3.2 Formation of the Ribbon-Like Swarm

$$u(x) = u^1(x) + u^2(x) + u^3(x) + u^4(x)$$
$$u^i(x_0) = 0 (i = 1, 2, 3, 4)$$
$$\lim_{x \to \infty} u^1(x) = \omega y \sin \kappa(t)$$
$$\lim_{x \to \infty} u^2(x) = -\omega x \sin \kappa(t)$$
$$\lim_{x \to \infty} u^3(x) = \omega z \cos \kappa(t)$$
$$\lim_{x \to \infty} u^4(x) = -\omega y \cos \kappa(t)$$
(3.18)

The simulation results of the flow field generated by a single oscillating particle chain (simplified to an ellipsoid) are shown in Fig. 3.10. The red arrows indicate the oscillation direction, and the white arrows indicate the induced flow. The size of the arrow indicates the magnitude of the velocity. From 0 to 0.01 s, the flow field is significantly enhanced due to the faster angular velocity of the chain oscillation. After 0.01 s, the velocity field becomes weaker with the decrease of the angular velocity. When the oscillation direction of the chain is changed (0.005 s), vortices are induced at both ends of the chain.

By considering the ribbon-like swarm as an array of oscillating nanoparticle chains, simulation results of the flow field induced by the swarm can be obtained, as shown in Fig. 3.11. When $t = 0$ s, a strong flow field is generated around the chain due to the maximum angular velocity. Then the strong flow field gradually disappears, and the flow field around the swarm is generated, exerting a repulsive force on surrounding non-magnetic objects along the long side of the swarm. Therefore, the swarm always avoids contact with external objects, enhancing the stability of the ribbon-like pattern as an entity. Figure 3.12 shows the behavior of 30 non-magnetic free particles in the flow field induced by the ribbon-like swarm. The moving trajectories show that these particles are drawn into the swarm first and then thrown out at the ends of the swarm. The hydrodynamic and magnetic interactions between particle chains inside the swarm balance each other, which enables the swarm pattern to remain dynamically equilibrium.

Particle chains in the swarm are constantly reconfigured under the actuation of the dynamic magnetic field, as shown in Fig. 3.13. Figure 3.13a shows the morphology of the swarm tip. The ribbon-like swarm can be considered as a series of parallel nanoparticle chains that consist of equal numbers of nanoparticles aligned along the red dashed lines (Fig. 3.13b). Nanoparticles located on the left and right sides of the swarm are indicated by green and blue regions, respectively. As the strength of the magnetic field decreases and the angular velocity increases, the nanoparticle chains cannot maintain their stability, and particles at the ends of the chains (green and blue regions) break away from their original chains (Fig. 3.13c). Subsequently, the disassembled nanoparticles are attracted to new chains again due to attractive magnetic force, as indicated by the orange dashed arrows. Take the particle p in Fig. 3.13 as an example to analyze the reconfiguration behavior of nanoparticles at the swarm tip. First, the particle p is expelled from the swarm due to the repulsive fluid force and then attracted by the magnetic force. The motion trajectory of the particle is

84   3 Ribbon-Like Magnetic Colloid Microswarm

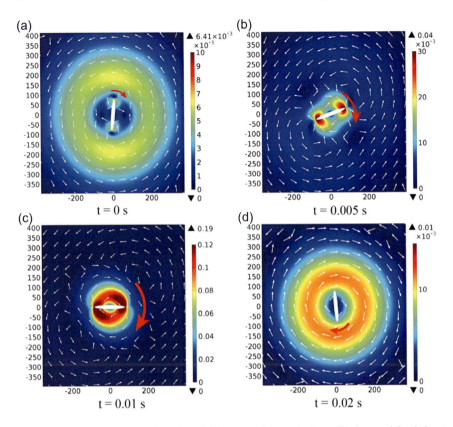

Fig. 3.10 Simulation results of the flow field generated by a single oscillating particle chain at **a** $t = 0$ s, **b** $t = 0.005$ s, **c** $t = 0.01$ s, and **d** $t = 0.02$ s

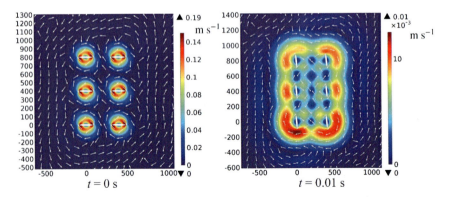

Fig. 3.11 Simulation results of the flow field generated by an array of oscillating particle chains at $t = 0$ s and $t = 0.01$ s. The field ratio is 3, and the oscillating frequency is 20 Hz

3.2 Formation of the Ribbon-Like Swarm

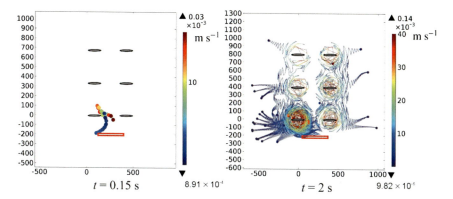

Fig. 3.12 Simulation results of moving trajectories of 30 free particles in the swarm-induced flow field. The particles with a diameter of 10 µm are released at the tip of the chain array (indicated by the red rectangle) at $t = 0$ s

shown by the curve in Fig. 3.13d, where the green part indicates that the particle is mainly affected by the repulsive fluid force, and the red part indicates that the attractive magnetic force is dominant. Therefore, due to the strong magnetic interaction, the positions of particles in region A are relatively fixed. In region B, particle chains continuously perform reconfiguration, showing a relatively loose structure. In addition, in the middle part of the swarm, the gaps generated by the chain's reconfiguration are immediately filled by surrounding particles, and the overall structure is relatively tight and stable.

The directional locomotion of the swarm as an entity may have an impact on the stability of the swarm pattern. As shown in Fig. 3.14, the particle p at the tip of the swarm is constantly moving around the edge under the influence of hydrodynamic and magnetic forces. When the swarm moves in a specific direction (as indicated by the yellow arrow in the figure), the distance between the particle p and the main pattern of the swarm may become larger due to the displacement of the swarm, resulting in insufficient magnetic attraction force to attract the particle p back again into the swarm. In this case, particles at the edge of the swarm will be lost continuously

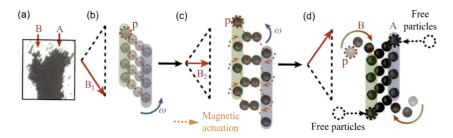

Fig. 3.13 Schematic diagram of dynamic reconfiguration of particles at the tip of the ribbon-like swarm

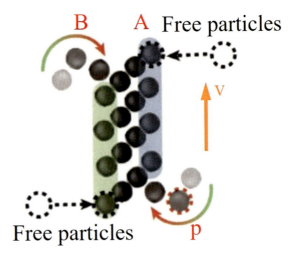

Fig. 3.14 Schematic illustration of particle loss at the tips of the ribbon-like swarm due to locomotion

during the locomotion, especially when the moving velocity is fast. Therefore, in order to avoid this, the moving velocity of the swarm should not be too large, that is, the magnetic field driving the movement of the swarm should have a suitable pitch angle to keep the swarm pattern stable.

3.2.5 *Pattern Formation Using Particles of Different Sizes and at Different Particle Concentrations*

The dynamic balance of magnetic and hydrodynamic interactions between nanoparticle chains is the key factor for maintaining the stability of the ribbon-like swarm. The magnetic interaction force is dominant during the formation process. When the magnetic field strength weakens and the angular velocity of the chain increases, the fluid drag causes the disassembly of the chains. When the field strength increases again, the chains assemble again. If the size of particles composing the swarm is too small, other interactions between particles, such as electrostatic, van der Waals, and capillary forces, may become dominant. In this case, the disassembly of nanoparticle chains cannot be induced, which may affect the formation of the ribbon-like swarm. Figure 3.15 shows the experimental results of generating ribbon-like swarms using two types of Fe_3O_4 nanoparticles with diameters of 100 nm and 250 nm, respectively. Particles with a diameter of 100 nm take a longer time to generate the swarm than particles with a diameter of 250 nm, which is caused by the decrease in magnetic force with particle size. Therefore, for magnetic nanoparticles with diameters larger than 100 nm, the oscillating magnetic field can effectively generate ribbon-like swarms.

The initial concentration of particles also has an effect on the formation of the ribbon-like swarm. Figure 3.16 shows the formation process when the initial particle concentrations are 2.1 and 6.5 μg/mm². When the particle concentration is lower

3.2 Formation of the Ribbon-Like Swarm

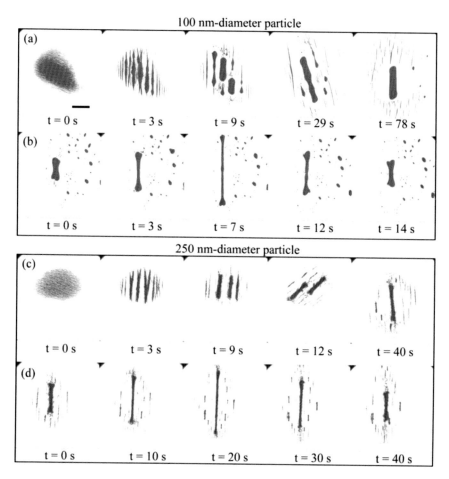

Fig. 3.15 Generating and reconfiguration of the ribbon-like swarm using Fe_3O_4 nanoparticles with diameters of 100 and 250 nm

(2.1 µg/mm^2), a larger number of sub-swarms are generated first. These sub-swarms are far away from each other, resulting in a longer merging time. When the particle concentration is higher (6.5 µg/mm^2), the generated sub-swarms are smaller in number and closer to each other, and thus, they can merge more quickly. The number of sub-swarms as a function of time is shown in Fig. 3.17. A single stable ribbon-like swarm is successfully generated in both particle concentrations. The final swarm area increases with initial particle concentration (Fig. 3.18). When more doses of nanoparticles are used, the swarm area also becomes larger, as shown by the comparison of the blue and red dashed curves in Fig. 3.18.

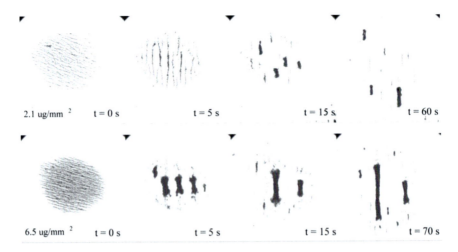

Fig. 3.16 Formation process of the ribbon-like swarm when the initial particle concentrations are 2.1 and 6.5 µg/mm^2

Fig. 3.17 Number of sub-swarms as a function of time during the pattern formation process under different initial particle concentrations

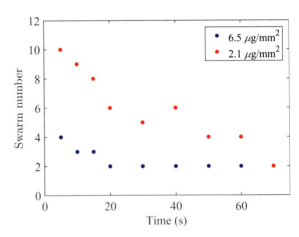

Fig. 3.18 Relationship between the final swarm area and initial particle concentrations

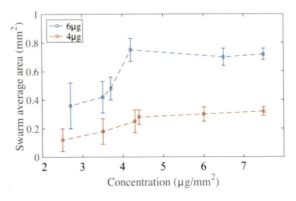

3.3 Multimodal Reconfiguration of the Swarm Pattern

3.3.1 Pattern Reconfiguration of the Ribbon-Like Swarm

The ribbon-like swarm has a flexible morphology that is entirely different from the single microrobot with a rigid body (Xu et al. 2015; Steager et al. 2013). A unique feature of such microswarm is the ability to perform reversible elongation and shrinkage along the long axis of the pattern, as shown in Fig. 3.19a. Here, the aspect ratio α is used to describe the shape of the swarm pattern. At the initial moment ($t = 0$ s), a ribbon-like swarm with an aspect ratio of around 4 is generated under the oscillating field (the field ratio $\gamma = 3$). The swarm can perform pattern elongation by increasing γ. When $t = 3$ s, γ is increased to 5, inducing the elongation of the swarm along the direction of the green arrows. After elongating for 14 s ($t = 17$ s), a slender swarm pattern with an aspect ratio of around 22 is obtained. Then, γ is tuned back to 3, and the elongated swarm gradually shrinks back to the initial state ($t = 36$ s).

During the formation process of the ribbon-like swarm, multiple sub-swarms may not be able to merge into a single pattern if the attractive force between them is too weak. This usually occurs when the initial particle concentration is low, and the individual sub-swarms are too far away from each other. The reversible pattern elongation and shrinkage ability enable the sub-swarms to perform active merging, as shown in Fig. 3.19b. Initially, two independent swarms are generated under low particle concentration ($\gamma = 3$). The two swarms are far apart and do not tend to approach each other for more than one minute. Therefore, it can be considered that they cannot spontaneously merge through hydrodynamic or magnetic attraction. At $t = 5$ s, the long axes of the two swarms are aligned by adjusting the magnetic field direction. The swarms are then elongated along the long axis by adjusting γ from 3 to 5. When the swarms are long enough, their ends touch each other and merge into one single pattern ($t = 17$ s). The merged swarm has an elongated morphology and can be transformed back to the contracted state by adjusting γ back to 3. Finally, the active merging of two independent swarms is realized by using the reversible elongation ability of the ribbon-like swarm.

In addition to active merging, the splitting of the ribbon-like swarm can be achieved by adjusting the magnetic field, as shown in Fig. 3.19c. First, an elongated swarm pattern is obtained after formation ($\gamma = 5$). The oscillating plane of the magnetic field is then adjusted to be perpendicular to the substrate ($t = 1$ s); i.e., the pitch angle is adjusted from 0° to 90°, as shown in the inset of Fig. 3.19c. At this time, the magnetic interaction in the horizontal direction between particle chains inside the swarm is repulsive, which makes the swarm pattern collapse and expand significantly. By adjusting the oscillating plane back to horizontal, multiple small swarm patterns can be obtained, indicating that the active splitting of the ribbon-like swarm is successfully completed.

The splitting process can also be conducted in a controlled manner by combining the oscillating field with the rotating field. When rotating the oscillating magnetic

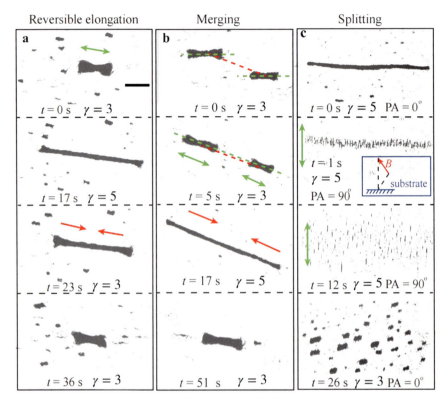

Fig. 3.19 Pattern reconfiguration of the ribbon-like swarm. **a** Reversible elongation and shrinkage of the swarm pattern by tuning the field ratio. The green arrow indicates the direction of pattern elongation, and the red arrow indicates the direction of pattern shrinkage. **b** Active merging of two independent ribbon-like swarms. The green dashed line indicates the long axis of the swarm pattern, and the red dashed line indicates the linking between the centers of the swarms. **c** Active splitting of a ribbon-like swarm. PA represents the pitch angle. The scale bar is 600 μm

field in the plane, the swarm pattern will rotate accordingly and be subjected to the fluid drag as an entity. The fluid drag exerted on the swarm is proportional to the angular velocity of rotation and the aspect ratio of the swarm, and the increase in fluid drag during rotation can cause the splitting of the swarm along the short axis direction. Therefore, a single ribbon-like swarm can be split into several sub-swarm by increasing the rotating frequency or field ratio γ. The magnetic and the fluid torque of the split sub-swarms balance each other. The number of split sub-swarms can be controlled by adjusting the rotating frequency of the swarm. Figure 3.20 shows the controllable splitting of a swarm into two to six sub-swarms which are nearly identical. The sub-swarms formed by the ends of the initial swarm are sometimes longer than the sub-swarms formed by the middle part, which is due to the fact that both ends of the swarm contain a higher number of particles. The phase diagram (Fig. 3.21) shows the relationship between the number of split sub-swarms and the

3.3 Multimodal Reconfiguration of the Swarm Pattern

field ratio and rotating frequency of the magnetic field. The higher rotating frequency and larger field ratio generate larger fluid drag and longer swarm patterns, resulting in a larger number of split sub-swarms. A single swarm is able to be stably split into 2, 3, 4, or 5 sub-swarms. When the field ratio and rotating frequency continue to increase, the number of split sub-swarms may have some uncertainty. For example, rotating an oscillating magnetic field with a field ratio of 4 at 0.35 Hz will split the swarm into 5 or 6 sub-swarms.

The aspect ratio of the ribbon-like swarm can be adjusted by tuning the field ratio γ of the magnetic field. The relationship between the aspect ratio α of the swarm pattern and the field ratio γ is shown in Fig. 3.22a. When the field ratio γ is small (< 4), the aspect ratio α of the swarm remains almost unchanged. When the value of γ

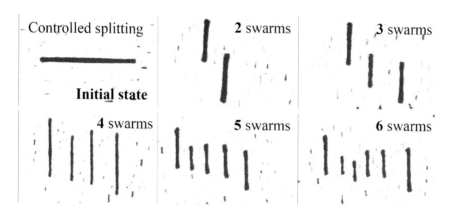

Fig. 3.20 Controlled splitting of the ribbon-like swarm

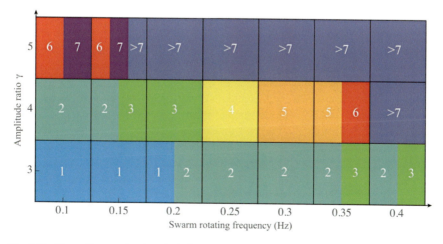

Fig. 3.21 Phase diagram showing the influence of swarm rotating frequency and field ratio on the number of split sub-swarms. The data show the average number of five independent experiments

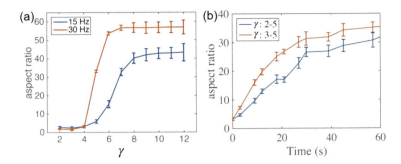

Fig. 3.22 Aspect ratio of the ribbon-like swarm. **a** Relationship between the aspect ratio α of the swarm pattern and the field ratio γ. **b** change in aspect ratio of the swarm during the pattern elongation process. The oscillation frequency is 30 Hz. Each data point represents the average of three trials. The error bar indicates the standard deviation

exceeds 4, the swarm starts to elongate. Then, the aspect ratio of the swarm increases with γ and gradually levels out. The aspect ratio has a maximum value that increases with the oscillating frequency, as shown by the blue and orange curves in Fig. 3.22a. During the elongation process, the change in α with time is shown in Fig. 3.22b. The aspect ratio first increases rapidly and then gradually becomes stable, indicating that the elongation is completed. It takes about one minute to elongate the aspect ratio of the swarm pattern from about 3 to more than 30.

3.3.2 Mechanism of Pattern Reconfiguration

The reversible pattern elongation and shrinkage of the ribbon-like swarm are mainly induced by the changes in the chain-chain magnetic interaction. The magnetic force between particle chains depends on their relative positions, which are influenced by the oscillating angle. Figure 3.23 shows the schematic diagram of two cases of relative positions between chains, that is, the oscillation angle changes from large (θ_L) to small (φ_S) and from small (θ_S) to large (φ_L). The red and blue parts in the figure represent the different magnetizations of the chains. A nanoparticle chain with a small initial oscillating angle is shown in Fig. 3.23a. From stage I to stage II, a single particle chain will split into two, and their center positions remain unchanged. The green line represents the overlap of the same magnetic dipoles of the two particle chains, and its length is $l_O = (L/2)\cos\alpha$, which determines the chain-chain magnetic interaction force. If $l_O < L/4$, the magnetic force is repulsive; if $l_O > L/4$, the magnetic force is attractive. The mechanism is the same for the case in Fig. 3.23b, where α can be expressed as:

$$\alpha = \pi - (\theta + \varphi), \tag{3.19}$$

3.3 Multimodal Reconfiguration of the Swarm Pattern

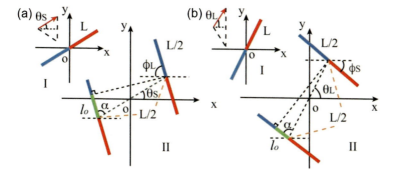

Fig. 3.23 Schematic diagram of two cases of relative positions between chains in the oscillating field. **a** Oscillation angle changes from large (θ_L) to small (φ_S). **b** Oscillation angle changes from small (θ_S) to large (φ_L)

where θ and φ denote the initial and current oscillating angle, respectively. Therefore, when $\theta + \varphi > 120°$, the magnetic force is attractive, and the chains tend to move close to each other; when $\theta + \varphi < 120°$, the magnetic force is repulsive, and the chains tend to move away from each other. Actually, both the attractive and repulsive magnetic interactions are crucial for the reconfiguration of the swarm pattern. The magnetic attraction makes the chains bond more tightly, forming a ribbon-like pattern. Besides, the magnetic repulsion does not always make the chain-chain distance larger but also causes the chains to stagger along their long axis until the overlap of identical dipoles disappears. Then, the chain-chain magnetic interaction turns attractive, which also has a significant influence on the pattern reconfiguration.

The schematic diagram of the mechanism of pattern elongation is shown in Fig. 3.24a. Stage I indicates that the particle chains oscillate with a small oscillating angle θ_S of the magnetic field. Stage II indicates that the magnetic field strength is reduced to the minimum value, and the chains are correspondingly split into shorter ones. At this stage, the central positions of these short chains are considered unchanged due to the small field strength. The oscillating angle is increased to φ_L at this stage, and the particle chains will approach each other under the influence of the attractive magnetic force. The repulsive magnetic force tends to make the chains end to end with each other. Due to the small initial oscillating angle θ_S (γ is small), the chains in stage II have a relatively broad distribution in the x-axis direction. Stage III indicates that after increasing the oscillating angle to φ_L, the chain reaches the position of the maximum oscillating angle. Identical magnetic dipoles in different particle chains overlap with each other, and repulsive magnetic forces are induced, causing chains to stagger along their long axis (i.e., the direction of the magnetic field). Eventually, the chain-chain overlap in the x-axis direction decreases, and the swarm pattern performs elongation along the y-axis direction as an entity (stage IV). In the x-axis direction, the chain-chain distance decreases due to the magnetic attraction, and the width of the swarm decreases accordingly.

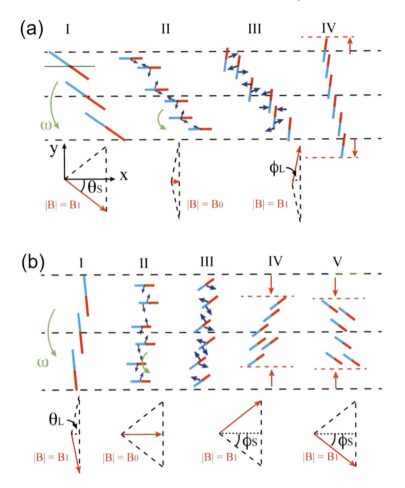

Fig. 3.24 Schematic diagram of the pattern transformation mechanism of the ribbon-like swarm. **a** Pattern elongation mechanism. **b** Pattern shrinkage mechanism. The red and blue parts of the rod indicate the different magnetizations of nanoparticle chains

The mechanism of pattern shrinkage is schematically presented in Fig. 3.24b. The initial oscillating angle (θ_L) of the magnetic field is relatively large (γ is large), and the swarm pattern is in an elongated state (stage I). Similarly, particle chains break into shorter ones when the magnetic field strength decreases, and their positions remain unchanged (stage II). Subsequently, the field oscillating angle becomes smaller (φ_S). In stage III, the chains reach the full oscillation amplitude, and the overlap between them is reduced by the repulsive magnetic force, which is similar to that in the elongation process. However, due to the smaller oscillating angle (φ_S) and the wider distribution of the chains along the y-axis, movements of the chains cause the swarm to contract along the y-axis and expand along the x-axis (stage IV). Eventually, the swarm performs pattern shrinkage along its long axis, and the aspect ratio decreases.

3.4 Navigated Locomotion

3.4.1 Two Locomotion Modes of the Ribbon-Like Swarm

The drag coefficient of an object will increase near a solid boundary (Sing et al. 2010). According to this principle, the motion symmetry of nanoparticle chains in the swarm can be broken by applying a pitch angle to the oscillating magnetic field. Therefore, the ribbon-like swarm is capable of performing locomotion in any direction on a two-dimensional plane as an entity, which is similar to the vortex-like swarm introduced in Chap. 2. Depending on how the pitch angle is applied to the oscillating field, the locomotion of the ribbon-like swarm is divided into two modes: longitudinal motion (mode I) and lateral motion (mode II), as shown in Fig. 3.25. The longitudinal motion refers to the movement of the swarm along the direction of its long axis (Fig. 3.25a). In this case, the trajectory of the magnetic field oscillation (the blue dashed line in Fig. 3.25a) is inclined to the substrate (the x–y plane); i.e., the alternating component B_A (Fig. 3.1) of the oscillating field is inclined to the substrate while the constant component B_C remains horizontal. When B_A is parallel to the substrate and B_C is inclined to the substrate, the oscillating trajectory of the magnetic field is also parallel to the x–y-plane. In this case, the swarm will move laterally along its short axis, as shown in Fig. 3.25b.

Take mode I (longitudinal motion) as an example to analyze the moving principle of the swarm, as shown in Fig. 3.26. The red rod in the figure represents the particle chain PQ, whose oscillation center is denoted by point O. When the pitch angle is added to the magnetic field, the oscillation plane of the chain is inclined to the substrate. The part of the chain close to the substrate experiences a larger drag force, causing the oscillation center to be closer to the bottom (denoted by $O1$, stage T1). Then, the chain rotates around $O1$. Phase T2 indicates that the chain is rotated to be parallel to the substrate. At this moment, the drag force on both ends of the chain is identical, and thus, the oscillation center is shifted to the geometric center

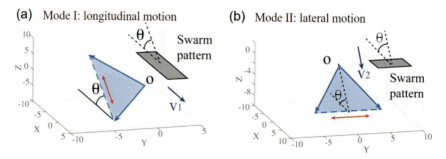

Fig. 3.25 Schematic diagram of two locomotion modes of the ribbon-like swarm. **a** Longitudinal motion (mode I) of the swarm. **b** Lateral motion (mode II) of the swarm. θ indicates the pitch angle. The blue area represents the oscillating plane

of the chain. The chain continues to rotate with the magnetic field until the end P touches the substrate (stage T3). The inset in Fig. 3.26 shows the force exerted on the end P at this time. Generally, the frictional force exerted by the substrate on the chain cannot keep the chain from slipping. Therefore, the horizontal component of the force (F_{torque}^P) will drive the chain to move forward. From stage T3 to T4, the oscillation plane of the chain also moves from plane A to plane B due to the net displacement. Particle chains in the swarm keep repeating this behavior under the actuation of the oscillating field, leading to the directional locomotion of the swarm along the long axis. In addition, under an appropriate pitch angle, the interaction force between particle chains can keep their relative positions stable, so even if the moving distance of the chains is inconsistent, the swarm can maintain a stable pattern during locomotion. The principle for mode II (lateral motion) is similar, except that the net displacement produced by the chain is along the short axis of the swarm. The above analysis theoretically explains how the ribbon-like swarm performs locomotion under an oscillating magnetic field with a pitch angle.

The magnitude of the pitch angle determines the moving velocity of the swarm and also has an influence on the pattern stability. Figure 3.27 shows the pattern of the ribbon-like swarm moving in two motion modes at different pitch angles. The green dashed rectangle represents the initial position of the swarm, and the red dashed rectangle represents the position where the swarm reaches. When the pitch angle is small (e.g., 8°), the swarms in both motion modes maintain pattern stability and integrity, indicating that the particle chains perform synchronized motion and the swarm moves as an entity. However, larger pitch angles can lead to unstable swarm patterns. For example, when the swarm is moving in mode I at a pitch angle of 20°, it will split into multiple sub-swarms with their long axes parallel to the moving direction. Similarly, actuating the swarm at a pitch angle of 30° in mode II

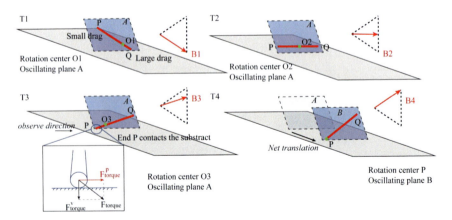

Fig. 3.26 Schematic diagram showing the mechanism of locomotion (mode I) of the ribbon-like swarm. The red rods indicate the particle chain, the blue area represents the oscillating plane, and the gray area represents the substrate. The magnetic field corresponding to each stage is shown on the right. The inset shows the force exerted on the end P when contacting the substrate

3.4 Navigated Locomotion

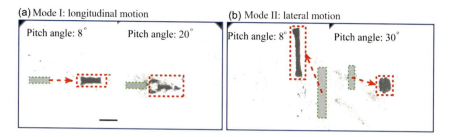

Fig. 3.27 Swarm morphologies during locomotion under different pitch angles. **a** Locomotion of the swarm in mode I. **b** Locomotion of the swarm in mode II. γ is 3, the field strength is 10 mT, and the oscillating frequency is 30 Hz. The green dashed rectangles represent the initial positions of the swarm, and the red dashed arrows indicate the moving direction. The scale bar is 800 μm

causes the swarm to split into sub-swarms whose long axes are perpendicular to the moving direction. A large pitch angle causes a larger net displacement of the particle chain per oscillation cycle, leading to a higher moving velocity of the swarm, but it also results in the change of relative position between particle chains of different lengths. If inner interactions of the swarm fail are not sufficient to maintain the relative positions between particle chains, the swarm will split. Therefore, a small pitch angle (generally less than 10°) is generally used to ensure good stability and controllability of the swarm during locomotion, which is crucial for the application of microswarms in complex environments.

The relationship between moving velocities of the ribbon-like swarm in both motion modes and the pitch angle is shown in Fig. 3.28. The moving velocity first increases with the pitch angle and then gradually decreases after the pitch angle exceeds a certain value. For example, the swarm moving in mode I (oscillating at 30 Hz) has a velocity that reaches a maximum at a pitch angle of 10° (Fig. 3.28a). It is worth noting that the maximum velocity that the ribbon-like swarm can achieve is affected by the field ratio γ, that is, increasing γ will reduce the maximum velocity, which may be caused by the reduction of the magnetic field strength. Furthermore, the oscillation frequency of the magnetic field has no significant effect on the moving velocity. A similar principle exists in swarms moving in mode II, but in this case, the drop in velocity after reaching a maximum is much more significant. The velocity fluctuations of swarms moving in mode II are greater than those in mode I. Therefore, mode I has advantages over mode II in terms of motion control and velocity optimization of the ribbon-like swarm.

3.4.2 Pattern Stability and Controlled Locomotion

The ribbon-like swarm can move as an entity at small pitch angles, which is very suitable for navigation in small channels. Therefore, the possible collision between the swarm and surrounding boundaries may influence the pattern stability, which

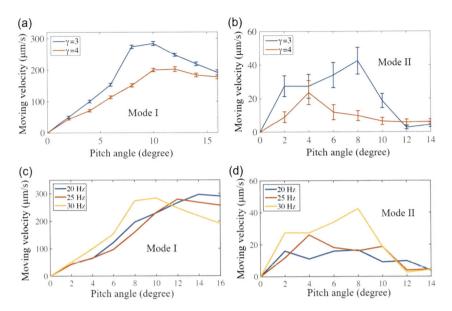

Fig. 3.28 Relationship between moving velocities of the ribbon-like swarm and the pitch angle. **a** Change in velocity of longitudinal motion (mode I) with pitch angles under different field ratio γ. **b** Change in velocity of lateral motion (mode II) with pitch angles under different field ratio γ. The oscillating frequency in **a** and **b** is 30 Hz. **c** Change in velocity of longitudinal motion (mode I) with pitch angles under different oscillating frequencies. **d** Change in velocity of lateral motion (mode II) with pitch angles under different oscillating frequencies. The field ratio in **c** and **d** is 3

needs to be considered. Figure 3.29 shows the stability of the ribbon-like swarm when it encounters various boundary conditions during locomotion. First, the swarm is directly driven toward a cylindrical obstacle (curved boundary) and a prismatic obstacle (planar boundary), as shown in Fig. 3.29a, b, respectively. When the swarm approaches the obstacle, it continues to move forward under the actuation of the external magnetic field (as shown by the red dotted arrow). Although the movement of the swarm is hindered by the obstacle, the swarm morphology remains intact. By changing the moving direction after a period of time, the swarm can be guided successfully away from the obstacle. The swarm pattern remains stable during the whole process, with only a small number of particles lost. Figure 3.29c shows that the swarm maintains pattern stability when moving toward a sharp boundary. When the ribbon-like swarm moves at an angle toward a solid boundary, it slips along the boundary (Fig. 3.29d). The moving direction without the boundary is indicated by the red dashed arrow, and the green arrow indicates the actual moving direction of the swarm under the influence of the boundary. Notably, the swarm is not in direct contact with the solid boundary due to the fluidic repulsion around the pattern. The above experiments show that ribbon-like swarms have excellent pattern stability near various types of boundaries, demonstrating great potential for applications in complex environments.

3.5 Preliminary Application Demonstration

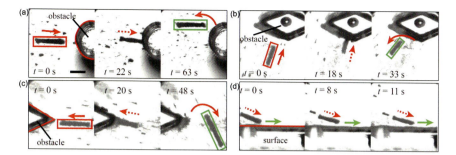

Fig. 3.29 Pattern stability of the ribbon-like swarm when contacting different boundaries during locomotion. **a** Curved boundary. **b** Planar boundary. **c** Sharp boundary. **d** Swarm moves along the direction tilted with a planar boundary. The scale bar is 800 μm

In addition to the directional locomotion, the multimodal reconfiguration property also enhances the ability of ribbon-like swarms to perform controlled navigation in complex environments. Figure 3.30a shows that a ribbon-like swarm reaches multiple target locations through a semicircular channel by using navigated locomotion and reversible pattern reconfiguration. The blue rectangles below the experimental images present the full view of the channel and the real-time location of the swarm (green rod). First, the swarm is navigated into the semicircular channel and then passes through it by adjusting moving direction. The swarm remains stable during this process, and most of the nanoparticles are trapped and pass through the channel with a high access rate. At $t = 113$ s, the field ratio is increased from 3.5 to 6 to induce pattern elongation, and the aspect ratio of the swarm is correspondingly increased from ~ 6 to ~ 28. By adjusting the magnetic field, the elongated swarm is split into multiple sub-swarms, and then they are successfully driven into three branch channels. Figure 3.30b shows that a ribbon-like swarm passes through a channel with two branches and then remerges. In the beginning, the swarm is wider than the channel and thus may get stuck and cannot move forward. The magnetic field is rotated at a frequency of 0.35 Hz to split the swarm into two sub-swarms with their long axis parallel to the channel. These two sub-swarms are then navigated through two channels to reach the target location, respectively. During movement, since the slide boundaries are not flat, the high stability of the ribbon-like swarm ensures that the swarm pattern does not collapse upon contact with the sidewalls. After passing through the narrow channels, the two sub-swarms remerge into a single pattern by adjusting the orientation and field ratio of the oscillating field.

3.5 Preliminary Application Demonstration

The ribbon-like swarm is able to reversibly transform morphology, perform controllable multimodal locomotion, maintain pattern stability, and generate strong fluid low, which can serve as a versatile tool for microscale applications. For example, the

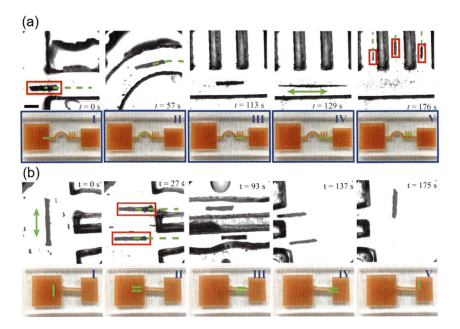

Fig. 3.30 Demonstration of the ribbon-like swarm navigating in confined channels. **a** Swarm passes through a semicircular channel and then enters three branched channels. **b** Swarm passes through a channel with two branches and then remerges into a single pattern. The swarm pattern is highlighted by the red rectangle. The figures below experimental images present the full view of the channel, and the green rods indicate real-time locations of the swarm. The scale bar is 800 μm

manipulation of tiny objects can be achieved by exploiting the motion controllability of the ribbon-like swarm. Figure 3.31 shows the non-contact manipulation of four polystyrene microbeads (approximately 200 μm in diameter) using the ribbon-like swarm as a robotic end effector. These microbeads are randomly distributed on the substrate at the beginning ($t = 0$ s), as indicated by the red circles in the figure. The swarm is navigated toward the microbead and then pushes it with the fluid flow generated by the tip after approaching. After pushing the microbead to the target position, the swarm is separated from the microbead by changing the moving direction ($t = 29$ s). Finally, through the sequential manipulation of the swarm, the four microbeads are lined up in a straight line ($t = 140$ s).

The induced flow field at the tip of the swarm is shown in Fig. 3.32. It is worth noting that particles smaller than 40 μm in diameter may be drawn into the swarm rather than being pushed by it. Thus, the ribbon-like swarm has a pumping effect, that is, tiny particles are sucked in from the tip and thrown away from other parts of the swarm pattern. Based on this property, the ribbon-like swarm can be used as a tool for sorting non-magnetic particles of different sizes. As shown in Fig. 3.33, there are three 70 μm-diameter beads and hundreds of 15 μm-diameter beads in a microchannel ($t = 0$ s). When actuating the swarm moving in the microchannel, 70 μm-diameter beads (indicated by red circles) are captured and pushed to move by the swarm.

3.5 Preliminary Application Demonstration

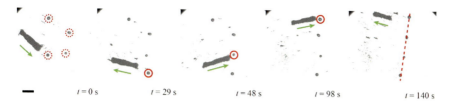

Fig. 3.31 Non-contact manipulation of four polystyrene microbeads using the ribbon-like swarm. The microbeads are highlighted by red circles. The green arrows indicate the moving direction of the swarm. The scale bar is 800 μm

Beads with a diameter of 15 μm are either pushed aside by the swarm-induced flow or are sucked into the swarm tip and thrown from the sides (as indicated by the green arrows). Finally, the swarm passes through the microchannel and selectively sorts the three larger beads from the numerous small beads ($t = 150$ s).

Besides, based on the fluid flow induced by the oscillation and reconfiguration of particle chains at the swarm tip, the ribbon-like swarm can also serve as a tool for manipulating the fluid in the microfluidic device. Figure 3.34 shows the antidiffusion ability of the ribbon-like swarm. In a microchannel system, the left chamber is filled with blue dye, and the nanoparticle chains are located in the right chamber ($t = 0$ s). After the ribbon-like swarm is generated under an oscillating magnetic field, the swarm is then navigated into the microchannel connecting the two chambers. The fluid flow induced by the swarm tip hinders the left-to-right diffusion of the dye ($t = 40$ s). After maintaining the swarm in the channel for approximately 140 s, the dye

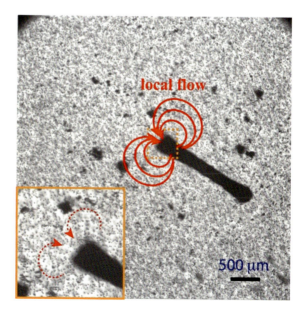

Fig. 3.32 Induced local flow at the tip of the ribbon-like swarm, which is labeled by tracer particles with an average diameter of 100 μm. The scale bar is 500 μm

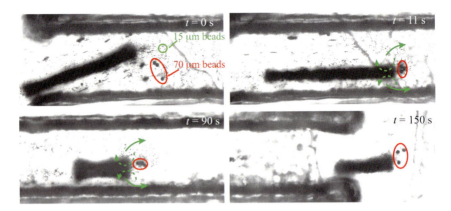

Fig. 3.33 Sorting of non-magnetic beads of different sizes in a microchannel using the ribbon-like swarm. The red and green circles indicate the 70 μm-diameter beads and 15 μm-diameter beads, respectively. The green arrows indicate that particles are sucked in and thrown out from the swarm pattern

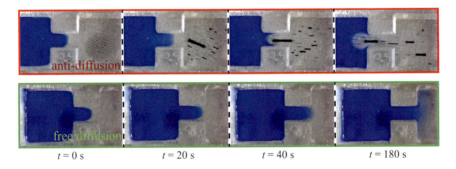

Fig. 3.34 Antidiffusion in a microchannel using the ribbon-like microswarm

remains confined within the left chamber. In the control group (Fig. 3.34, bottom), the dye diffuses spontaneously from left to right without the influence of the swarm.

3.6 Conclusion

This chapter introduces a ribbon-like paramagnetic nanoparticles swarm actuated by an oscillating magnetic field. The ribbon-like swarm has reversible elongation and shrinkage ability, and the aspect ratio of the swarm pattern can vary significantly by more than an order of magnitude. This pattern transformation behavior is very similar to some swarms in nature, such as the swarm of *B. paradoxa*. The pattern formation and transformation of the ribbon-like swarm mainly depend on the interactions between nanoparticle chains in the swarm, and our related in-depth analysis may pave the way for understanding the transformable morphology of living systems in nature. Based on the pattern transformation ability, the ribbon-like swarm is able to perform reversible merging and splitting, and the number of split sub-swarms can be controlled by adjusting the field parameters, which further improves controllability and flexibility. By adding a pitch angle to the oscillating magnetic field, the directional locomotion of the ribbon-like swarm can be achieved. The behavior of nanoparticle chains under the tilted oscillating field is analyzed in detail, providing a clear explanation for the moving mechanism of the swarm. The ribbon-like swarm has two motion modes, namely longitudinal motion along the long axis and lateral motion along the short axis, which facilitates motion control and path planning. Besides, the swarm pattern remains highly stable during locomotion; thus, the particle loss is less, and the access rate is high. In addition, the ribbon-like swarm can also maintain pattern stability when contacting different types of boundaries, demonstrating great potential for applications in complex environments. Various applications of the ribbon-like swarm are demonstrated, such as navigation through complex microchannels, manipulation of tiny objects, sorting particles of different sizes, and antidiffusion in microfluidic channels. The ribbon-like swarm can have more functions by functionalizing building blocks of the swarm or replacing them with other types of magnetic nanoparticles, which will be described in subsequent chapters. Thanks to good environmental adaptability, motion controllability, and pattern stability, the ribbon-like swarm is expected to be used in the biomedical field in the future as a targeted drug delivery tool in complex environments (such as blood vessels, etc.) in the human body.

References

Anderson C, Theraulaz G, Deneubourg J-L (2002) Self-assemblages in insect societies. Insectes Soc 49(2):99–110

Belkin M, Snezhko A, Aranson I, Kwok W-K (2007) Driven magnetic particles on a fluid surface: pattern assisted surface flows. Phys Rev Lett 99(15):158301

Cai J, Chen M, Wang Y, Pan J, Li A, Zhang D (2013) Culture and motion analysis of diatom *Bacillaria paradoxa* on a microfluidic platform. Curr Microbiol 67(6):652–658

Camassa R, Leiterman TJ, McLaughlin RM (2008) Trajectory and flow properties for a rod spinning in a viscous fluid. Part 1. An exact solution. J Fluid Mech 612:153–200

Chen Q, Bae SC, Granick S (2011) Directed self-assembly of a colloidal Kagome lattice. Nature 469(7330):381–384

Felfoul O, Mohammadi M, Taherkhani S, De Lanauze D, Zhong Xu Y, Loghin D, Essa S, Jancik S, Houle D, Lafleur M (2016) Magneto-aerotactic bacteria deliver drug-containing nanoliposomes to tumour hypoxic regions. Nat Nanotechnol 11(11):941–947

Gao Y, Beerens J, van Reenen A, Hulsen MA, de Jong AM, Prins MW, den Toonder JM (2015) Strong vortical flows generated by the collective motion of magnetic particle chains rotating in a fluid cell. Lab Chip 15(1):351–360

Garattoni L, Birattari M (2018) Autonomous task sequencing in a robot swarm. Sci Robot 3(20):eaat0430

Groß R, Bonani M, Mondada F, Dorigo M (2006) Autonomous self-assembly in swarm-bots. IEEE Trans Rob 22(6):1115–1130

Hemelrijk CK, Hildenbrandt H (2011) Some causes of the variable shape of flocks of birds. PLoS ONE 6(8):e22479

Hemelrijk CK, Hildenbrandt H (2012) Schools of fish and flocks of birds: their shape and internal structure by self-organization. Interface Focus 2(6):726–737

Hemelrijk CK, Hildenbrandt H, Reinders J, Stamhuis EJ (2010) Emergence of oblong school shape: models and empirical data of fish. Ethology 116(11):1099–1112

Jahn R, Schmid A-MM (2007) Revision of the brackish-freshwater diatom genus Bacillaria Gmelin (Bacillariophyta) with the description of a new variety and two new species. Eur J Phycol 42(3):295–312

Kapinga MR, Gordon R (1992) Cell motility rhythms in *Bacillaria paxillifer*. Diatom Res 7(2):221–225

Landau LD, Bell J, Kearsley M, Pitaevskii L, Lifshitz E, Sykes J (2013) Electrodynamics of continuous media, vol 8. Elsevier

Lee S-H, Pak H, Chon T-S (2006) Dynamics of prey-flock escaping behavior in response to predator's attack. J Theor Biol 240(2):250–259

Li S, Batra R, Brown D, Chang H-D, Ranganathan N, Hoberman C, Rus D, Lipson H (2019) Particle robotics based on statistical mechanics of loosely coupled components. Nature 567(7748):361–365

Ma F, Wang S, Wu DT, Wu N (2015) Electric-field-induced assembly and propulsion of chiral colloidal clusters. Proc Natl Acad Sci 112(20):6307–6312

Mao X, Chen Q, Granick S (2013) Entropy favours open colloidal lattices. Nat Mater 12(3):217–222

Martel S, Mohammadi M, Felfoul O, Lu Z, Pouponneau P (2009) Flagellated magnetotactic bacteria as controlled MRI-trackable propulsion and steering systems for medical nanorobots operating in the human microvasculature. Int J Robot Res 28(4):571–582

Mlot NJ, Tovey CA, Hu DL (2011) Fire ants self-assemble into waterproof rafts to survive floods. Proc Natl Acad Sci 108(19):7669–7673

Nagy M, Akos Z, Biro D, Vicsek T (2010) Hierarchical group dynamics in pigeon flocks. Nature 464(7290):890–893

Ozkan-Aydin Y, Goldman DI (2021) Self-reconfigurable multilegged robot swarms collectively accomplish challenging terradynamic tasks. Sci Robot 6(56):eabf1628

Palacci J, Sacanna S, Steinberg AP, Pine DJ, Chaikin PM (2013) Living crystals of light-activated colloidal surfers. Science 339(6122):936–940

Parrish JK, Edelstein-Keshet L (1999) Complexity, pattern, and evolutionary trade-offs in animal aggregation. Science 284(5411):99–101

Peleg O, Peters JM, Salcedo MK, Mahadevan L (2018) Collective mechanical adaptation of honeybee swarms. Nat Phys 14(12):1193–1198

Rubenstein M, Cornejo A, Nagpal R (2014) Programmable self-assembly in a thousand-robot swarm. Science 345(6198):795–799

Schill RA (2003) General relation for the vector magnetic field of a circular current loop: a closer look. IEEE Trans Magn 39(2):961–967

Sing CE, Schmid L, Schneider MF, Franke T, Alexander-Katz A (2010) Controlled surface-induced flows from the motion of self-assembled colloidal walkers. Proc Natl Acad Sci USA 107(2):535–540

Singh H, Laibinis PE, Hatton TA (2005) Rigid, superparamagnetic chains of permanently linked beads coated with magnetic nanoparticles. Synthesis and rotational dynamics under applied magnetic fields. Langmuir 21(24):11500–11509

Snezhko A, Aranson IS (2011) Magnetic manipulation of self-assembled colloidal asters. Nat Mater 10(9):698–703

Snezhko A, Belkin M, Aranson I, Kwok W-K (2009) Self-assembled magnetic surface swimmers. Phys Rev Lett 102(11):118103

Steager EB, Sakar MS, Magee C, Kennedy M, Cowley A, Kumar V (2013) Automated biomanipulation of single cells using magnetic microrobots. Int J Robot Res 32(3):346–359

References

Storms R, Carere C, Zoratto F, Hemelrijk C (2019) Complex patterns of collective escape in starling flocks under predation. Behav Ecol Sociobiol 73:10

Sumpter DJ (2010) Collective animal behavior. Princeton University Press

Tasci T, Herson P, Neeves K, Marr D (2016) Surface-enabled propulsion and control of colloidal microwheels. Nat Commun 7:10225

Tolley MT, Lipson H (2011) On-line assembly planning for stochastically reconfigurable systems. Int J Robot Res 30(13):1566–1584

Van Reenen A, de Jong AM, den Toonder JM, Prins MW (2014) Integrated lab-on-chip biosensing systems based on magnetic particle actuation–a comprehensive review. Lab Chip 14(12):1966–1986

Ward A, Webster M (2016) Sociality: the behaviour of group-living animals

Weimerskirch H, Martin J, Clerquin Y, Alexandre P, Jiraskova S (2001) Energy saving in flight formation. Nature 413(6857):697–698

Xu TT, Hwang G, Andreff N, Regnier S (2015) Planar path following of 3-D steering scaled-up helical microswimmers. IEEE Trans Rob 31(1):117–127

Yamaoka N, Suetomo Y, Yoshihisa T, Sonobe S (2016) Motion analysis and ultrastructural study of a colonial diatom, *Bacillaria paxillifer*. J Electron Microsc 65(3):211–221

Yan J, Bloom M, Bae SC, Luijten E, Granick S (2012) Linking synchronization to self-assembly using magnetic Janus colloids. Nature 491(7425):578–581

Yu J, Yang L, Zhang L (2018) Pattern generation and motion control of a vortex-like paramagnetic nanoparticle swarm. Int J Robot Res 37(8):912–930

Yu J, Yang L, Du X, Chen H, Xu T, Zhang L (2021) Adaptive pattern and motion control of magnetic microrobotic swarms. IEEE Trans Rob 38(3):1552–1570

Zhang J, Luijten E, Grzybowski BA, Granick S (2017) Active colloids with collective mobility status and research opportunities. Chem Soc Rev 46(18):5551–5569

Zhou Q, Petit T, Choi H, Nelson BJ, Zhang L (2017) Dumbbell fluidic tweezers for dynamical trapping and selective transport of microobjects. Adv Func Mater 27(1):1604571

Chapter 4
Heterogeneous Colloidal Microswarm with Multifunction

Abstract Heterogeneous microswarms consisting of different types of building blocks have richer functionality than homogeneous microswarms. This chapter introduces a heterogeneous colloidal microswarm with hierarchical and cooperative functionalities inspired by swarms with adaptive morphology and functional divisions in nature. The microswarm consists of core–shell structured magnetic nanoparticles, exhibiting a vortex-like or ribbon-like pattern driven by corresponding magnetic fields. The vortex-like pattern is conducive to rapid movement in a tortuous environment, while the ribbon-like pattern enables passing through narrow areas. Transformation between the two patterns can be completed quickly by changing the magnetic field, allowing the microswarm to adapt to complex environments. Three types of building blocks specialized by doxorubicin, glucose oxidase, and manganese ion, respectively, endow the heterogeneous microswarm with three different anticancer functions, i.e., chemotherapy, starvation therapy, and chemodynamic therapy. Different building blocks enable combination cancer therapy through domino reactions by cooperatively destroying the nucleus, reducing the energy supply, and generating toxic radicals.

Keywords Heterogeneous swarm · Domino reaction · Pattern transformation · Adaptability · Cancer therapy

4.1 Introduction

In nature, one of the most striking phenomena is that a variety of natural organisms can self-organize into a coordinated system to accomplish complex tasks that cannot be achieved by individuals. For example, the bird flocks fly in a V-formation to save energy during migration (Weimerskirch et al. 2001); the fish schools swim in a coordinated large group to forage and defend themselves from predators (Pitcher 2012); the insect colonies cooperatively construct structures with their bodies to traverse difficult terrains (Hölldobler and Wilson 1990; Sumpter 2010). Inspired by these swarm behaviors with group-level functionality, researchers create artificial swarm systems (Krieger et al. 2000; Werfel et al. 2014; Rubenstein et al. 2014; Garattoni and Birattari 2018; Li et al. 2019) with the collective abilities emerging from the local interactions between individual components, which are especially important for micro/nanorobotics where building blocks are relatively simple and unsophisticated compared to the large-scale counterparts (Wang and Pumera 2020). Due to the tiny size and active mobility, swarming micro/nanorobots are envisioned to be capable of accessing the hard-to-reach regions inside the human body, thus promising major benefits in revolutionizing biomedical fields (Li et al. 2017; Nelson et al. 2010; Sitti et al. 2015; Gao et al. 2021).

Generally, a group of micro/nanorobots is needed to perform tasks as a swarm system because the individuals only possess limited capabilities (Erkoc et al. 2019). Especially in biomedical fields, microswarm has several advantages which cannot or can hardly be realized by using individual tiny agents. First, the payload on individual micro/nanoscale components is commonly limited. To perform drug/energy delivery with a reasonably higher dose or concentration, using swarm systems with thousands or even millions of agents, which are capable of performing targeted delivery with high precision, are highly required. Second, the swarming pattern as an entity may provide much higher imaging contrast than that formed by the individual agents. Third, a reconfigurable swarm may be able to adjust its morphology to adapt itself to the applied complex environment; hence, a high access rate for targeted therapy might be guaranteed. In this respect, a variety of strategies have been proposed to actuate the active generation and navigation of swarming micro/nanorobots, including chemotaxis (Felfoul et al. 2016; Ji et al. 2019), magnetic field (Servant et al. 2015; Wu et al. 2018; Yan et al. 2017), acoustic field (Wang et al. 2012), and so on (Solovev et al. 2013; Wang et al. 2015; Tang et al. 2020).

Although considerable progress has been made in the development of colloidal swarming systems in recent years, there still exists a substantial gap between the conceptual design and real applications. First, achieving synchronized manipulation of microscale swarm systems with environmental adaptability is quite challenging. This may result in inconsistent spatiotemporal distributions of building blocks when passing through complex environments, which deteriorates the targeting efficiency of microswarms. Besides, to realize their applications, the tiny components in a swarm usually need to be able to perform a wide spectrum of tasks, which, if taking targeted therapy for instance, range from sensing and responding to surrounding environment, to storing and releasing molecules or cells when stimulated by physical cues or other chemicals. However, it is quite challenging to integrate all desired functionalities as well as active propulsion capability in a micro/nanosized colloidal particle. Therefore, developing a synchronized microswarm whose building blocks possess multiple specialized functions and are capable of adapting to diverse environments is highly desired for further applications (Liang et al. 2020).

In this regard, nature provides a variety of inspirations. For example, as shown in Fig. 4.1, during long-term migration, goose team can self-organize into a synchronized formation that is adaptively adjusted according to the dynamic changing of surrounding environments, leading to the improvement of locomotion efficiency. Emulating such collective behaviors in micro/nanorobotic swam systems may

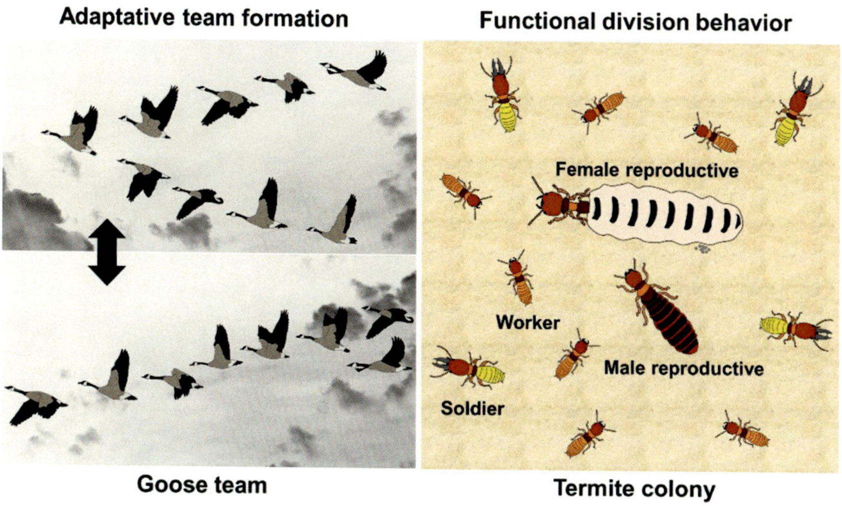

4.1 Introduction

Fig. 4.1 Schematic of the adaptive morphology and functional division behaviors in nature. The bird flocks actively adjust their team formations in response to changing environments (e.g., wind direction and velocity) to migrate over long distances, and the division of labor in insect colonies improves the multitasking efficiency and facilitates their survival

promise major benefits in the enhancement of targeting efficiency. While functional division is also a ubiquitous behavior in natural insect colonies to perform multi-tasks. In the society of termites, specialized castes are created according to the division of labor, among which, a fertile female and male known as queen and king are responsible for reproduction, soldier group defends the nest with highly developed mandibles, and worker caste takes charge of nest construction, foraging, feeding and so on.

In this chapter, inspired from the adaptive morphology and functional division behavior in nature, we present a heterogeneous colloidal microswarm with hierarchical and cooperative functionalities, which is demonstrated by performing targeted drug delivery as a proof-of-concept task (Fig. 4.2). The microswarm consists of millions of building blocks, which are core–shell structured nanoparticles-based drug delivery vehicles capable of magnetic control and on-demand cargo loading and release. Under the wireless actuation of programmed magnetic fields, two kinds of synchronized swarm behaviors are achieved using the system, i.e., vortex-like microswarm under a rotating magnetic field (introduced in Chap. 2) and ribbon-like microswarm under an oscillating magnetic field (introduced in Chap. 3). The vortex-like microswarm has an isotropic circular pattern and is able to move dexterously and rapidly in tortuous environments, while the ribbon-like microswarm has an anisotropic slender pattern and can pass through narrow environments. Through changing the types of applied magnetic fields, microswarm is reconfigured and switched between two formations within seconds, which thus can adapt to highly complex environment. Besides, the cooperative tasking capability is integrated into microswarm by the specialization and cooperation between different building blocks. The anticancer function is divided into three subgroups, i.e., chemotherapy, starvation therapy, and chemodynamic therapy by specializing three groups of building blocks with doxorubicin, glucose oxidase, and manganese ion, respectively, aiming to demonstrate the synergistic effect among microswarms and achieve combination therapy via domino reaction (Tietze 1996; Chen et al. 2020). Therefore, when microswarm is navigated to targeted cancer cells under the guidance of magnetic field, different specialized building blocks can inhibit their growth by cooperatively destroying the nucleus, reducing the energy supply, and generating toxic radicals. Such swarm behavior emulates the division and cooperation of labor in natural social animals, which we hope to serve as a new drug delivery platform with efficient targeting and therapeutic performances for practical applications in the future.

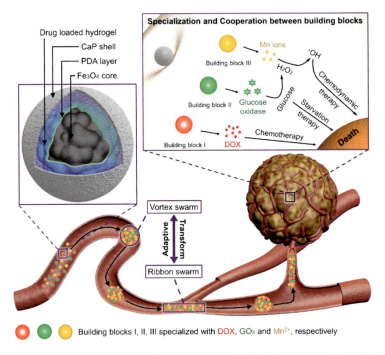

Fig. 4.2 Schematic of the adaptive and heterogeneous colloidal microswarm with multifunction for targeted delivery

4.2 Design and Preparation of Building Blocks as Drug Delivery Vehicles

To achieve the desired functions, the building block of swarm system is designed as core–shell structured, in which, magnetite (Fe$_3$O$_4$) nanoparticle serves as the core, an acrylic acid-based hydrogel is the middle layer to load drugs, and pH-responsive calcium phosphate (CaP) is coated as the outer shell (Fig. 4.3) (Habraken et al. 2016; Rim et al. 2011). Fe$_3$O$_4$ nanoparticle with a spherical shape is first synthesized using the solvothermal method (Deng et al. 2005), followed by an in situ polymerization process of a thin polydopamine (PDA) layer on the surface (Lee et al. 2007). Then the nanoparticle is treated by 3-methacryloxypropyltrimethoxysilane (MPS) through silanization reaction to immobilize carbon–carbon double bonds on the nanoparticle (Fe$_3$O$_4$@PDA-MPS), which provides crosslinking sites for the next-step hydrogel coating. Subsequently, using methacrylic acid (MAAc) as the monomer, a hydrogel layer is anchored via precipitation polymerization process, thus obtaining the nanoparticle (Fe$_3$O$_4$@PDA@PMAAc) for on-demand drug encapsulation (Li et al. 2013). At last, a biocompatible calcium phosphate (CaP) shell is further coated using biomineralization method to obtain the building block (Fe$_3$O$_4$@PDA@PMAAc@CaP) (Qi et al. 2018). The shell shows pH-responsive degradability, which is stable in neutral solution but dissociates under weakly acidic conditions, and therefore can be used to prevent premature leakage and realize on-demand drug release.

The transmission electron microscope (TEM) images are first provided in Fig. 4.4, indicating the morphology of nanoparticles during each preparation stage. It can be found that the pristine Fe$_3$O$_4$ nanoparticle is a cluster of smaller particles and possesses a relatively rough surface. After performing PDA and MPS functionalization, a layer as thin as ~6 nm is uniformly formed on the surface. This step facilitates the immobilization of carbon–carbon double bonds and paves the foundation for precipitation polymerization, which is achieved by immersing nanoparticles, MAAc (monomer), N,N'-methylenebisacrylamide (MBA, crosslinker) and azobisisobutyronitrile (AIBN, thermoinitiator) into acetonitrile, followed by reflux heating to initiate the polymerization reaction between MAAc and MBA. In the TEM image of corresponding Fe$_3$O$_4$@PDA@PMAAc nanoparticle, a polymer layer with a thickness of ~26 nm is found to entirely coat the magnetic nanoparticle core. The last TEM image shows the morphology of CaP shell on the synthesized building block (Fe$_3$O$_4$@PDA@PMAAc@CaP), which is achieved by the electrostatic absorption of Ca^{2+} ions in PMAAc and subsequent biomineralization deposition via the gradual addition of HPO$_4^{2-}$. To characterize the structure of building block more clearly, energy-dispersive X-ray (EDX) analysis is conducted to verify its elemental distribution as shown in Fig. 4.4b, suggesting the existence of Fe element in the central part, and CaP shell on the surface of the nanoparticle.

The surface functional groups of building blocks during the fabrication process are tested by Fourier-transform infrared (FT-IR) analysis (Fig. 4.5a). The peaks at 588, 1650, and 3450 cm^{-1} indicate Fe–O stretching vibration, absorbed water, and

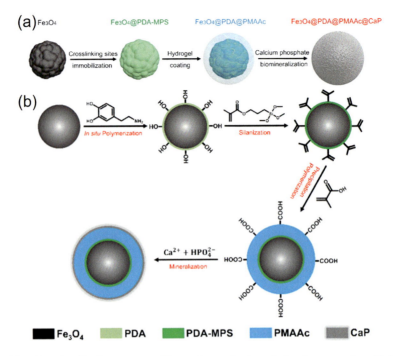

Fig. 4.3 **a** Functionalization process and **b** reaction mechanisms during the preparation of building blocks

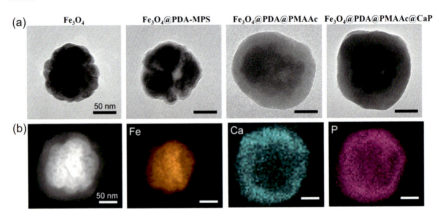

Fig. 4.4 **a** TEM images of the functionalized nanoparticle at each fabrication stage. **b** EDX analysis of the building block, indicating the distribution of Fe, Ca, and P elements

4.3 Magnetic Control of the Synchronized and Adaptive Microswarm

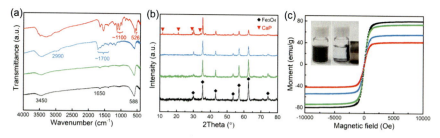

Fig. 4.5 **a** FT-IR spectra, **b** XRD patterns, and **c** magnetic hysteresis loops of the nanoparticles during fabrication process. The black, green, blue, and red curves indicate Fe_3O_4, Fe_3O_4@PDA-MPS, Fe_3O_4@PDA@PMAAc, and Fe_3O_4@PDA@PMAAc@CaP (building block), respectively

hydroxyl groups, respectively, demonstrating the successful fabrication of iron oxide nanoparticles. While the peaks at ~1700 and 2990 cm^{-1}, which correspond to the stretching vibration of C = O and C-H groups, verify the successful encapsulation of hydrogel layer. There exist new absorption peaks of P-O group at 526 and ~1100 cm^{-1} for building blocks, demonstrating the coating of phosphate composites after biomineralization deposition. Besides, X-ray powder diffraction (XRD) patterns are provided in Fig. 4.5b to characterize the crystal structure of nanoparticles at each fabrication stage. The phase of iron oxide Fe_3O_4 exists in all samples, yet no obvious difference is found between Fe_3O_4, Fe_3O_4@PDA-MPS, and Fe_3O_4@PDA@PMAAc nanoparticles due to the amorphous nature of PDA and PMAAc. In contrast, the building blocks exhibit feature peaks corresponding to $CaHPO_4$, which is consistent with the deposition of CaP shell. Moreover, a vibrating sample magnetometer (VSM) is conducted to characterize the magnetic properties of nanoparticles (Fig. 4.5c). The hysteresis loops reveal paramagnetic property of nanoparticles, while the saturation magnetization values decrease from 80, 74, 55 to 42 emu/g as fabrication process goes on, as the proportion of magnetic material gradually decreases. In the inset, it can be found that the prepared building blocks exhibit a strong response to a permanent magnet, facilitating the next-step magnetic control.

4.3 Magnetic Control of the Synchronized and Adaptive Microswarm

Two kinds of synchronized swarm patterns, including vortex-like and ribbon-like patterns, are achieved using the same colloidal system by programing external magnetic fields (Yu et al. 2018a, b), including rotating magnetic field ($\boldsymbol{B_R}$) and oscillating magnetic field ($\boldsymbol{B_O}$) depicted as follows:

$$\boldsymbol{B_R} = A_R\left[\sin(2\pi ft)\boldsymbol{B_x} + \cos(2\pi ft)\boldsymbol{B_y}\right], \qquad (4.1)$$

where A_R is the amplitude of magnetic field, f is the alternating frequency, and t is the time. $\boldsymbol{B_x}$ and $\boldsymbol{B_y}$ correspond to the unit vectors of magnetic field along x- and y-axes, respectively.

$$\boldsymbol{B_o} = [A_o\sin(2\pi f t)\cos\theta + C_o\sin\theta]\boldsymbol{B_x} + [A_o\sin(2\pi f t)\sin\theta + C_o\cos\theta]\boldsymbol{B_y}, \quad (4.2)$$

where A_O and C_O are the amplitude of the alternating and constant components, respectively. θ is the angle between constant component and y-axis.

One more magnetic field is applied to disassemble the swarm patterns of colloidal system and makes the building blocks distributed uniformly. The corresponding disassembly magnetic field ($\boldsymbol{B_D}$) is composed of an oscillating component along z-axis and a rotating component on x–y plane as follows:

$$\boldsymbol{B_D} = A_{xy}[\sin(2\pi f_{xy}t)\boldsymbol{B_x} + \cos(2\pi f_{xy}t)\boldsymbol{B_y}] + A_z\sin(2\pi f_z t)\boldsymbol{B_z}, \quad (4.3)$$

where A_{xy} and A_z are the amplitude of the rotating and oscillating components, respectively. f_{xy} is rotating frequency on x–y plane, f_z is oscillating frequency along z-axis, and $\boldsymbol{B_z}$ is the unit vector of magnetic field along z-axis (Wang et al. 2020; Yu et al. 2017). The mechanisms of different swarm behaviors and corresponding external magnetic fields are shown in Fig. 4.6 with the experimental results. Through switching the types of imposed magnetic fields (rotating, oscillating, and disassembly magnetic fields), three swarm patterns (i.e., vortex, ribbon, and dispersed state) are reconfigured between each other rapidly using a same colloidal system.

Active and synchronized navigation is actuated by lifting the alternating plane of magnetic fields after generating stable vortex- and ribbon-like patterns, and the locomotion velocity can be actively controlled by lifting the pitch angle. The effect of pitch angle on the translational speed of microswarm in blood plasma-simulated

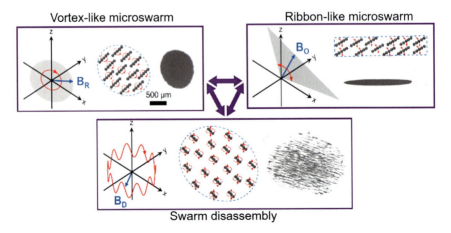

Fig. 4.6 Three kinds of transformable swarm behaviors, including vortex-like microswarm, ribbon-like swarm, and swarm disassembly actuated by magnetic fields $\boldsymbol{B_R}$, $\boldsymbol{B_O}$, and $\boldsymbol{B_D}$, respectively

4.3 Magnetic Control of the Synchronized and Adaptive Microswarm

fluid is shown in Fig. 4.7a. When pitch angle is set as 1°, the speed of vortex-like microswarm reaches ~20.6 μm/s, but the ribbon-like microswarm only shows a speed of ~6.2 μm/s. If pitch angle is increased to 5°, a speed of ~256.1 μm/s is realized by vortex-like microswarm, approximately 5.6 times higher than that of the ribbon-like microswarm (~38.55 μm/s). Further increasing the pitch angle would destroy the pattern of the ribbon-like microswarm, while the vortex-like microswarm still takes effect with a speed as high as 333.5 μm/s, indicating the relatively stronger locomotion capability of the vortex-like microswarm compared to the ribbon-like counterpart. Figure 4.7b demonstrates that the ribbon-like microswarm can reach the target location faster by transforming into the vortex-like microswarm during locomotion. Furthermore, by fully taking advantage of the specific pattern and controlled mobility, the colloidal system with vortex- and ribbon-like bimodal formations can adapt to highly complex environments by active transformation. As shown in Fig. 4.7c, for the identical group of building blocks, a vortex-like microswarm with an isotropic pattern can easily turn direction when encountering a sharp corner. In contrast, the ribbon-like microswarm needs to turn its body direction at first, which is hard to achieve in a complex branched environment. However, the ribbon-like microswarm with a slender and stable pattern can pass through narrow channels impossible for the vortex-like microswarm (Fig. 4.8d). Therefore, through actively adjusting the formations of microswarm, the colloidal system can traverse terrains with various configurations, thus holding excellent adaptability to complex surrounding environments.

To further demonstrate the adaptability of the colloidal system, a microfluidic chip with various branched and confined channels is used to conduct magnetic control experiments. As shown in Fig. 4.8, in region I with a tortuous path, the vortex-like swarm is generated and navigated by a rotating magnetic field until entering region II with an inaccessible narrowed channel. Then the external magnetic field is switched to the oscillating magnetic field, and the transformation from vortex-like to ribbon-like swarm occurs. After forming the stable slender pattern within 1 s, a pitch angle is applied to drive the locomotion of the swarm again, resulting in smooth navigation. Subsequently, the microswarm is recovered to a vortex-like formation again and then moves along the next branch because of the robust and rapid locomotion capability. In this manner, the colloidal system finally arrives at the targeted position via active transformation, followed by the employment of a disassembly magnetic field, aiming to enhance the interaction between building blocks and surrounding environments. It can be found that the majority of building blocks access this position, paving the foundation for targeted delivery in complex environments.

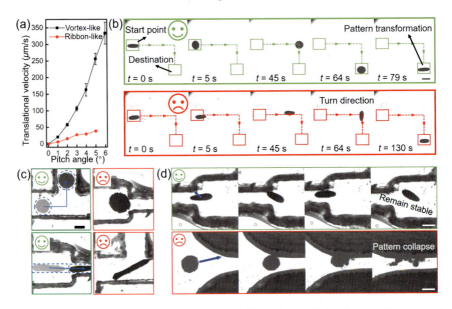

Fig. 4.7 a Effect of the pitch angle of magnetic fields on the translational speed of vortex- and ribbon-like microswarms. The error bars represent the standard deviation ($n = 3$). **b** The ribbon-like microswarm moves to the targeted position along the predetermined trajectory directly (bottom) or by transforming into a vortex-like microswarm (top). **c, d** Vortex-like microswarm capable of moving in tortuous channel and ribbon-like microswarm capable of passing through confined channel. The scale bars are 500 μm

4.4 Domino Reaction Encoded Cooperative Function of Heterogeneous Microswarm

The capability to perform the cooperative function in microswarm is demonstrated by taking targeted drug delivery as a proof-of-concept task. Doxorubicin (DOX) is used as a small molecular anticancer model drug to evaluate the loading and release capability of building blocks. Through mixing with PMAAc-coated nanoparticles in CaCl$_2$ solution, DOX is loaded inside the building blocks, thus obtaining the drug delivery nanovehicles (denoted as NV-DOX) with a DOX content of ~4.2%. Figure 4.9a depicts the scheme of the building blocks for pH-responsive drug release. In the neutral environment, the formation of a CaP shell prevents the undesired drug leakage, which would collapse in the acidic condition and thus induce drug release. Phosphate buffer solutions (PBS) with the pH of 7.2 and 5.5 are utilized to simulate the neutral physiological condition and acidic tumor cellular environment (Mura et al. 2013), respectively. As shown in Fig. 4.9b, the nanoparticles without CaP shell exhibit an obvious DOX leakage of ~38.7% within 12 h at pH 7.2. When CaP is deposited, this value decreases to ~10.5%, suggesting the gate effect of the CaP shell. In comparison, if in a weakly acidic condition with a pH value of ~5.5, a notable DOX amount calculated as ~77.9% is released from the nanovehicles, which can also be

4.4 Domino Reaction Encoded Cooperative Function of Heterogeneous ...

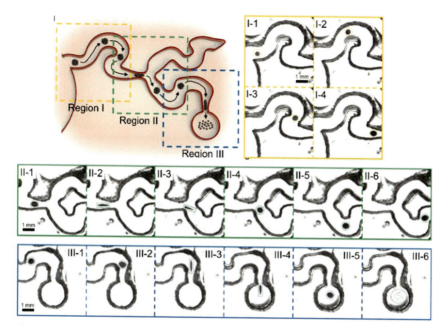

Fig. 4.8 Demonstration of synchronized and adaptive microswarm in complex environment through magnetic controlled active transformation

judged by the color change of the medium (Fig. 4.9c). The discharged profile of DOX at different pH values after 12 h is provided in Fig. 4.9d. Besides, it can be found that without a CaP shell, the nanoparticles also exhibit a pH-dependent release profile, mainly due to the existence of abundant pendant carboxylic groups in the hydrogel matrix (Qiu and Park 2001). The hydrodynamic diameter of Fe_3O_4@PDA@PMAAc nanoparticles at different pH values is shown in Fig. 4.9e, which clearly shows the size shrinking of nanoparticles under acidic conditions. Therefore, the PMAAc hydrogel layer of the building blocks can be regarded as a pH-responsive reservoir for on-demand cargo release in an acidic tumor environment.

Except for DOX, a variety of functional drugs can be encapsulated into building blocks with ease, facilitating the design and customization of specialized building blocks. So, a domino reaction encoded cooperative function is performed by employing glucose oxidase (GO_X) and manganese ions (Mn^{2+}) as drugs, which are expected to perform tasks as shown in Fig. 4.10. After internalized by cancer cells, the decrease of environmental pH to ~4–6 (Mura et al. 2013) would trigger the dissolution of CaP shell and subsequent release of GO_X and Mn^{2+}. GO_X can catalyze the decomposition of glucose to generate acidic glucono-1,5-lactone and H_2O_2, thus cutting off the energy supply of cancer cells (Fu et al. 2018). Besides, on the one hand, the corresponding acidic product can promote CaP shell to break and facilitate drug release in turn. On the other hand, in the existence of Mn^{2+}, another product H_2O_2 can be further converted to highly toxic hydroxyl radicals ·OH via a

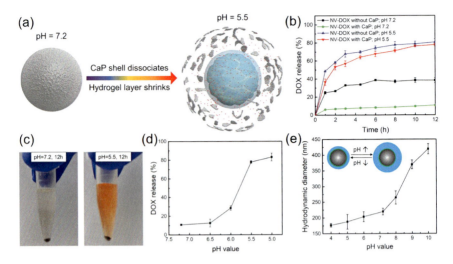

Fig. 4.9 a Schematic of the pH-responsive building block for drug loading and release. b The temporal release profiles of DOX from different nanovehicles. c Nanoparticle solution after 12 h immersion of NV-DOX in PBS with pH 7.2 and 5.5, respectively. d DOX release profiles from NV-DOX at different pH values after 12 h. e Hydrodynamic diameter of Fe_3O_4@PDA@PMAAc nanoparticles as a function of pH value. The error bars represent the standard deviation ($n = 3$)

Fenton-like reaction (Tang et al. 2019; Ranji-Burachaloo et al. 2018), thus achieving a domino reaction-based combination therapy.

The actual performance is then evaluated in Fig. 4.11, in which the GO_X loading in nanovehicles (NV-GO_X) is presented by measuring the change of environmental pH in Fig. 4.11a. As the GO_X-catalyzed reaction would produce acidic glucono-1,5-lactone, the variation of pH can be used to justify the release behaviors of GO_X from nanovehicles. After adding NV-GO_X into glucose solution at 37 °C for 12 h, it is found that the pH decreases from 5.5 to ~2.8, while the incubation with pure glucose solution or pure NV-GO_X solution at an initial pH of 7.2 does not possess any obvious pH change, indicating the effective loading and pH-responsive release

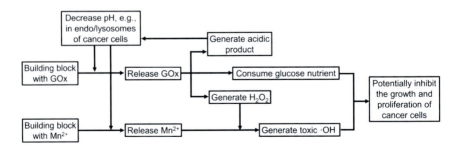

Fig. 4.10 Domino reaction encoded cooperative functions between two subgroups of building blocks specialized with GOX and Mn^{2+}, respectively

4.5 Adaptive and Heterogeneous Colloidal Swarm for Anticancer Targeted ...

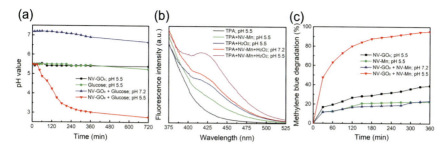

Fig. 4.11 **a** Change in glucose solution pH during incubation with NV-GOX. **b** Fluorescence of TPA solution after incubation with NV-Mn. **c** Degradation profile of methylene blue after treatment with drug-loaded nanovehicles, indicating the enhanced generation of ·OH via domino reaction

behaviors. Similar results are obtained when investigating Mn^{2+} as a drug. During the dissolution of the CaP shell in an acidic environment, Mn^{2+} is released to trigger the Fenton-like reaction with H$_2$O$_2$, thus producing ·OH. Terephthalate (TPA) is used here to confirm whether radicals are generated, as it would change from a non-fluorescent state to fluorescent 2-hydroxyterephthalic acid when encountering ·OH (Nosaka and Nosaka 2017). Therefore, in Fig. 4.11b, the fluorescence detected in a pH-dependent manner when incubating TPA and H$_2$O$_2$ are incubated with NV-Mn clearly proves the effective loading and release behaviors of Mn^{2+} in building blocks, which is consistent with the other drugs. Moreover, methylene blue (MB) degradation experiments are performed to investigate the domino reaction encoded by the specialization and cooperation between heterogeneous building blocks. As shown in Fig. 4.11c, adding 100 μg/mL NV-GO$_X$ or NV-Mn alone only decreases a small amount of MB in solution after 6 h, mainly due to the physical absorption of MB on the building blocks. In contrast, a significant increase in degradation efficiency is found when GO$_X$- and Mn^{2+}-loaded building blocks are incubated together with the same total amount of 100 μg/mL (mass ratio 1:1) under an acidic condition. More than 95% of the MB is degraded at this time, suggesting the domino reaction-enhanced production of active ·OH.

4.5 Adaptive and Heterogeneous Colloidal Swarm for Anticancer Targeted Drug Delivery

The anticancer performances of drug-loaded building blocks on liver cancer HepG2 cells are evaluated at first. As shown in Fig. 4.12a, the biocompatible empty building blocks are nontoxic to cancer cells even at a concentration of 250 μg/mL. When DOX is loaded, ~80% of cancer cells are killed after 48 h when the concentration of building blocks is 250 μg/mL, indicating the cytotoxic effect of DOX-loaded nanovehicles (NV-DOX). Such performance is even better than that achieved by pure DOX, which is probably due to the enrichment of drugs inside HepG2 cells enabled by

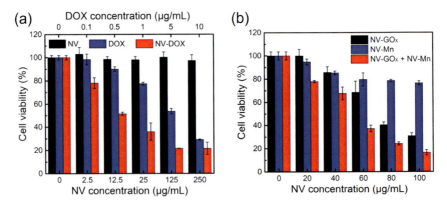

Fig. 4.12 **a** HepG2 cell viabilities after 48 h incubation with empty nanovehicles (building blocks), free DOX, and NV-DOX. **b** HepG2 cell viabilities after 24 h incubation with NV-GOX alone, NV-Mn alone, and the combination of NV-GOX and NV-Mn. The error bars represent the standard deviation ($n = 3$)

the endocytosis process (Fig. 4.13). The domino reaction-enhanced anticancer effects of heterogeneous NV-GO$_X$ and NV-Mn building blocks are further confirmed. As shown in Fig. 4.12b, the addition of NV-GO$_X$ inhibits the growth and proliferation of HepG2 cells, which becomes more obvious at higher NV-GO$_X$ concentrations. While for NV-Mn, the cell viability becomes lower at first and then keeps stable if increasing NV-Mn concentration, which may be ascribed to the limited amount of inherent H_2O_2 in tumor cellular environment (Fan et al. 2019). So, when adding NV-GO$_X$ and NV-Mn together with the same mass ratio, extra H_2O_2 can be provided via the GO$_X$-mediated catalytic reaction, which subsequently promotes ·OH production via a Fenton-like reaction. Therefore, the anticancer performance achieved by the combination of NV-GO$_X$ and NV-Mn is much better than that realized by single nanovehicles, suggesting the therapeutic efficiency can be improved by the domino reaction between heterogeneous building blocks.

Finally, the multitasking capability of the synchronized and heterogeneous colloidal swarm is investigated in a biochip (Fig. 4.14a). After adding three subgroups of building blocks (NV-DOX, NV-GO$_X$, and NV-Mn with a mass ratio of 2:1:1) in the left channel, the swarm system is navigated to the right targeted area where HepG2 cancer cells are incubated in a distributed or synchronized formation. The former would lose most of the building blocks during the locomotion as the building blocks move independently, making the corresponding access rate only ~5%. While for the synchronized swarm control strategy developed in this chapter, most of the building blocks (>90%) are successfully guided to the targeted position (Fig. 4.14b). Then the survival states of HepG2 cancer cells after 24 h delivery are investigated via live/dead staining experiments, in which, the live and dead cells possess green and red fluorescence, respectively (Fig. 4.14c). Compared to the control experiment, the passive drug diffusion and distributed swarm only have a limited effect on the growth and proliferation of HepG2 cells. In comparison, the synchronized and heterogeneous

4.6 Experimental Section

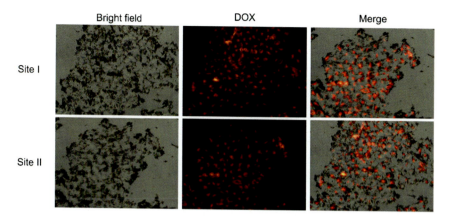

Fig. 4.13 Bright field and fluorescence images of HepG2 cells after 3 h incubation with 125 μg/mL NV-DOX. Strong red fluorescence of DOX could be detected in the cells at two random sites, indicating the cellular uptake and intracellular release behaviors of NV-DOX

microswarm with domino reaction encoded multiple functions is capable of killing almost all the cancer cells, indicating superior targeting and therapeutic efficiency.

4.6 Experimental Section

4.6.1 Nanoparticle Preparation

Fe$_3$O$_4$ nanoparticles were prepared via a solvothermal method. 1.35 g FeCl$_3$·6H$_2$O, 3.6 g NaAc, and 1 g polyethylene glycol ($M_w = 2000$) are sequentially dissolved in 40 mL ethylene glycol by magnetic stirring, leading to the formation of a russet mixture after 12 h vigorous stirring. Next, the obtained mixture is sealed in a 50 mL autoclave and transferred to an electronic oven, followed by heating at 200 °C for 10 h. After the autoclave naturally cooled to room temperature, magnetic nanoparticles are obtained, washed with deionized (DI) water, and collected by a permanent magnet. Finally, the nanoparticles are dispersed in DI water with a concentration of 50 mg/mL for further use.

The next step is to fabricate Fe$_3$O$_4$@PDA-MPS nanoparticles. The PDA layer was first coated on the surface of Fe$_3$O$_4$ via an in situ polymerization process. 0.12 g tris(hydroxymethyl)aminomethane is dissolved in 100 mL DI water, and the pH value is adjusted to 8.5 by dropwise adding 1 M HCl. Next, 100 mg Fe$_3$O$_4$ nanoparticle is added to the buffer solution, followed by mechanical stirring and sonication for 15 min, aiming to completely disperse the magnetic nanoparticles in solution. Afterward, 0.02 g dopamine hydrochloride is added to trigger the polymerization process, which is allowed to react for 5 h in an ice bath. The functionalized nanoparticles

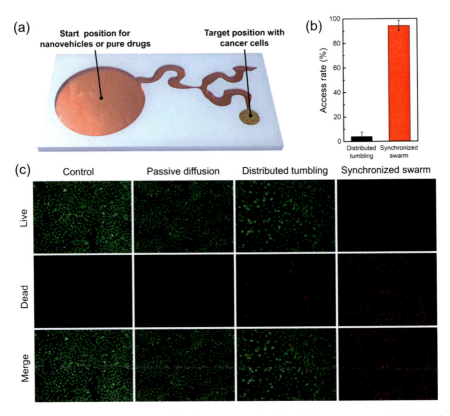

Fig. 4.14 **a** Schematic of the colloidal microswarm in biochip. **b** The access rates of distributed microswarm and adaptive microswarm in biochip. The error bars represent the standard deviation ($n = 3$). **c** Live/dead staining results of HepG2 cancer cells by different treatments. The green and red fluorescence indicate the live and dead cells, respectively

are washed with DI water and collected by a permanent magnet. Subsequently, the nanoparticles are dispersed in 160 mL ethanol, 40 mL DI water, and 2 mL ammonia solution, followed by mechanical stirring for 15 min and the addition of 3 mL MPS. Then the mixture is allowed to react for 24 h, thus obtaining Fe_3O_4@PDA-MPS nanoparticles. Finally, the product is washed with ethanol and acetonitrile with a permanent magnet and stored in acetonitrile with a concentration of 10 mg/mL for further use.

Then, Fe_3O_4@PDA@PMAAc nanoparticles were prepared. A precipitation polymerization process is used to coat the hydrogel layer on the surface of nanoparticles. 15 mg Fe_3O_4@PDA-MPS nanoparticles in 100 mL acetonitrile is mechanically stirred and treated by ultrasonication for 30 min under the argon atmosphere, followed by the addition of 1.5 mL MAAc, 0.1 g MBA, and 0.015 g azobisisobutyronitrile. The mixture is quickly heated to boiling state and refluxed for 90 min, thus obtaining Fe_3O_4@PDA@PMAAc nanoparticles. Finally, the product is washed with ethanol and DI water with a permanent magnet and stored in DI water with a concentration of 5 mg/mL for further use.

The last step is to fabricate Fe_3O_4@PDA@PMAAc@CaP nanoparticles that serve as the building blocks. A biomimetic biomineralization process is conducted to deposit CaP shell on the nanoparticles. 15 mg Fe_3O_4@PDA@PMAAc nanoparticles in 2 mL 30 mM $CaCl_2$ solution is firstly mechanically stirred overnight, followed by the injection of 4 mL 20 mM Na_2HPO_4 solution with a rate of 1 mL/h. The mixture undergoes further mechanical stirring for 2 h, thus obtaining the building blocks. Finally, the product is washed with DI water with a permanent magnet and stored in DI water with a concentration of 5 mg/mL for further use. Different kinds of drugs are simply loaded by adding DOX (1 mg/mL), GO_X (1 mg/mL), or $MnCl_2$ (10 mM) into the $CaCl_2$ solution.

4.6.2 *Cellular Experiments and Characterization Techniques*

HepG2 cells are seeded in 96-well plate with a density of 2000 cells/well, and then incubated in 100 μL EMEM with 10% FBS for 12 h. Then the medium is removed, and 100 μL fresh medium containing drug-loaded building blocks are added. After incubation for different time, MTS assay is performed to evaluate the cell viability. 10 μL MTS solution is added to each well and aged for 2 h. Then the building blocks are gathered to the bottom with a permanent magnet and the supernatant solution is used for absorbance detection at 490 nm with a microplate reader. All of the tests were repeated for three times. For live/dead cell staining experiments, HepG2 cells with a density of 2000 cells/well are first seeded in biochip and incubated for 12 h. After performing delivery task, the cells are further cultured with different building blocks for 24 h. Subsequently, 5 μM calcein-AM and 10 μM PI are added and aged for 30 min. The cells are then transferred to an inverted fluorescence microscope (Nikon Eclipse Ti), and the fluorescence signal is observed. For drug uptake experiments, HepG2 cells with a density of 2000 cells/well are seeded in 96-well plate and incubated in 100 μL EMEM with 10% FBS for 12 h. NV-DOX is then added with a concentration of 125 μg/mL, followed by incubation of 3 h, and observation with the fluorescence microscope.

Scanning electron microscope (SEM, JEOL 7800F) and transmission electron microscope (TEM, Tecnai F20) are used to characterize the morphology and composition of nanoparticles. Fourier-transform infrared spectrometer (FT-IR, Thermo Nicolet Nexus 670) and X-ray powder diffractometer (XRD, Rigaku SmartLab) are used to test the surface functional groups and crystal structure of nanoparticles. The magnetic properties are investigated via a PPMS Model 6000 Quantum Design VSM, and the concentrations of DOX and MB are measured by UV/VIS spectrophotometer (Hitachi U2910). Zeta Sizer (Malvern Nano ZS) is employed to characterize the hydrodynamic diameter of nanoparticles, while microplate reader (Tecan Infinite M Plex) is used to obtain the fluorescence spectra for cell viability evaluation.

4.7 Conclusion

In conclusion, a synchronized colloidal swarm system consisting of heterogeneous building blocks is introduced in this chapter. Each building block is a core–shell structured nanoparticle, which enables both magnetic control and function customization. Through applying different programmable magnetic fields, the colloidal system is organized into two formations, i.e., vortex- and ribbon-like microswarms, and allows controllable, reversible, and rapid transformation between different formations. In this manner, the colloidal swarm can adapt to highly complex environments by fully taking advantage of the unique patterns and locomotion capabilities of different formations, and an access rate more than 90% is achieved when performing drug delivery tasks in a biochip with various branched and narrow channels. Furthermore, cooperative tasking capability is divided and integrated into heterogeneous building blocks to construct a multifunctional swarm system. Through designing the loaded drugs, a domino reaction encoded combination therapy is accomplished to significantly improve the killing efficiency of HepG2 cancer cells. Therefore, the colloidal system introduced in this chapter may stimulate the future development of multifunctional swarming systems and promote the realization of their applications in targeted therapy, verifying how science and technology can be inspired by nature.

References

Chen J, Zhu Y, Wu C, Shi J (2020) Nanoplatform-based cascade engineering for cancer therapy. Chem Soc Rev 49(24):9057–9094

Deng H, Li X, Peng Q, Wang X, Chen J, Li Y (2005) Monodisperse magnetic single-crystal ferrite microspheres. Angew Chem 117(18):2842–2845

Erkoc P, Yasa IC, Ceylan H, Yasa O, Alapan Y, Sitti M (2019) Mobile microrobots for active therapeutic delivery. Adv Therapeutics 2(1):1800064

Fan JX, Peng MY, Wang H, Zheng HR, Liu ZL, Li CX, Wang XN, Liu XH, Cheng SX, Zhang XZ (2019) Engineered bacterial bioreactor for tumor therapy via Fenton-like reaction with localized H_2O_2 generation. Adv Mater 31(16):1808278

Felfoul O, Mohammadi M, Taherkhani S, De Lanauze D, Xu YZ, Loghin D, Essa S, Jancik S, Houle D, Lafleur M (2016) Magneto-aerotactic bacteria deliver drug-containing nanoliposomes to tumour hypoxic regions. Nat Nanotechnol 11(11):941–947

Fu L-H, Qi C, Lin J, Huang P (2018) Catalytic chemistry of glucose oxidase in cancer diagnosis and treatment. Chem Soc Rev 47(17):6454–6472

Gao C, Wang Y, Ye Z, Lin Z, Ma X, He Q (2021) Biomedical micro-/nanomotors: From overcoming biological barriers to in vivo imaging. Adv Mater 33(6):2000512

Garattoni L, Birattari M (2018) Autonomous task sequencing in a robot swarm. Sci Robotics 3(20):eaat0430

Habraken W, Habibovic P, Epple M, Bohner M (2016) Calcium phosphates in biomedical applications: materials for the future? Mater Today 19(2):69–87

Hölldobler B, Wilson EO (1990) The ants. Harvard University Press

Ji Y, Lin X, Wu Z, Wu Y, Gao W, He Q (2019) Macroscale chemotaxis from a swarm of bacteria-mimicking nanoswimmers. Angew Chem Int Ed 58(35):12200–12205

References

Krieger MJ, Billeter J-B, Keller L (2000) Ant-like task allocation and recruitment in cooperative robots. Nature 406(6799):992–995

Lee H, Dellatore SM, Miller WM, Messersmith PB (2007) Mussel-inspired surface chemistry for multifunctional coatings. Science 318(5849):426–430

Li GL, Möhwald H, Shchukin DG (2013) Precipitation polymerization for fabrication of complex core-shell hybrid particles and hollow structures. Chem Soc Rev 42(8):3628–3646

Li S, Batra R, Brown D, Chang H-D, Ranganathan N, Hoberman C, Rus D, Lipson H (2019) Particle robotics based on statistical mechanics of loosely coupled components. Nature 567(7748):361–365

Li J, de Ávila BE-F, Gao W, Zhang L, Wang J (2017) Micro/nanorobots for biomedicine: Delivery, surgery, sensing, and detoxification. Sci Robotics 2(4):eaam6431

Liang X, Mou F, Huang Z, Zhang J, You M, Xu L, Luo M, Guan J (2020) Hierarchical microswarms with leader-follower-like structures: electrohydrodynamic self-organization and multimode collective photoresponses. Adv Func Mater 30(16):1908602

Mura S, Nicolas J, Couvreur P (2013) Stimuli-responsive nanocarriers for drug delivery. Nat Mater 12(11):991–1003

Nelson BJ, Kaliakatsos IK, Abbott JJ (2010) Microrobots for minimally invasive medicine. Annu Rev Biomed Eng 12:55–85

Nosaka Y, Nosaka AY (2017) Generation and detection of reactive oxygen species in photocatalysis. Chem Rev 117(17):11302–11336

Pitcher TJ (2012). The behaviour of teleost fishes. Springer Science & Business Media

Qi C, Lin J, Fu L-H, Huang P (2018) Calcium-based biomaterials for diagnosis, treatment, and theranostics. Chem Soc Rev 47(2):357–403

Qiu Y, Park K (2001) Responsive polymeric delivery systems. Adv Drug Deliv Rev 53:321–339

Ranji-Burachaloo H, Gurr PA, Dunstan DE, Qiao GG (2018) Cancer treatment through nanoparticle-facilitated fenton reaction. ACS Nano 12(12):11819–11837

Rim HP, Min KH, Lee HJ, Jeong SY, Lee SC (2011) pH-tunable calcium phosphate covered mesoporous silica nanocontainers for intracellular controlled release of guest drugs. Angew Chem Int Ed 50(38):8853–8857

Rubenstein M, Cornejo A, Nagpal R (2014) Programmable self-assembly in a thousand-robot swarm. Science 345(6198):795–799

Servant A, Qiu F, Mazza M, Kostarelos K, Nelson BJ (2015) Controlled in vivo swimming of a swarm of bacteria-like microrobotic flagella. Adv Mater 27(19):2981–2988

Sitti M, Ceylan H, Hu W, Giltinan J, Turan M, Yim S, Diller E (2015) Biomedical applications of untethered mobile milli/microrobots. Proc IEEE 103(2):205–224

Solovev AA, Sanchez S, Schmidt OG (2013) Collective behaviour of self-propelled catalytic micromotors. Nanoscale 5(4):1284–1293

Sumpter DJ (2010) Collective animal behavior. Princeton University Press

Tang Z, Liu Y, He M, Bu W (2019) Chemodynamic therapy: tumour microenvironment-mediated Fenton and Fenton-like reactions. Angew Chem Int Ed 58(4):946–956

Tang S, Zhang F, Gong H, Wei F, Zhuang J, Karshalev E, de Ávila BE-F, Huang C, Zhou Z, Li Z 2020 Enzyme-powered Janus platelet cell robots for active and targeted drug delivery. Sci Robotics 5(43):eaba6137

Tietze LF (1996) Domino reactions in organic synthesis. Chem Rev 96(1):115–136

Wang H, Pumera M (2020) Coordinated behaviors of artificial micro/nanomachines: from mutual interactions to interactions with the environment. Chem Soc Rev 49(10):3211–3230

Wang W, Castro LA, Hoyos M, Mallouk TE (2012) Autonomous motion of metallic microrods propelled by ultrasound. ACS Nano 6(7):6122–6132

Wang W, Duan W, Ahmed S, Sen A, Mallouk TE (2015) From one to many: dynamic assembly and collective behavior of self-propelled colloidal motors. Acc Chem Res 48(7):1938–1946

Wang Q, Yu J, Yuan K, Yang L, Jin D, Zhang L (2020) Disassembly and spreading of magnetic nanoparticle clusters on uneven surfaces. Appl Mater Today 18:100489

Weimerskirch H, Martin J, Clerquin Y, Alexandre P, Jiraskova S (2001) Energy saving in flight formation. Nature 413(6857):697–698

Werfel J, Petersen K, Nagpal R (2014) Designing collective behavior in a termite-inspired robot construction team. Science 343(6172):754–758

Wu Z, Troll J, Jeong H-H, Wei Q, Stang M, Ziemssen F, Wang Z, Dong M, Schnichels S, Qiu T, Fischer P, 2018 A swarm of slippery micropropellers penetrates the vitreous body of the eye. Sci Adv 4(11):eaat4388

Yan X, Zhou Q, Vincent M, Deng Y, Yu J, Xu J, Xu T, Tang T, Bian L, Wang Y-XJ (2017). Multifunctional biohybrid magnetite microrobots for imaging-guided therapy. Sci Robotics 2(12):eaaq1155

Yu J, Xu T, Lu Z, Vong CI, Zhang L (2017) On-Demand disassembly of paramagnetic nanoparticle chains for microrobotic cargo delivery. IEEE Trans Rob 33(5):1213–1225

Yu J, Yang L, Zhang L (2018a) Pattern generation and motion control of a vortex-like paramagnetic nanoparticle swarm. Int J Robotics Res 37(8):912–930

Yu J, Wang B, Du X, Wang Q, Zhang L (2018b) Ultra-extensible ribbon-like magnetic microswarm. Nat Comm 9:3260

Chapter 5
Three-Dimensional Structure and Independent Control of Micro/Nanorobot Swarms

Abstract Developing microswarms with three-dimensional structures and realizing independent control of microswarms are two significant challenges in the field of micro/nanorobot swarms. This chapter first presents a strategy to construct a tornado-like microswarm with a three-dimensional structure through the combination of light and magnetic fields. The two-dimensional vortex-like paramagnetic nanoparticle swarm driven by the rotating magnetic field generates convection under illumination due to photothermal effects. The induced convection causes the swarm pattern to contract further and gradually rise vertically, forming a tornado-like 3D pattern that can be maintained for controlling chemical reactions. This chapter then introduces an independent control strategy of microswarms based on different magnetic responses of building blocks. The variation of the aspect ratio of the ribbon-like swarm formed by nickel nanorods with the field ratio of the oscillating field is very different from that formed by Fe_3O_4 nanoparticles. A theoretical model based on dipole–dipole interaction is constructed and clarifies that the main reason for this difference is attributed to the magnetic anisotropy of nanorods. Thus, independent control of the swarm pattern under the same input is achieved by using the nickel nanorod swarm and the Fe_3O_4 nanoparticle swarm.

Keywords 3D pattern · Light vertical motion · Independent control · Nanorods · Nanoparticles

5.1 Introduction

Natural swarms consisting of living organisms provide inspiration for the design of swarming systems in the field of micro/nanorobotics (Jin and Zhang 2022; Wang and Pumera 2020; Yang et al. 2022). To date, various micro/nanorobot swarms have been developed using a variety of materials as the building blocks, and they are actuated by many different energy sources, such as magnetic fields (Sun et al. 2021; Yigit et al. 2019; Snezhko and Aranson 2011), electric fields (Sapozhnikov et al. 2003; Yan et al. 2016), acoustic field (Zhou et al. 2018; Ahmed et al. 2017), and light (Ibele et al.

2009; Mou et al. 2019; Hernàndez-Navarro et al. 2014). For example, magnetic field-actuated peanut-like hematite particles can perform multimode swarming behaviors (Xie et al. 2019). The acoustic field-driven liquid metal colloidal particle swarm has a dandelion-like pattern (Li et al. 2020). Collective behaviors of TiO$_2$ particles are induced by light due to electrolyte diffusion electrophoresis (Mou et al. 2019). In addition, two representative magnetic nanoparticle swarms are highlighted in previous chapters, namely the vortex-like swarm and ribbon-like swarm (Yu et al. 2018a, 2021, 2018b). Although these micro/nanorobot swarm systems have overcome many difficult issues in this field, such as swarm formation, transformation, locomotion, imaging, and navigation, more challenges still need to be addressed to facilitate the development of microswarms. This chapter will focus on two important challenges in the research of micro/nanorobot swarms: the three-dimensional structure and independent control of microswarms.

In nature, animal swarms such as fish schools and bird flocks generally have three-dimensional (3D) structures (Nagy et al. 2010; Morse 1963; Vicsek and Zafeiris 2012; Sumpter 2010; Isaeva 2012). However, most artificial micro/nanorobot swarms are generally two-dimensional (2D), that is, their collective behaviors are confined to the vicinity of an interface (usually a horizontal interface). For instance, swarming systems, including living bacteria colonies (Chen et al. 2017; Felfoul et al. 2016; Loghin et al. 2017) and abiotic particles (Ibele et al. 2009; Palacci et al. 2013), have been implemented in 2D space. Current artificial microswarms can perform pattern formation, transformation, and locomotion on the solid substrate (Kudrolli et al. 2008; Narayan et al. 2007) or liquid–air interface (Snezhko et al. 2009; Lv et al. 2018; Dong and Sitti 2020; Wang et al. 2019a). Dependence on the interface usually makes it difficult for the building blocks of the microswarm to overcome gravity, resulting in the difficulty for them to break away from the interface to form structures distributed in three-dimensional space. The 2D swarms have relatively limited functions, especially in the vertical direction. For example, it is difficult for a 2D swarm to induce fluid flow in the entire 3D space in a liquid environment, and the effect in enhancing catalysis is limited (Wang et al. 2019b). Mimicking 3D structures of natural swarms has received less attention in the field of micro/nanorobot swarms. Microswarms with three-dimensional structures can manipulate objects vertically without being bound by boundaries, which may provide methods for applications in complex biological environments, such as 3D cell culture systems (Guo et al. 2019; Edmondson et al. 2014; Lee et al. 2019). Many attempts have been made in research to make tiny agents to overcome gravity. Inspired by the antigravity movement of zooplankton in nature toward the light, scientists have utilized light-induced self-diffusiophoresis for lifting Janus particles above the boundary (Singh et al. 2018). Using the thermal gradient generated by radiation, comet-like colloidal particles can self-assemble in the vertical direction (Cohen and Golestanian 2014). Metal–organic framework particles can utilize buoyancy to achieve ascension and descension based on pH-sensitivity (Lee et al. 2019). Polystyrene particles with metal patches are capable of performing three-dimensional motion under an alternating electric field (Guo et al. 2019). However, maintaining the dynamic stability of a large number of

5.1 Introduction

particles in 3D space and forming reconfigurable 3D swarms is still one of the main challenges.

In the micro/nanorobotic system, the collaboration between individual robots is the key to expanding the capabilities of single agents, which makes independent control of robots important (Johnson et al. 2020; Wang et al. 2018). Among various actuation methods of micro/nanorobots, the magnetic field has good tunability, which makes it possible for the magnetic-driven micro/nanorobot swarms to realize independent control. At present, many studies have proposed and verified independent control strategies for micro/nanorobots. For example, multiple microrobots can be driven independently using magnetic torques or gradient forces generated by specially designed magnetic drives (Rahmer et al. 2017). Magnetic fields can also enable independent control of micro/nanorobots with different geometrical characteristics (Diller et al. 2013; Tottori et al. 2013; Howell et al. 2018; Khalil et al. 2018). In addition to the morphology, the magnetic properties of each micro/nanorobots, such as the response to magnetic fields and agent-agent interactions, have also been proven to be utilized for independent control (Diller et al. 2012; Salehizadeh and Diller 2020; Becker et al. 2014). However, these studies mainly investigate robotic systems with a small number of individuals, and the independent control of microswarms consisting of millions of agents remains challenging, especially in generating swarm patterns in the same applied fields and external environments. Achieving independent control can enrich the functionality of microswarms and improve the ability of microswarms to perform a variety of complex tasks. In addition, research on the independent control of microswarms may deepen the understanding of the fundamental principles of collective behavior and facilitate the further development of micro/nanorobot swarms.

Based on the previously introduced vortex-like and ribbon-like swarms (Yu et al. 2018a, 2021, 2018b), this chapter introduces new pattern formation strategies to solve the above two key challenges in the field of micro/nanorobot swarms. The first is generating a tornado-like paramagnetic $Fe_3O_4@SiO_2$ nanoparticles swarm in the aqueous solutions by integrating magnetic fields and light, which is based on the vortex-like swarm driven by rotating magnetic fields. The tornado-like swarm can perform collective behavior in the three-dimensional space, including four steps, $i.e.$, rising, hovering, oscillation, and landing (Fig. 5.1). The first step is to generate a two-dimensional vortex-like swarm under the actuation of a conical rotating magnetic field. Then, the swarm is exposed to light, and the convection low induced by the illumination shrinks the swarm pattern (Weinert and Braun 2008), as shown in Fig. 5.1b. Under the influence of illumination and the rotation of the swarm itself, strong induced convection makes the particles in the swarm gradually rise vertically, generating a 3D swarm pattern that is similar to a tornado (Fig. 5.1c). The tornado-like swarm can maintain its 3D structure (in a hovering state) and can be used to control the rate of chemical reactions, such as the degradation of methylene blue and ascorbic acid. The swarm can speed up the reaction while confining reactants inside the pattern. This effect is similar to the "calmness" inside a tornado, that is, there is also a relatively stable cylindrical void inside the 3D pattern. By turning the magnetic field into an oscillating field, the swarm will oscillate in situ (Fig. 5.1e, f),

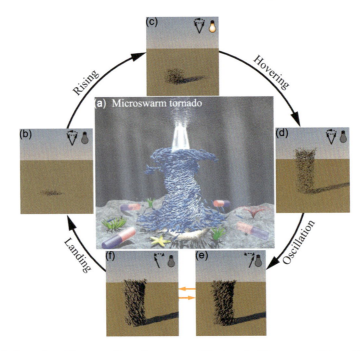

Fig. 5.1 Reconfigurable pattern formation process of the tornado-like swarm. **a** Schematic diagram of a school of fish swimming in the sea with a tornado-like 3D structure. **b** Two-dimensional vortex-like swarm on the substrate. (**c**) Rising the swarm by using light. **d** The tornado-like swarm maintains its 3D structure. **e, f** The oscillation of the 3D swarm pattern in the longitudinal plane

facilitating the manipulation of trapped reactants and switchable control (Wei et al. 2010). Finally, the particles are deposited onto the substrate by gravity, and the swarm return to its original 2D state. These four stages can be cycled, so the swarm has a reconfigurable 3D pattern, based on which vertical mass transportation and reaction control can be realized.

In terms of the independent control of micro/nanorobot swarms, the independent pattern formation and transformation can be achieved by using the previously introduced oscillating magnetic field (Fig. 3.1) and two different magnetic materials, *i.e.*, nickel nanorods and Fe_3O_4 nanoparticles. Nickel nanorods can act like Fe_3O_4 nanoparticles as building blocks for generating the ribbon-like swarm. However, the relationship between the aspect ratio of the ribbon-like nickel nanorod swarm (nanorod swarm) and the field ratio γ of the oscillating field is nonlinear, which is different from pattern transformation of the ribbon-like Fe_3O_4 nanoparticle swarm (nanoparticle swarms) introduced in Chap. 3. Here, a theoretical model based on dipole–dipole interactions of nanorods is first described, which can elucidate the pattern transformation mechanism of the nanorod swarm. This model takes the magnetic anisotropy of the nanorods into account, which can well describe the magnetization of the nanorods under the oscillating magnetic field and is in good agreement with the experimental results. Compared with the nanoparticle swarm, it

5.2 Two-Dimensional Swarm Under the Conical Rotating Field

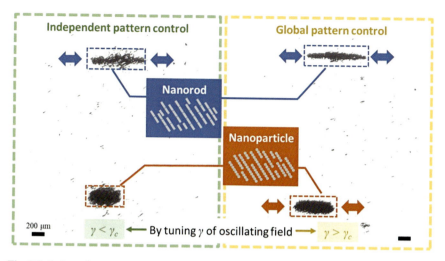

Fig. 5.2 Independent pattern transformation of the nanorod and nanoparticle swarms under the same oscillating magnetic field

is found that the nanorod chains are longer than the nanoparticle chains at a small γ value under the same oscillation field. This phenomenon results in the presence of anomalous regions (AR) in the variation of the aspect ratio of the nanorod swarm with γ, i.e., the increase in γ leads to the decrease in the aspect ratio. When the value of γ is large, pattern transformation of the nanorod swarm is in the normal region (NR). The magnetic anisotropy of building blocks that make up the swarm is the key to the different pattern transformation behavior of the swarm. Therefore, independent control of two swarm patterns can be achieved using nickel nanorods and Fe_3O_4 nanoparticles under the same input oscillating magnetic field (Fig. 5.2).

5.2 Two-Dimensional Swarm Under the Conical Rotating Field

5.2.1 *Building Blocks and the Magnetic and Optical Control System*

The building blocks of the tornado-like swarm are $Fe_3O_4@SiO_2$ nanoparticles with a diameter of about 160 nm, as shown in the transmission electron microscope image in Fig. 5.3. The core part of the nanoparticle is Fe_3O_4, and it is coated by a thin shell of SiO_2, as indicated by the translucent layer in Fig. 5.3. The formation of a tornado-like swarm requires a combination of magnetic fields and lights, so magnetic and optical actuation systems need to be designed to avoid interfering with each other. The magnetic actuation device employed here is called Magdisk, which

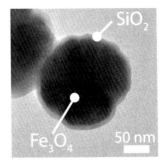

Fig. 5.3 Transmission electron microscope image of Fe$_3$O$_4$@SiO$_2$ nanoparticles. The average diameter is about 160 nm. The black central area is composed of Fe$_3$O$_4$, and the transparent outer shell is SiO$_2$

consists of five electromagnetic coils (four in the horizontal plane and one along the vertical direction) and the control system. The Magdisk can continuously generate dynamic magnetic fields in three-dimensional space by programming LabVIEW code to control the coils. In addition, it is smaller than general three-axis Helmholtz coils, which facilitates integration with the optical actuation device. Another reason for using Magdisk is that the Fe$_3$O$_4$@SiO$_2$ nanoparticles have a small size, and the stronger magnetic field (about 30 mT) generated by Magdisk is beneficial to the actuation of these nanoparticles. The light used to generate the swarm is provided by a laser with a wavelength of 808 nm, and its optical path configuration is shown in Fig. 5.4a. Light is transmitted through an infrared/visible (IR/Vis) beam splitter and objective lens to produce a spot with a radius of approximately 125 μm. By integrating the magnetic actuation and light generating devices, a 3D swarm generating system is constructed, and the overall schematic diagram is shown in Fig. 5.4b.

5.2.2 Formation of the 2D Swarm Pattern

Generating a two-dimensional pattern on the substrate is a precondition for the formation of the three-dimensional swarm. First, a drop of Fe$_3$O$_4$@SiO$_2$ nanoparticle solution (0.15 mg/mL) is added to an open container, and a permanent magnet is used to preliminarily aggregate the particles. After applying the magnetic field, Fe$_3$O$_4$@SiO$_2$ nanoparticles form chain-like structures due to the magnetic dipole–dipole interaction, which is consistent with the behavior of Fe$_3$O$_4$ nanoparticles (Reddy et al. 2012). Under the rotating magnetic field, the nanoparticle chains will form the vortex-like swarm, and the formation mechanism is described in Chap. 2. Briefly, the rotating nanoparticle chains are subjected to three forces, *i.e.*, the magnetic and hydrodynamic interaction forces between the chains and the inward force exerted by the main vortex (Fig. 5.5). Their resultant force acts as a centrifugal force, confining nanoparticle chains within the vortex-like swarm.

In order to transform the 2D swarm pattern into three-dimensional, the key problem is to maintain the stability of the swarm while lifting the particles inside it. In complex environments, the rotating frequency of the nanoparticle chains may

5.2 Two-Dimensional Swarm Under the Conical Rotating Field

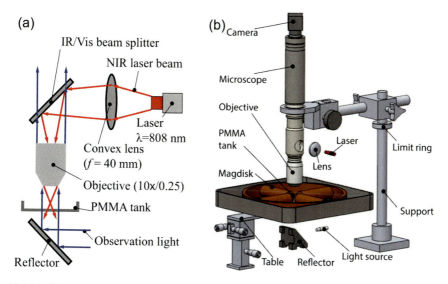

Fig. 5.4 3D swarm generating system that integrates magnetic and optical actuation devices. **a** Schematic diagram of the light path of the optical control device. The red lines indicate the laser beam, and the blue lines indicate the observation light. **b** The overall schematic diagram of the 3D swarm generating system

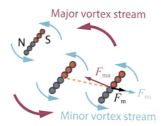

Fig. 5.5 Schematic diagram of forces acting on nanoparticle chains in the vortex-like swarm. F_m is the chain-chain magnetic force, F_{mi} is the hydrodynamic force between particle chains, and F_{ma} is the inward force induced by the main vortex

be inconsistent with the rotating frequency of the field, which may easily lead to the instability of the swarm pattern. Therefore, a conical rotating magnetic field is considered to be applied to lift the nanoparticle chains by an angle relative to the substrate, which may facilitate the formation of the 3D swarming structure. The pattern formation process using the $Fe_3O_4@SiO_2$ nanoparticles in a conical rotating magnetic field is shown in Fig. 5.6. In the generated swarm pattern ($t = 15$ s), S_1 denotes the core area of the swarm, in which the density of particles is high, and their combination is relatively tight. S_2 denotes the whole influence area of the swarm, that is, the distribution area of particles affected by the swarm. The successful formation of a 2D swarm pattern requires the following criteria to be met. First, there should

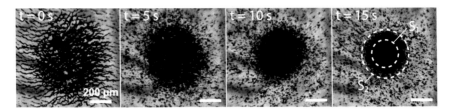

Fig. 5.6 Formation process of a two-dimensional vortex-like swarm on the substrate. S_1 is the core region, and S_2 denotes the whole affected region of the swarm. The magnetic field strength is 28 mT, the frequency is 9 Hz, and the conical angle is 80°

be no large voids in the core region of the swarm, *i.e.*, no obvious white pixels in the optical image. Second, the formation process needs to be completed within a certain period of time. Finally, all particles in the swarm should rotate with the pattern, and particles that are not affected by the swarm are considered ineffective particles in the pattern formation process. The core ratio κ is defined as the ratio of S_1 and S_2, *i.e.*, $\kappa = S_1/S_2$, which is used to quantitatively characterize the quality of the swarm. S_1/t is the generation rate used to describe the swarm formation process.

Measurements of these two areas are conducted by modifying the image threshold to cover the selected particles. The experimental image is converted to an 8-bit grayscale picture with the threshold range set as 0–111, which makes the particles appear black and the background appear white (Fig. 5.7). An ellipse with four anchor points is used to mark the core area S_1 of the 2D swarm pattern, which covers the largest area without any white pixels. Another ellipse is used to label the influence area of the swarm (S_2), which contains all particles that can rotate with the swarm and allows the existence of white pixels. The ellipse of S_2 is considered the whole swarm area, whose boundaries distinguish the particles in the swarm from the dispersed particles.

The parameters of the conical magnetic field, such as the conical angle α, frequency f, and strength B, have a significant influence on the swarm formation, so it is necessary to determine appropriate parameters for increasing the stability of the swarm. For example, low rotating frequencies (*e.g.*, $f \leq 2$ Hz) cause nanoparticle chains to be large in size, which cannot induce the vortex for swarm formation (Fig. 5.8a). High rotating frequencies (*e.g.*, $f \geq 17$ Hz) cause the particle chains to be too dispersed to form a tightly bound pattern (Fig. 5.8b). Figure 5.8c shows the swarm pattern generated at an appropriate rotating frequency ($f = 7$ Hz). The relationship between the conical angle and the generation rate (S_1/t) is shown in Fig. 5.9. When the conical angle is 90°, the fluctuation of the swarm formation process is violent, and the 2D pattern may drift. The small conical angle ($\alpha \leq 60°$) will prevent particle chains from forming tight swarm patterns. The swarm formation process is stable when the conical angle is 80°, and the rotating frequency range is between 7 and 13 Hz. In addition, Fig. 5.10 shows the influence of magnetic field strength on the core ratio κ of the swarm. When the field strength is low ($B \leq 16$ mT), no obvious swarm core region exists, so it can be considered that the swarm is not successfully

5.2 Two-Dimensional Swarm Under the Conical Rotating Field

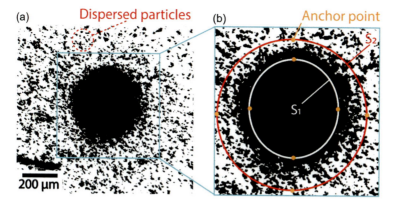

Fig. 5.7 Image processing for swarm area measurement. **a** Adjust the image threshold to cover all the selected particles. **b** The magnified image of the central area of **a**, the swarm core area S_1 is inside the white ellipse, and the influence area of the swarm is inside the red ellipse. The orange dots indicate the anchor points

generated. The strong magnetic field ($B \geq 32$ mT) enhances the repulsion between nanoparticle chains, causing them to spread apart and fail to generate the swarm. By decreasing the field strength or increasing the rotating frequency, the core ratio can be improved. Therefore, a conical rotating magnetic field (conical angle: 80°) with suitable rotating frequency and magnetic field mentioned above is selected to generate the 2D swarm for further forming the 3D swarming pattern.

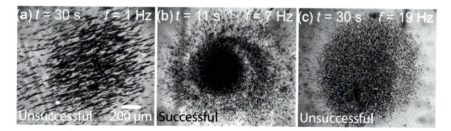

Fig. 5.8 2D swarm pattern generated under the conical rotating magnetic field with different frequencies. **a** Large nanoparticle chains are formed when $f = 1$ Hz, and no swarm pattern is generated. **b** The successful formation of a 2D swarm at a proper rotating frequency of 7 Hz. **c** The high frequency (19 Hz) causes nanoparticle chains to disperse, and the swarm formation is unsuccessful. The magnetic field strength is 28 mT, and the conical angle is 70°

Fig. 5.9 Relationship between the generation rate of the 2D swarm and the frequency of the conical rotating field under different conical angles. The inset shows the definition of the conical angle α

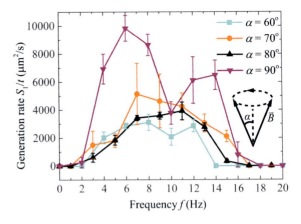

Fig. 5.10 Core ratio of the 2D swarm pattern is influenced by the strength and frequency of the magnetic field. The conical angle is 70°

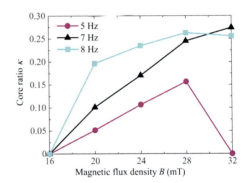

5.3 The Tornado-Like Swarm with a 3D Structure

5.3.1 Light-Driven Hovering of the Swarm

Under the irradiation of the laser beam, the nanoparticle chains are further aggregated due to the flow generated by the thermal effect and rise from the substrate, and thus, the transformation from a circular 2D pattern to a 3D tornado-like pattern can be realized. Figure 5.11 demonstrates the light-induced aggregation of nanoparticle chains, where the initial distribution area of the chains is denoted by S_0. The measurement of swarm area in the presence of light is similar to the previous method. The initial state is set as the time after both magnetic field and light are applied. Since the illumination increases the overall brightness of the image, the threshold in image processing is adjusted to around 0–135 to eliminate measurement errors. By comparing the swarm actuated by magnetic field only and that actuated by both magnetic field and light, it can be found that the swarm under light illumination has a smaller effective area (S_2). This indicates that the flow induced by high rotating frequency and strong magnetic field strength causes the swarm pattern to interact with surrounding environments

5.3 The Tornado-Like Swarm with a 3D Structure

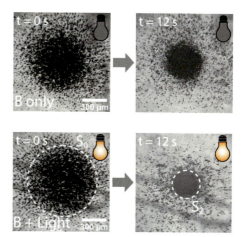

Fig. 5.11 Comparison of swarm formation when only magnetic field is applied and when both magnetic field and light are applied

frequently and strongly, while illumination can make the nanoparticle chains in the swarm bind more tightly.

The enhanced aggregation of particles under illumination is due to light-induced convection, and the finite element simulation is used to elucidate this effect. Three modules are coupled in the simulation, namely heat transfer in fluids, laminar flow, and particle tracing for fluid flow. In the simulation, a cylindrical laser beam is set as the light source according to the actual experimental conditions. Besides, the diameter of the nanoparticles is 160 nm, the fluid viscosity is 30 mPa·s (the viscosity of the polyvinylpyrrolidone solution used in the experiment), the thermal coefficient is 0.27 J/(kg K), the heat capacity is 1.74 kJ (kg K), and the density is 1064 kg/m^3 (Korycka et al. 2018; Xie et al. 2016; Westrum and Grønvold 1969). The number of particles is set as 1000, and they are randomly distributed in a circular area with a radius of 500 μm. The heat generated by the light is transferred in the fluid and nanoparticles, resulting in non-isothermal flow. The fluid then flows and creates convection for heat exchange, resulting in the nanoparticles flowing with the fluid. The momentum equation is expressed as (Ji et al. 2018)

$$\rho \frac{\partial \boldsymbol{u}}{\partial t} = \boldsymbol{F} + \nabla \cdot \left(-p\boldsymbol{I} + \mu \left(\nabla \boldsymbol{u} + (\nabla \boldsymbol{u})^\mathrm{T} \right) \right) \tag{5.1}$$

where $\boldsymbol{u}$ is the flow velocity, $\boldsymbol{F}$ is the body force, ρ is the fluid density, p is the pressure, $\boldsymbol{I}$ is the identity tensor, and μ is the dynamic viscosity. The motion of nanoparticles is caused by the convection, so the Stokes equation is expressed as

$$\boldsymbol{F}_\mathrm{d} = \frac{18\mu}{\rho_\mathrm{p} d^2} m(\boldsymbol{u} - \boldsymbol{v}) \tag{5.2}$$

where $\boldsymbol{F}_\mathrm{d}$ is the drag force, d is the diameter, m is the mass, ρ_p is the density, and $\boldsymbol{v}$ is the velocity of the particles, respectively. The heat generated by the light is transferred between the fluid and particles, and the thermal conduction is governed by the following equation:

$$mc_p \frac{dT_p}{dt} = hA_p(T - T_p) \tag{5.3}$$

where c_p is the specific heat capacity, T_p is the particle temperature, T is the fluid temperature, h is the heat transfer coefficient, A_p is the surface area of particles, and t is the time. The temperature of the fluid depends on the continuous irradiation of the laser and the heat conduction between the particles, which can be obtained from the following equation (Bird et al. 2002):

$$\rho c_f \left(\frac{\partial T}{\partial t} + \mathbf{u} \cdot \nabla T \right) - k \nabla^2 T = Q \tag{5.4}$$

where c_f is the specific heat capacity of fluid at constant pressure, k is the thermal conductivity of fluid, and Q is the heat source. By solving this 3D model with a time-dependent solver, the thermal and flow field of this system can be obtained, as shown in Fig. 5.12.

On the substrate, light-induced flow promotes the aggregation of nanoparticles. In the vertical direction, the illumination changes the temperature distribution of the cylindrical space above the swarm, and the temperature increases by about 2.9 K after the pattern formation. After the nanoparticles aggregate in the center of the light beam, they rise with the upward convection (Fig. 5.12, upper right). During this process, the light-driven particles move at a speed of about 7 μm/s. Overall, the light-induced flow causes particles to gather toward the center and then be lifted upward (Fig. 5.12, bottom). The aggregation ratio of nanoparticles is defined as $\gamma = S_2/S_0$. γ is used to indicate the effect of light on enhancing the aggregation of nanoparticles, and the smaller the γ, the better the aggregation effect and the tighter the swarm. Figure 5.13a shows that γ decreases as the light power increases,

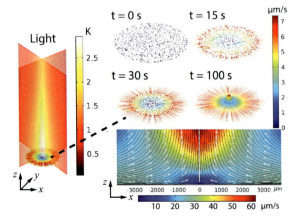

Fig. 5.12 Simulation results showing light-induced temperature changes and liquid flow. In the temperature profile, two cross planes are presented to show the thermal effect of the Gaussian beam. The red lines indicate the trajectories of the particles

5.3 The Tornado-Like Swarm with a 3D Structure

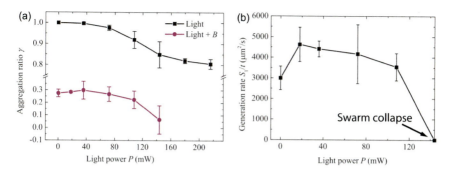

Fig. 5.13 Influence of the light power on swarm formation. **a** The relationship between the aggregation ratio and the light power. **b** The relationship between the swarm generation rate and the light power. Each error bar denotes the standard deviation from three trials

indicating that the stronger illumination results in tighter particle aggregation. By comparing the purple (light and magnetic field applied simultaneously) and black curves (light applied only) in Fig. 5.13a, it can be found that just applying light also causes particles to aggregate, but not as well as the combined effect of light and magnetic field. This is because the vortex induced by the rotating magnetic field also has the effect of aggregating particles, as described earlier. Figure 5.13b shows the relationship between swarm generation rate and light power. The results show that increasing the power from 36 to 110 mW can accelerate swarm formation. It is worth noting that the strong fluid flow generated by strong light may cause the swarm to collapse, so the light power should be kept below 140 mW. Therefore, high light power favors the formation of the 3D swarm pattern, and light with a power of about 110 mW works best.

5.3.2 Reconfigurable Formation of the Tornado-Like Swarm

The formation of a three-dimensional swarm pattern is realized based on particle aggregation caused by rotating fields and illumination and the vertical rise of particles caused by light-induced convection. The 2D pattern transforms to a tornado-like 3D pattern under light and can maintain the 3D structure for extended periods of time (over 5 min) (Fig. 5.14). The 3D pattern may drift or even collapse when subjected to external disturbances (Fig. 5.15). By turning the conical rotating magnetic field into an oscillating field, the swarm can perform 3D oscillation in the y–z plane ($t = 87$ s) or the x–z plane ($t = 114$ s). The lifted particles gradually land back on the substrate due to gravity and form the same two-dimensional swarm as the initial state. This process is repeatable, indicating that the tornado-like swarm is reconfigurable.

In the hovering stage after the rising of particles, the weak light power is beneficial to maintain the tornado-like pattern for a long time. For example, global convection induced by light with a power of 40 mW can prolong the lifetime of a tornado-like

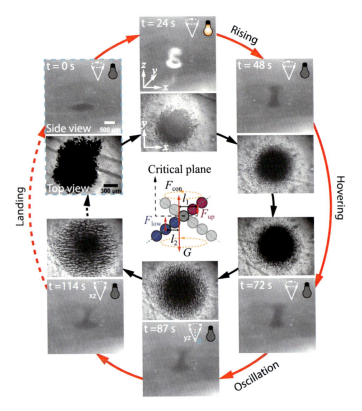

Fig. 5.14 Repeatable formation process of the tornado-like swarm under the actuation of light and magnetic field. The side view has a viewing angle of 25°. The inset in the middle shows the force analysis of particle chains in the hovering stage. The light power is 110 mW. The field strength is 16 mT, the frequency is 8 Hz, and the conical angle is 80°. In the oscillating stage, the frequency is 1 Hz, and the oscillating angle is 90°

swarm. In the analysis of the hovering state, the nanoparticle chains are divided into upper and lower parts according to the center of rotation, and the plane dividing these two parts is defined as the critical plane. The nanoparticle chain in the swarm is dynamically stable, so the exerted forces can be expressed as

$$F_{\text{con}} + F_{\text{low}} = F_{\text{up}} + G \tag{5.5}$$

where F_{con} is the force of the global convection, F_{low} is the force of the lower rotating part, F_{up} is the force of the upper rotating part, and G is gravity. According to the balance of torque, $F_{\text{up}} l_1 = F_{\text{low}} l_2$, where l_1 and l_2 are the effective distance of F_{up} and F_{low} from the center of gravity, respectively. Generally, $l_1 > l_2$ and $F_{\text{up}} < F_{\text{low}}$ due to the upward convection. Therefore, the magnetic field should increase the pattern stability compared to the rotation without force and torque balance, without the requirements of uniform fields and smooth boundaries.

5.3 The Tornado-Like Swarm with a 3D Structure

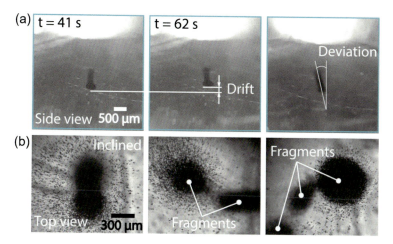

Fig. 5.15 Drift and collapse of the tornado-like swarm. **a** The tornado-like swarm drifts on the substrate and deviates from the z-axis. **b** The 3D pattern of the tornado-like swarm collapses and breaks into several fragments

During the formation of the 3D tornado-like swarm, the mass transfer in the vertical direction can be estimated according to particle area on the substrate. The specific bottom area is defined as the ratio of the bottom particle area to that at $t = 50$ s (hovering state), and its variation over time during the cycle formation process is shown in Fig. 5.16a. First, particles gather toward the swarm center under the actuation of light and the magnetic field. Then, the rise of particles with convection causes the specific bottom area to decrease. When the light power is 110 mW, the rise of particles is completed within 20 s. During the subsequent hovering stage, the specific bottom area remains almost unchanged. This is due to the balance of force and torque of particle chains, as shown in the previous analysis. When the particle chains oscillate in y–z and x–z planes by changing the magnetic field, the particles gradually fall back to the substrate, and the specific bottom area increases. Reapplying the light again can cause the specific bottom area to reduce again, which shows that this process can be repeated. The vertical displacement of particles during the formation of the tornado-like swarm is shown in Fig. 5.16b. In the initial swarm shrinkage stage, the particles gather together and stack vertically, resulting in a slight increase in vertical displacement. During the rising stage, the velocity of particles can reach approximately 42 μm/s, and the vertical displacement increases substantially. In the hovering stage, there is only a small increase in the vertical displacement of the particles due to the stable 3D structure of the tornado-like swarm.

The tornado-like swarm in the hovering state generates an annular flow around the pattern, which is analyzed using numerical simulations. The radius and height of the tornado-like swarm are set as 250 μm and 800 μm, respectively. The swarm rotates at a frequency of 6 Hz, and its bottom is located at the non-slip boundary. In low Reynolds number environments, the effect of inertia is ignored. Therefore, for the incompressible fluid, the momentum equation (Eq. 5.1) is also satisfied. The

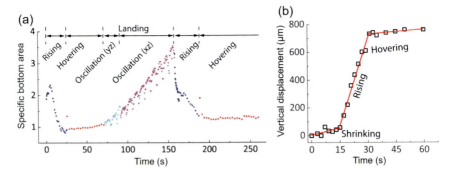

Fig. 5.16 Changes of the specific bottom area and vertical displacement over time in the pattern formation process. **a** Periodic variation of the specific bottom area during the cyclic formation process. **b** Vertical displacement of particles in the shrinking, rising, and hovering stage

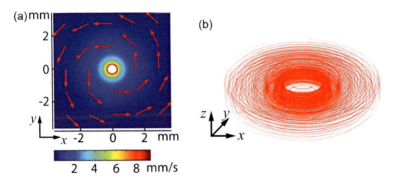

Fig. 5.17 Simulation results of the flow induced by the rotating of the tornado-like swarm. **a** The flow field of the cross section at the center of the swarm. **b** Three-dimensional distribution of streamlines around the swarm at $t = 28$ s

simulation results are shown in Fig. 5.17. Streamlines are densely distributed near the swarm and evenly distributed in regions farther away from the swarm. The velocity of the induced circling flow around the swarm is about 9 mm/s. Fast flow is only induced in a small area around the swarm, so the surrounding environment of the swarm can be divided into a near effective area and a far stable area.

5.3.3 Reaction Rate Control Using the Tornado-Like Swarm

The fluid flow induced by the tornado-like swarm enables chemical reactions to be accelerated outside it and retarded inside it. Here, the degradation reaction of methylene blue with ascorbic acid is used to demonstrate the reaction rate control ability of the tornado-like swarm. Normally, methylene blue will degrade after mixing

5.3 The Tornado-Like Swarm with a 3D Structure

with ascorbic acid for 2 min, as indicated by the lightening of the color of the solution (Fig. 5.18a1, a2). When the tornado-like swarm is generated in the solution, the strong fluid flow induced by the 3D swarm pattern significantly accelerates the degradation of methylene blue, resulting in the solution in most areas of the open container becoming transparent (Fig. 5.18a3 and a4). But the solution in the area close to the swarm appears blue, indicating that methylene blue is trapped and still present in this region.

The swarm-induced fluid flow can enhance the interaction between the chemical species, so the reaction rate in distal regions of the swarm is accelerated. According to Bernoulli's principle (Bernoulli and Bernoulli 1968), there is a region of low pressure in the vicinity of the swarm due to the high-flow velocity. Line velocities are highest at the swarm boundary, and the swarm can trap methylene blue nearby (Fig. 5.18b). The tornado-like swarm rotates as an entity, so the solution inside is hardly mixed, leading to a slow degradation rate inside it. To further investigate the effect of reaction rate control, ultraviolet–visible (UV–vis) spectroscopy experiments are conducted in the proximal and distal regions of the swarm. The absorption peak of methylene blue is around 660 nm. The results in Fig. 5.19 show that methylene blue in both regions is significantly degraded, but it always maintains a higher concentration in the proximal region. Quantitative comparison with the control group also leads to a similar conclusion (Fig. 5.20a) that the tornado-like swarm can speed up the rate of chemical reactions, and the reaction area is divided into the fast-reaction zone far from the swarm and the slow-reaction zone close to the swarm. The hovering state of the tornado-like swarm can be transformed into an oscillating state by adjusting the magnetic field, which can be used to degrade methylene blue in the proximal region. In the control group, the magnetic field is turned off after the swarm is generated and maintained for one or two minutes. By comparison, it is found that the swarm in the oscillating state is capable of speeding up the reaction rate, especially when the swarm remains in the hovering state for two minutes and then oscillates for one minute (Fig. 5.20b). In conclusion, the use of tornado-like swarms accelerates the chemical reaction in a step-by-step manner, enabling control of reaction rates.

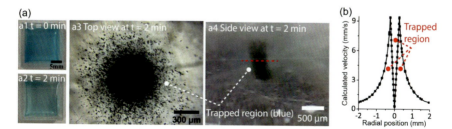

Fig. 5.18 Reaction rate control using the tornado-like swarm. **a** Degradation of the methylene blue. (a1), (a2) Degradation of methylene blue after mixing with ascorbic acid. (a3), (a4) Tornado-like swarm-assisted degradation. **b** Flow velocities at different radial locations in the cross plane in the middle of the swarm, *i.e.*, along the red line in (a4)

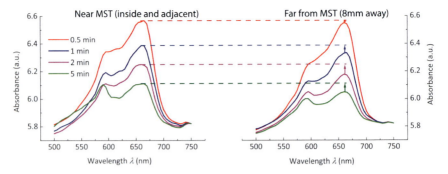

Fig. 5.19 UV–vis absorption spectra of the methylene blue near the swarm and at 8 mm from the swarm

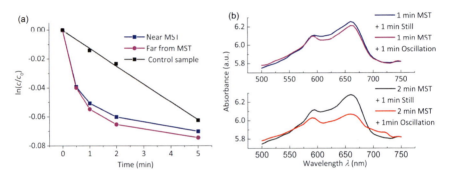

Fig. 5.20 Further characterization of methylene blue degradation. **a** Changes in $\ln(c/c_0)$ over time in regions near and far from the swarm. The slope of the curve can approximate the degradation rate. **b** The UV–vis absorption spectra of methylene blue after maintaining the hovering state for a period of time and then applying the oscillating field or resting

5.4 The Nickel Nanorod Swarm

5.4.1 Modeling of the Magnetized Nickel Nanorod

The rest of this chapter will introduce the independent pattern formation and transformation based on the nickel nanorod swarm. The nickel nanorods are synthesized by electrochemical deposition in the anodic aluminum oxide template with a length of about 1 μm and a diameter of about 300 nm (Fig. 5.21a). Actuated by the oscillating magnetic field (Fig. 3.1), nickel nanorods can form the ribbon-like swarm, which is called nanorod swarm. To investigate the behaviors of the nanorod swarm, the analysis of interactions between agents inside the swarm is required. Unlike the modeling of swarms presented previously, the analysis here starts with individual nanorods. In the magnetic field **B**, the magnetic moment of the nanorod is expressed as

5.4 The Nickel Nanorod Swarm

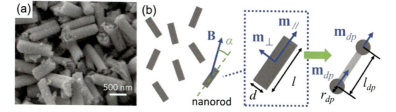

Fig. 5.21 Nickel nanorod. **a** Scanning electron microscope image of nickel nanorods. **b** The two-dipole model of the magnetized nanorod. The nanorod is considered two rigidly connected magnetic dipoles

$$\mathbf{m} = \iiint \mathbf{M} dV_r = \frac{V_r \chi_s}{\mu_0} \mathbf{B} \tag{5.6}$$

where μ_0 is the magnetic permeability of the vacuum and V_r is the volume of the nanorod. As shown in Fig. 5.21b, the nanorod is cylindrical, so its axial magnetization $\mathbf{m}_\parallel$ is different from its radial magnetization $\mathbf{m}_\perp$. This can be denoted by using the magnetic tensor χ_s, which is defined as (Singh et al. 2005)

$$\chi_s = \begin{bmatrix} \frac{\chi}{1+n_\parallel \chi} & 0 & 0 \\ 0 & \frac{\chi}{1+n_\perp \chi} & 0 \\ 0 & 0 & \frac{\chi}{1+n_\perp \chi} \end{bmatrix} \tag{5.7}$$

where $n_\parallel$ and $n_\perp$ are the demagnetizing factors along the axial direction and radial direction of the nanorod, respectively. The demagnetization factor of a nanorod of diameter d and length l can be calculated as (Prozorov and Kogan 2018)

$$\begin{cases} n_\parallel \approx \frac{d}{d+1.6l} \\ n_\parallel + 2n_\perp = 1 \end{cases} \tag{5.8}$$

The axial and radial magnetization of the nanorods can be calculated according to the above equations. The angle between the direction of the magnetic field and the long axis of the nanorod is defined as α (Fig. 5.21b). For a nanorod with a length of 1 μm and an aspect ratio of 3 ($l/d = 3$), the variation of $\mathbf{m}_\parallel$ and $\mathbf{m}_\perp$ with α is shown in Fig. 5.22a. $\mathbf{m}_\parallel$ is larger than $\mathbf{m}_\perp$ in most cases, so the nanorods are more inclined to rotate to make their long axis parallel to the direction of the external magnetic field, and then form chain-like structures through end-to-end connection. In the modeling of nanoparticle chains, each nanoparticle is treated as a single magnetic dipole. The shape of the nanorod affects properties of the nanorod chain, so the nanorod cannot be simply considered as a single magnetic dipole using previous modeling methods. Here, a two-dipole model is introduced for analyzing the magnetic interaction between nanorods, thereby clarifying the difference from nanoparticles. In this model, the nanorod is considered two magnetic dipoles ($\mathbf{m}_{dp}$) connected by a non-magnetic rigid rod. The two magnetic dipoles are both along the long axis of

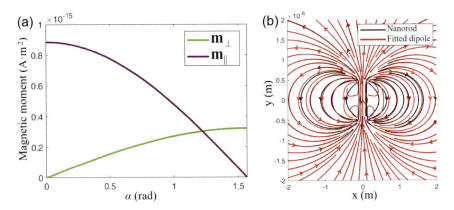

Fig. 5.22 Magnetization of the nanorod. **a** The axial magnetization $\mathbf{m}_\parallel$ and radial magnetization $\mathbf{m}_\perp$ vary with the angle α. **b** Mapping results between the finite element results and the proposed two-dipole model of the magnetic field generated by a magnetized nanorod

the nanorod (Fig. 5.21b). The magnetic field generated by each magnetic dipole at position **p** can be expressed as

$$\widehat{\mathbf{B}}(m, \mathbf{p}) = \frac{\mu_0 m}{4\pi |\mathbf{p}|^3} \left(\frac{3(\widehat{\mathbf{m}} \cdot \mathbf{p})\mathbf{p}}{|\mathbf{p}|^2} - \widehat{\mathbf{m}} \right) \quad (5.9)$$

where $m = \|\mathbf{m}\|$, $\widehat{\mathbf{m}}$ is the unit vector along the direction of **m**. To better describe the magnetic field distribution around the magnetized nanorod, a fitting process is performed to match the model to the results of the finite element method, calculated as (Derby and Olbert 2010)

$$(m_{dp}, l_{dp}) = \mathrm{argmin}\left[\widehat{\mathbf{B}}(m_{dp}, \mathbf{p}_1) + \widehat{\mathbf{B}}(m_{dp}, \mathbf{p}_2) - \mathbf{B}_r(\mathbf{p})\right] \quad (5.10)$$

where $\mathbf{p}_1$ is the position vector pointing from the first magnetic dipole to the position of **p**. $\boldsymbol{l}_{dp}$ is the vector from the second dipole to the first dipole, and $l_{dp} = \|\boldsymbol{l}_{dp}\|$. The position vector $\mathbf{p}_2$ pointing from the second dipole to point **p** can be expressed as $\mathbf{p}_2 = \mathbf{p}_1 + \boldsymbol{l}_{dp}$. $\widehat{\mathbf{B}}(m_{dp}, \mathbf{p}_1)$ and $\widehat{\mathbf{B}}(m_{dp}, \mathbf{p}_2)$ denote the magnetic field calculated using Eq. (5.9). $\mathbf{B}_r(\mathbf{p})$ is the magnetic flux density at point **p** calculated by the finite element method. The fitted region of interest is $|y| > l/2$. When the strength of the applied magnetic field is 10 mT, $l_{dp} = 0.57\ l$ and $\|\boldsymbol{m}_{dp}\| = 1.12 \times 10^{-14}$ A·m², and the results are shown in Fig. 5.22b. The average error on the norm of the magnetic flux density is 2.9%, and the average angular error is 7.8×10^{-3} rad. This model can describe the magnetic field distribution around the magnetized nanorods well.

5.4.2 Formation Mechanism of the Nanorod Swarm

Under the actuation of the oscillating magnetic field introduced earlier, the nanorods can form the ribbon-like swarm. The aspect ratio of the swarm pattern can be changed by adjusting the field ratio γ as well. Therefore, in order to understand the behavior of nanorod swarms, we further analyze interactions between nanorod chains. The schematic diagram of the nanorod chain magnetized by an oscillating magnetic field is shown in Fig. 5.23a. Similar to nanoparticle chains, the length of nanorod chains is also determined by the balance between magnetic torque and fluid drag torque. For a nanorod chain composed of $2N + 1$ nanorods, the middle nanorod is located at the origin, and the other nanorods are labeled from $-N$ to $+N$ in turn. The magnetic torque exerted on the nanorod chain can be expressed as

$$\Gamma_m = 2\sum_{k=1}^{N}(F_k r_k) = \frac{3\mu_0||\mathbf{m}||^2}{4\pi}\sin(2\varphi)\sum_{k=1}^{N}\left(2r_k \sum_{\substack{j=-N \\ j \neq k}}^{N} \frac{1}{r_{jk}^4}\right) \quad (5.11)$$

where r_k is the distance from the origin to the kth nanorod and r_{jk} is the distance between the jth and kth nanorods. F_k is the tangential component of the magnetic force exerted on the nanorod chain. φ is the angle between the direction of the magnetic field and the long axis of the chain. Here, only considering the interaction between adjacent magnetic dipoles, the above equation can be simplified as

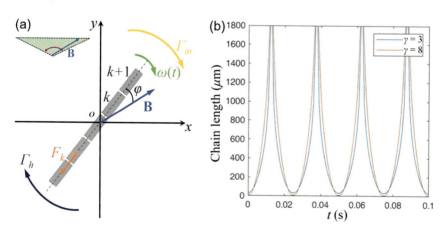

Fig. 5.23 Nanorod chain model. **a** Schematic diagram of a nanorod chain. **b** Length of nanorod chains as a function of time under the oscillating field

$$\Gamma_m = \frac{3\mu_0 \mathbf{m}_{dp}^2}{2\pi} \sin(2\varphi) \frac{Nl}{(2r_{dp})^4} \tag{5.12}$$

where $r_{dp} = (l - l_{dp})/2$ is the distance between the magnetic dipole $\mathbf{m}_{dp}$ and the end of the nanorod, as shown in the two-dipole model (Fig. 5.21). Considering the cylindrical shape of the nanorods, the fluidic drag torque exerted by the surrounding fluid on the nanorod chain is calculated as (Wilhelm et al. 2003)

$$\Gamma_h = 4\pi d^2 l \frac{N^3}{\ln(N) + \frac{1.2}{N}} \eta\omega(t) \tag{5.13}$$

where $\eta = 1.09 \times 10^{-3}$ Pa·s is the fluid viscosity and $\omega(t)$ is the angular velocity of the chain. The length of the nanorod chain can be estimated according to the Mason number $R_T = \Gamma_h / \Gamma_m$. $R_T > 1$ indicates that the drag torque is larger than the magnetic torque, causing the chain to be unstable and break. $R_T < 1$ means that the chain can exist stably. By setting $R_T = 1$ and calculating the value of N, the chain length can be estimated. Figure 5.23b shows the chain length as a function of time under an oscillating magnetic field with a field strength of 10 mT and an oscillation frequency of 20 Hz. When the angular velocity reaches a maximum, the chain length becomes the shortest. When the angular velocity tends to zero, the chain length is the longest.

Pattern formation of the ribbon-like swarm using nickel nanorods under an oscillating magnetic field is shown in Fig. 5.24a. Initially, the nanorods are randomly distributed on the substrate, and several small sub-swarms are generated after applying the oscillating magnetic field. Then the sub-swarms merge with each other by adjusting their direction. Finally, a stable ribbon-like nanorod swarm is generated. In addition, by adding a pitch angle (Fig. 5.24b) to the oscillating field, the nanorod swarm is able to perform directional motion with controllable moving direction (Fig. 5.24c), and the mechanism is the same as that described in Sect. 5.4.1. The behavior of nanorod chains during the swarm formation process is shown in Fig. 5.25. First, the nanorods form chain-like structures under the influence of an oscillating magnetic field (Fig. 5.25, stage I). When the magnetic field strength is maximized, the angular velocity of the chain is minimized, and the chain length is increased to the longest (Fig. 5.25, stage II). At this point, nanorod chains are actuated by chain-chain interactions to the proper position, resulting in a balance of magnetic and hydrodynamic forces. Then, the field strength decreases to a minimum, the angular velocity becomes maximum, and the chain breaks (Fig. 5.25, stage III). In another half-cycle, the field strength increases again, and the nanorod chains reassemble (Fig. 5.25, stage VI). After multiple cycles, the nanorod chains form a stable ribbon-like swarm pattern from a macroscopic view.

5.4 The Nickel Nanorod Swarm

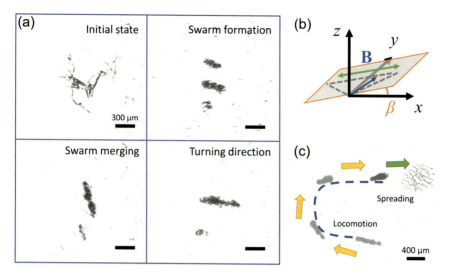

Fig. 5.24 Pattern formation and locomotion of the ribbon-like nanorod swarm. **a** Formation of the ribbon-like nanorod swarm under the oscillating magnetic field ($\gamma = 12$, the frequency is 20 Hz, and the field strength is 5 mT). **b** Schematic diagram of the oscillating magnetic field with a pitch angle. **c** Locomotion of the nanorod swarm

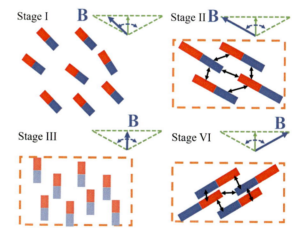

Fig. 5.25 Schematic diagram of the formation mechanism of the nanorod swarm

5.4.3 Pattern Transformation of the Nanorod Swarm

The aspect ratio of the nanorod swarm can be changed by adjusting the field ratio γ of the oscillating field. The pattern transformation is based on the interactions inside the swarm, so the interactions between nanorod chains are first analyzed. In the analysis, the magnetic interaction between the chains is mainly studied when the magnetic field oscillates to one side (*i.e.*, the magnetic field strength is maximum), because the chain-chain magnetic force is dominant and reaches the maximum value at this time in the oscillation period. For a chain consisting of $2N + 1$ nanorods, it contains $4N + 2$ magnetic dipoles. After applying the magnetic field, the magnetic force between every two dipoles in different chains can be calculated as

$$\mathbf{F}(j, k) = \frac{3\mu_0}{4\pi} (\mathbf{m}_j \cdot \nabla) \left(\frac{\mathbf{m}_k \cdot \mathbf{p}_{jk}}{\mathbf{p}_{jk}^4} \hat{p}_{jk} - \frac{\mathbf{m}_k}{\mathbf{p}_{jk}^3} \right) \quad (5.14)$$

where $\mathbf{m}_j$ and $\mathbf{m}_k$ are the magnetic moments of the jth and kth dipoles, respectively. $\mathbf{p}_{jk}$ is the vector pointing from the jth dipole j to the kth dipole, and $\hat{p}_{jk}$ is the unit vector in the direction of $\mathbf{p}_{jk}$. The magnetic potential between these two dipoles can be expressed as (Buyevich and Ivanov 1992)

$$U(j, k) = -\left(3 \frac{(\mathbf{m}_j \cdot \mathbf{p}_{jk})(\mathbf{m}_k \cdot \mathbf{p}_{jk})}{\mathbf{p}_{jk}^5} - \frac{\mathbf{m}_j \cdot \mathbf{m}_k}{\mathbf{p}_{jk}^3} \right) \quad (5.15)$$

Assuming that the chains are parallel to each other, the total magnetic potential between the two chains can be expressed as

$$W_m = \sum_{j=1}^{4N+2} \sum_{k=1}^{4N+2} U(j, k) = -m^2 \sum_{j=1}^{4N+2} \sum_{k=1}^{4N+2} \left(3 \frac{\varepsilon_{jk}^2}{\left(h^2 + \varepsilon_{jk}^2\right)^{5/2}} - \frac{1}{\left(h^2 + \varepsilon_{jk}^2\right)^{3/2}} \right) \quad (5.16)$$

where $h = d_c \cos(\cot^{-1}\gamma)$ is the shortest distance between two parallel chains and ε_{jk} is the distance between the jth and kth dipoles along the long axis of the chain, which can be calculated as $\varepsilon_{jk} = h\gamma + 2r(k-j)$, where r is the radius of each magnetic dipole and $r \approx r_{dp}$. The variable ΔW_m ($\Delta W_m = W_m(h + \Delta h) - W_m(h)$, $\Delta h > 0$) is defined to denote the tendency of the chain's movement. When $\Delta W_m < 0$, the magnetic interaction between the chains is repulsive; $\Delta W_m > 0$ means that there will be a mutually attractive magnetic force between the chains. Changing γ results in the variation of ΔW_m, as shown in Fig. 5.26. For nanorod chains with a length of 200 μm, ΔW_m gradually changes from negative to positive with increasing γ (Fig. 5.26a), indicating that the chain-chain magnetic interaction changes from repulsive to attractive. With the increase of γ, ΔW_m reaches a maximum value and then gradually decreases, which shows that the magnetic attraction is weakened. The magnetic interaction between chains of different lengths is further calculated, as shown in Fig. 5.26b. When the chain length increases (500 μm), ΔW_m becomes negative, and the chains repel each other; when the chain length decreases (160 μm), ΔW_m is positive, and the chain-chain magnetic interaction is attractive. Figure 5.27 shows the finite element simulation results of the magnetic field distribution around two nanorod chains. When γ is small, the two chains tend to repel each other. When γ becomes large, the two chains will attract each other.

5.4 The Nickel Nanorod Swarm

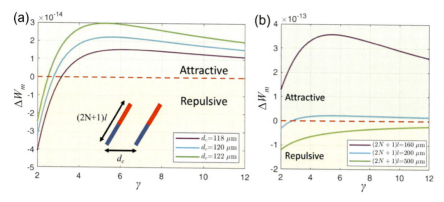

Fig. 5.26 Variation of ΔW_m with γ. **a** The relationship between γ and ΔW_m when the chain lengths are 118 µm, 120 µm, and 122 µm, respectively. **b** The relationship between γ and ΔW_m when the chain lengths are 160 µm, 200 µm, and 500 µm, respectively

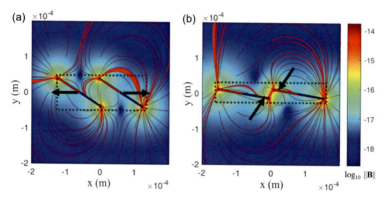

Fig. 5.27 Simulation results of the magnetic field distribution around nanorod chains in **a** anomalous regions and **b** normal regions

Inside the swarm pattern, the relative positions between nanorod chains need to ensure the balance of chain-chain magnetic and hydrodynamic interactions. According to the change of ΔW_m when γ is increased, the pattern transformation of nanorod swarms by adjusting γ can be divided into two regions. When γ increases from smaller values, the chain-chain magnetic interaction will change from repulsive to attractive, so the swarm will shrink along its long axis (x-axis). As γ keeps increasing, the magnetic attraction decreases, leading to the elongation of the swarm along the x-axis and the decrease in its width. Therefore, these two regions are named anomalous region (AR) and normal region (NR), respectively, and the demarcation value between them is γ_c. In the AR ($\gamma < \gamma_c$), the aspect ratio of the swarm pattern becomes smaller with increasing γ, which is contrary to the pattern transformation behavior of the ribbon-like nanoparticle swarm introduced in Chapter 3. While in the

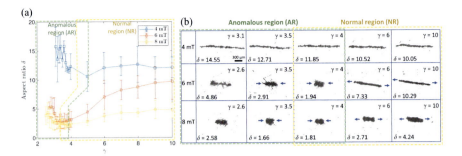

Fig. 5.28 Anomalous region (AR) and normal region (NR) of nanorod swarm's pattern transformation. **a** Aspect ratio δ of nanorod swarms as a function of field ratio γ under different magnetic field strengths. Each error bar denotes the standard deviation from four trials. **b** Experimental images of nanorod swarms during pattern transformation

NR ($\gamma > \gamma_c$), the aspect ratio becomes larger with increasing γ, which is consistent with the phenomenon observed in the nanoparticle swarm.

Figure 5.28a shows the relationship between the aspect ratio δ and γ when the nanorod swarm performs pattern transformation in two regions. The aspect ratio decreases with increasing magnetic field strength. Under different magnetic field strengths, there is a clear distinction between anomalous and normal regions. For the field strength of 4 mT, the demarcation value γ_c between AR and NR is about 4–6. For magnetic field strengths of 6 mT and 8 mT, γ_c is about 3–4. In the AR, the aspect ratio can be increased to 2.28, 2.01, and 1.49 times the initial state at magnetic field strengths of 8 mT, 6 mT, and 4 mT, respectively. In the NR, when the magnetic field strength is 8 mT, 6 mT, and 4 mT, the aspect ratio can be increased to a maximum of 2.56, 3.66, and 1.20 times, respectively. Figure 5.28b shows experimental images of pattern transformation of the nanorod swarm, from which the difference between the two regions is evident. The oscillation of nanorod chains in the AR is more vigorous than that in the NR, resulting in a looser swarm pattern and lower stability.

The oscillating frequency of the magnetic field also has an effect on the morphology of the nanorod swarm. As the oscillating frequency increases, the phase lag between the nanorod chain and the magnetic field increases accordingly, as shown in Fig. 5.29a. When the magnetic field changes from $\mathbf{B_1}$ to $\mathbf{B_2}$, the oscillation angle of the nanorod chain changes from θ_1 to θ_2. Subsequently, the magnetic field oscillates to the other side ($\mathbf{B_3}$), and the chain also turns to the other side with the magnetic field before reaching the position of $\mathbf{B_2}$, and its oscillation angle becomes θ_3. Therefore, the oscillation range of the chain is smaller than that of the magnetic field and decreases as the input oscillating frequency increases. The oscillating frequency of the magnetic field affects the relaxation time of the nanorod chains, which causes changes in the length of the chains as well as the chain-chain interactions. Since this effect is difficult to quantitatively characterize using an accurate mathematical model, the effect of the input oscillating frequency on the behavior of the nanorod swarm is experimentally demonstrated. When the field ratio γ keeps constant, increasing the

5.5 Independent Control of Micro/Nanorobot Swarms

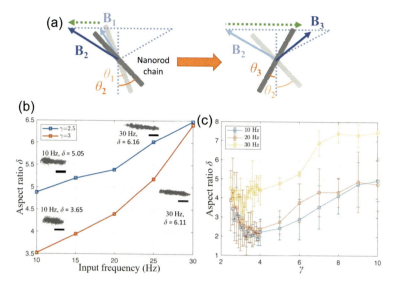

Fig. 5.29 Influence of input oscillating frequency on pattern transformation of the nanorod swarm. **a** Schematic diagram of the phase lag of nanorod chains under an oscillating magnetic field. **b** The relationship between aspect ratio and oscillating frequency in the anomalous region. The scale bar in the inset is 400 μm. **c** variation of aspect ratio with γ at different oscillating frequencies. Each data point is the mean of two experiments

input frequency causes the swarm to elongate, and the aspect ratio becomes larger (Fig. 5.29b). Pattern transformation of the swarm is conducted under three different frequencies, respectively, and the changes in aspect ratio are shown in Fig. 5.29c. Although increasing the frequency will increase the aspect ratio of the swarm, it can still be observed that the aspect ratio first decreases and then increases with the increase of γ.

5.5 Independent Control of Micro/Nanorobot Swarms

5.5.1 Comparison Between Nanorod and Nanoparticle Swarms

Pattern transformation behaviors of the nanorod swarm and the nanoparticle swarm are different in some cases, which is mainly caused by the difference between chain-chain magnetic interactions of the nanorod chain and nanoparticle chain. Due to the different shapes and magnetic susceptibility of materials, the nickel nanorod chains are generally longer than Fe_3O_4 nanoparticle chains under the same oscillating magnetic field. The experimentally measured and mathematical model calculated lengths of nanorod chains and nanoparticle chains are shown in Fig. 5.30a, showing

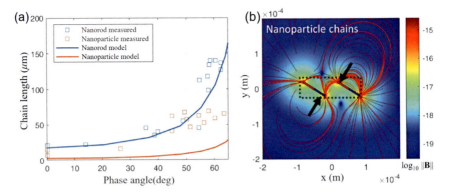

Fig. 5.30 Comparison of nanorod chains and nanoparticle chains in swarms. **a** Experimental measurements and theoretical calculations of chain lengths in the nanorod swarm and nanoparticle swarm. The strength is 6 mT, the frequency is 20 Hz, and $\gamma = 3$. The nanorods are 300 nm in diameter and 1 μm in length, and the Fe$_3$O$_4$ nanoparticles are 100 nm in diameter. **b** Simulation results of the magnetic field distribution around the magnetized nanoparticle chains

that the length of nanorod chains is at least twice as long as that of nanoparticle chains. The chain length affects the chain-chain magnetic interaction, especially when the γ value of the oscillating field is small. According to Eq. (5.15), Eq. (5.16), and the simulation results in Figs. 5.27a and 5.30b, longer nanorod chains produce repulsive magnetic forces at smaller values of γ, rather than attractive magnetic forces between shorter nanoparticle chains. The relationship between ΔW_m and chain length is shown in Fig. 5.31. In the calculation, $m = \|\mathbf{m}_{dp}\|$, $r = r_{dp}$, $h = 22$ μm, $\Delta h = 1$ μm, $\gamma = 2.5$, and the chain length is denoted as $(4N + 2)r_{dp}$. The results show that the chain-chain magnetic interaction changes from attractive to repulsive with increasing chain length when the chain-chain distance is kept constant. For nanoparticle chains, the length is generally shorter than 100 μm, so the chains tend to attract each other; for nanorod chains whose length is generally longer than 100 μm, the chains tend to repel each other.

The flow induced by oscillating nanorod and nanoparticle chains is studied by using finite element simulation, and the results are shown in Fig. 5.32. The hydrodynamic interaction force between the oscillating chains is repulsive, and increasing the chain length results in higher velocity flow and stronger chain-chain hydrodynamic repulsion. This hydrodynamic repulsion, together with the magnetic repulsion, will lead to a tendency for the swarms to elongate, especially for chains with long lengths. Overall, the different lengths of nanorod and nanoparticle chains lead to large differences in chain-chain interactions. Therefore, in the anomalous region, the pattern transformation behavior of the nanorod swarm is distinct from that of the nanoparticle swarm.

Pattern transformation of the nickel nanorod swarm and the Fe$_3$O$_4$ nanoparticle swarm under different magnetic field strengths (6 mT and 8 mT) is shown in Fig. 5.33. Figure 5.33a shows patterns of the two swarms in AR. It can be seen

5.5 Independent Control of Micro/Nanorobot Swarms

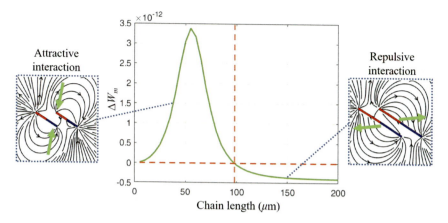

Fig. 5.31 Relationship between ΔW_m and the length of chains. The chain-chain magnetic interaction changes from attractive to repulsive with increasing chain lengths

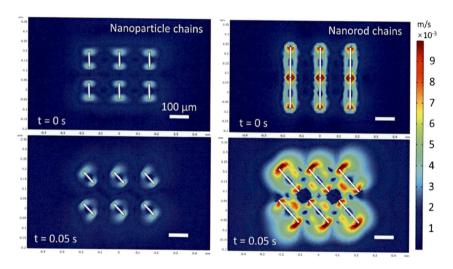

Fig. 5.32 Simulation results of flow induced by oscillating nanorod and nanoparticle chains

that the nanorod swarm in AR shrinks with the increase of γ value. For nanoparticle swarms, increasing the value of γ does not result in significant elongation of the swarm pattern, as described in Sect. 3.3.1. In the experiment, the aspect ratio is measured after the swarm pattern remains unchanged for more than 15 s, and the measured results of the nanorod and nanoparticle swarms at different γ values are shown in Fig. 5.33b, c. The results show that in the normal region, both the nanoparticle swarm and the nanorod swarm elongate with the increase of γ and shrink with the decrease of γ, which is consistent with the model prediction. In the anomalous region, the aspect ratio of the nanoparticle swarm remains almost unchanged. When

$\gamma < 4$, the aspect ratio fluctuation of the nanoparticle swarm is within 15.7% when the strength is 6 mT and within 20.5% when the strength is 8 mT. However, the aspect ratio of the nanorod swarm can still be adjusted in the AR, and it increases with the decrease of γ. In addition, the interaction between nanorod chains is more violent than that of nanoparticle chains, which leads to less stability of nanorod swarm, thus leading to larger errors in the measured aspect ratios.

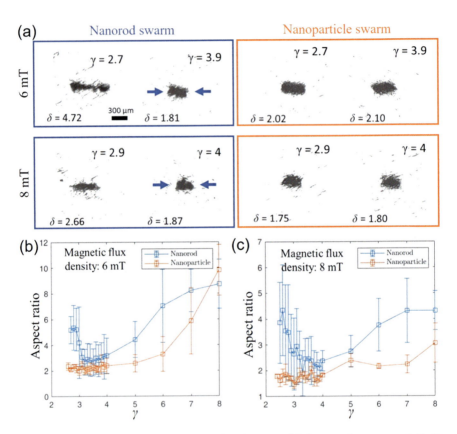

Fig. 5.33 Comparison of pattern transformation between nickel nanorod swarms and Fe$_3$O$_4$ nanoparticle swarms. **a** Patterns of the two swarms in the anomalous region when the magnetic field strengths are 6 mT and 8 mT, respectively. **b** The relationship between the swarm aspect ratio and the field ratio γ when the magnetic field strength is 6 mT. **c** The relationship between the swarm aspect ratio and the field ratio γ when the magnetic field strength is 8 mT

5.5.2 Independent Pattern Formation and Transformation of Nanorod and Nanoparticle Swarms

Taking advantage of the different behaviors of nickel nanorod swarms and Fe_3O_4 nanoparticle swarms, the independent pattern formation and transformation of microswarms under the same field input and surrounding environment can be achieved, and this process is simply demonstrated by experiments here. First, two magnets with a distance of about three millimeters are placed under the open tank, and then nanorod and nanoparticle solutions are added into the tank at the positions above the two magnets, respectively. Due to the generated local field gradient at the tips of the magnets, the two magnetic materials are concentrated at different locations and do not mix. The tank is then placed into the workspace of a three-axis Helmholtz coil, and an oscillating field (the strength is 6 mT and $\gamma = 12$) is applied. After two stable ribbon-like swarm patterns are generated, γ is adjusted to induce the pattern transformation of the swarms.

The experimental results are shown in Fig. 5.34. Two kinds of swarms are successfully generated in the same external field and environment, and they have different pattern transformation behaviors. In AR, the nanorod swarm can perform pattern elongation by decreasing γ or pattern shrinkage by increasing γ, while the pattern of nanoparticle swarm is basically unchanged. In NR, the nanorod and nanoparticle swarms change their patterns synchronously, i.e., they simultaneously elongate as γ increases and shrink as γ decreases. Besides, the nanorod swarm is less stable in AR than the nanoparticle swarm, while in NR patterns of both the two swarms are stable.

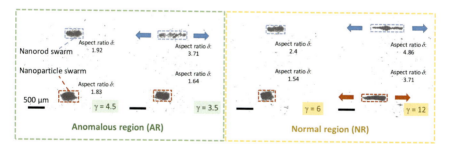

Fig. 5.34 Independent pattern formation and transformation of two swarms under the same field input and environment. The left shows that in AR, the nanorod swarm can perform reversible pattern transformation by changing γ, while the nanoparticle swarm remains largely unchanged. The right side shows that in NR, the patterns of nanorod and nanoparticle swarms can elongate or shrink in synchrony with changing γ

5.6 Conclusion

This chapter introduces two of the research challenges in the development of micro/nanorobot swarms, *i.e.*, the three-dimensional structure and independent control of micro/nanorobot swarms. First, due to the lack of effective methods to generate vertical displacements of tiny objects in the swarm, micro/nanorobot swarms in most research have two-dimensional patterns and rely on boundaries. The construction of three-dimensional patterns can provide a new dimension for micro/nanorobot swarms, which is one of the major challenges in this field. On the other hand, realizing the independent control of micro/nanorobot swarms can significantly enhance the function of the swarms. However, since the control strategies of microswarms are very different from those of individual micro/nanorobots, special control methods and structure designs of microswarms are required to realize the independent control. This chapter presents preliminary strategies to address these two challenges.

In order to build 3D microswarm, a strategy of integrating magnetic fields and light is used to generate a tornado-like swarm. $Fe_3O_4@SiO_2$ particles are used as building blocks in the swarm to form three-dimensional patterns in water. The swarm formation process is divided into four stages, namely rinsing, hovering, oscillating, and landing. The formation starts with forming a circular 2D swarm on the substrate under a conical rotating magnetic field, followed by applying a laser beam to raise particles in the swarm. The light-induced convection is the main reason for the vertical upward displacement of particles against gravity. The conical field, on the other hand, creates an inclination between the nanoparticle chains and the substrate, making it easier to rise with light-induced flow. As particles in the swarm continue to rise, the swarm gradually appears to a three-dimensional tornado-like structure. The swarm can maintain this 3D structure for a long time, so it is in the hovering stage. By switching the conical rotating field to a vertical oscillating field, particle chains in the swarm oscillate in situ, which makes the swarm into the oscillating stage. Finally, in the landing stage, the rising particles sink back onto the substrate due to gravity. The above stages are repeatable, which means that the tornado-like swarm is reconfigurable. The tornado-like swarm causes fluid flow throughout three-dimensional space, which can be used to control the rate of reactions, such as enhancing the degradation of methylene blue. Experimental results show that a tornado-like swarm in the hovering stage confines methylene blue within or near it, resulting in a lower reaction rate in the proximal region than in the distal region of the swarm. After methylene blue at the distal end of the swarm has been degraded, the swarm is turned into oscillating stage, which allows rapid degradation of remained methylene blue. Overall, such tornado-like swarm provides a method for constructing 3D swarming structures in the vertical direction. Building blocks in the swarm can be further medicated or functionalized, demonstrating great potential for applications in biomedicine and micro/nanoengineering. This hybrid actuation method combining magnetic field and light also broadens the ideas for the design of other types of micro/nanorobot swarms. Furthermore, the study of such tornado-like swarm also deepens

5.6 Conclusion

the understanding of ubiquitous 3D swarms in nature, providing a new perspective for mimicking natural collective behaviors.

The independent control strategy for micro/nanorobot swarms is based on the nickel nanorod swarm and the Fe_3O_4 nanoparticle swarm, both of which are driven by the oscillating magnetic field and have ribbon-like patterns. The nanoparticle swarm has been introduced in detail in Chap. 3, so this chapter focuses on the nanorod swarm and compares the difference between them. Due to the cylindrical shape and magnetic anisotropy of nanorods, the chains composed of nanorods are longer than those composed of nanoparticles. When the γ value of the oscillating magnetic field is small, longer nanorod chains tend to repel each other, while shorter nanoparticle chains attract each other. By quantitatively analyzing the change of the magnetic potential between chains with γ, it can be found that pattern transformation of the nanorod swarm is divided into two regions, namely the anomalous regions when γ is small and the normal region when γ is large. In the anomalous region, the nanorod swarm shrinks with increasing γ and elongates with decreasing γ. In the normal region, the pattern transformation behavior of the nanorod swarm is consistent with that of the nanoparticle swarm. These phenomena are experimentally verified, and independent pattern formation and transformation of the two swarms under the same external input and environment are demonstrated. Different magnetic properties of tiny building blocks lead to differences in behaviors of swarms in millimeter or submillimeter scale, based on which the independent control of micro/nanorobot swarms is initially realized.

In general, the two strategies introduced in this chapter have only initially addressed the corresponding challenges, and further researches are needed. For example, although the tornado-like swarm has a three-dimensional structure, it is fixed on the substrate and cannot perform locomotion as an entity, which limits its application to a large extent. In addition, although some of the particles in the tornado-like swarm are lifted from the substrate, the formation and maintenance of the 3D pattern still need to rely on the boundary. It is still a major challenge to achieve the movement of the microswarm as an entity in three-dimensional space, especially in the vertical direction. In terms of independent control of micro/nanorobot swarms, the strategy introduced here only realizes the independent pattern formation and transformation of the two types of swarms. The limitation is that the adjustable range of independent control is small, and the pattern transformation behavior is relatively simple, so it is difficult to make each swarm form the required pattern. In the future work, the closed-loop control algorithm can be considered to improve the performance of independent control of microswarms. In addition to the independent transformation of morphology, the independent control of microswarms should also include independent motion control, which requires further in-depth exploration.

References

Ahmed D, Baasch T, Blondel N, Laubli N, Dual J, Nelson BJ (2017) Neutrophil-inspired propulsion in a combined acoustic and magnetic field. Nat Commun 8:770
Becker A, Onyuksel C, Bretl T, McLurkin J (2014) Controlling many differential-drive robots with uniform control inputs. Int J Robot Res 33(13):1626–1644
Bernoulli D, Bernoulli J (1968) Hydrodynamics and hydraulics. Dover Publications, New York, pp 289–313
Bird RB, Stewart WE, Lightfoot EN (2002) Transport phenomena, 2nd edn. Wiley, New York, pp 336–338
Buyevich YA, Ivanov AO (1992) Equilibrium properties of ferrocolloids. Physica A 190(3–4):276–294
Chen C, Liu S, Shi XQ, Chate H, Wu YL (2017) Weak synchronization and large-scale collective oscillation in dense bacterial suspensions. Nature 542(7640):210–214
Cohen JA, Golestanian R (2014) Emergent cometlike swarming of optically driven thermally active colloids. Phys Rev Lett 112(6):068302
Derby N, Olbert S (2010) Cylindrical magnets and ideal solenoids. Am J Phys 78(3):229–235
Diller E, Floyd S, Pawashe C, Sitti M (2012) Control of multiple heterogeneous magnetic microrobots in two dimensions on nonspecialized surfaces. IEEE Trans Rob 28(1):172–182
Diller E, Giltinan J, Sitti M (2013) Independent control of multiple magnetic microrobots in three dimensions. Int J Robot Res 32(5):614–631
Dong X, Sitti M (2020) Controlling two-dimensional collective formation and cooperative behavior of magnetic microrobot swarms. Int J Robot Res 39(5):617–638
Edmondson R, Broglie JJ, Adcock AF, Yang LJ (2014) Three-dimensional cell culture systems and their applications in drug discovery and cell-based biosensors. Assay Drug Dev Technol 12(4):207–218
Felfoul O, Mohammadi M, Taherkhani S, de Lanauze D, Xu YZ, Loghin D, Essa S, Jancik S, Houle D, Lafleur M, Gaboury L, Tabrizian M, Kaou N, Atkin M, Vuong T, Batist G, Beauchemin N, Radzioch D, Martel S (2016) Magneto-aerotactic bacteria deliver drug-containing nanoliposomes to tumour hypoxic regions. Nat Nanotechnol 11(11):941–947
Guo ZY, Wang T, Rawal A, Hou JW, Cao ZB, Zhang H, Xu JT, Gu Z, Chen V, Liang K (2019) Biocatalytic self-propelled submarine-like metal-organic framework microparticles with pH-triggered buoyancy control for directional vertical motion. Mater Today 28:10–16
Hernàndez-Navarro S, Tierno P, Farrera JA, Ignés-Mullol J, Sagués F (2014) Reconfigurable swarms of nematic colloids controlled by photoactivated surface patterns. Angew Chem Int Ed 53(40):10696–10700
Howell TA, Osting B, Abbott JJ (2018) Sorting rotating micromachines by variations in their magnetic properties. Phys Rev Appl 9(5):054021
Ibele M, Mallouk TE, Sen A (2009) Schooling behavior of light-powered autonomous micromotors in water. Angew Chem 121(18):3358–3362
Isaeva V (2012) Self-organization in biological systems. Biol Bull 39(2):110–118
Ji FT, Zhou DK, Zhang GY, Li LQ (2018) Numerical analysis of visible light driven gold/Ferric oxide nanomotors. IEEE Trans Nanotechnol 17(4):692–696
Jin D, Zhang L (2022) Collective behaviors of magnetic active matter: recent progress toward reconfigurable, adaptive, and multifunctional swarming micro/nanorobots. Acc Chem Res 55(1):98–109
Johnson BV, Chowdhury S, Cappelleri DJ (2020) Local magnetic field design and characterization for independent closed-loop control of multiple mobile microrobots. IEEE-ASME Trans Mech 25(2):526–534
Khalil IS, Tabak AF, Hamed Y, Tawakol M, Klingner A, El Gohary N, Mizaikoff B, Sitti M (2018) Independent actuation of two-tailed microrobots. IEEE Robot Auto Lett 3(3):1703–1710

Korycka P, Mirek A, Kramek-Romanowska K, Grzeczkowicz M, Lewinska D (2018) Effect of electrospinning process variables on the size of polymer fibers and bead-on-string structures established with a 2(3) factorial design. Beilstein J Nanotechnol 9:2466–2478

Kudrolli A, Lumay G, Volfson D, Tsimring LS (2008) Swarming and swirling in self-propelled polar granular rods. Phys Rev Lett 100(5):058001

Lee JG, Brooks AM, Shelton WA, Bishop KJM, Bharti B (2019) Directed propulsion of spherical particles along three dimensional helical trajectories. Nat Commun 10:2575

Li Z, Zhang H, Wang D, Gao C, Sun M, Wu Z, He Q (2020) Reconfigurable assembly of active liquid metal colloidal cluster. Angew Chem Int Ed 59(45):19884–19888

Loghin D, Tremblay C, Mohammadi M, Martel S (2017) Exploiting the responses of magnetotactic bacteria robotic agents to enhance displacement control and swarm formation for drug delivery platforms. The International Journal of Robotics Research 36(11):1195–1210

Lv CJ, Varanakkottu SN, Baier T, Hardt S (2018) Controlling the trajectories of nano/micro particles using light-actuated marangoni flow. Nano Lett 18(11):6924–6930

Morse RA (1963) Swarm orientation in honeybees. Science 141(3578):357–358

Mou F, Zhang J, Wu Z, Du S, Zhang Z, Xu L, Guan J (2019) Phototactic Flocking of Photochemical Micromotors. Iscience 19:415–424

Nagy M, Akos Z, Biro D, Vicsek T (2010) Hierarchical group dynamics in pigeon flocks. Nature 464(7290):890–893

Narayan V, Ramaswamy S, Menon N (2007) Long-lived giant number fluctuations in a swarming granular nematic. Science 317(5834):105–108

Palacci J, Sacanna S, Steinberg AP, Pine DJ, Chaikin PM (2013) Living crystals of light-activated colloidal surfers. Science 339(6122):936–940

Prozorov R, Kogan VG (2018) Effective demagnetizing factors of diamagnetic samples of various shapes. Phys Rev Appl 10(1):014030

Rahmer J, Stehning C, Gleich B (2017) Spatially selective remote magnetic actuation of identical helical micromachines. Sci Robot 2(3):eaal2845

Reddy LH, Arias JL, Nicolas J, Couvreur P (2012) Magnetic nanoparticles: design and characterization, toxicity and biocompatibility, pharmaceutical and biomedical applications. Chem Rev 112(11):5818–5878

Salehizadeh M, Diller E (2020) Three-dimensional independent control of multiple magnetic microrobots via inter-agent forces. Int J Robot Res 39(12):1377–1396

Sapozhnikov MV, Tolmachev YV, Aranson IS, Kwok WK (2003) Dynamic self-assembly and patterns in electrostatically driven granular media. Phys Rev Lett 90(11):114301

Singh DP, Uspal WE, Popescu MN, Wilson LG, Fischer P (2018) Photogravitactic microswimmers. Adv Func Mater 28(25):1706660

Singh H, Laibinis PE, Hatton TA (2005) Rigid, superparamagnetic chains of permanently linked beads coated with magnetic nanoparticles. Synthesis and rotational dynamics under applied magnetic fields. Langmuir 21(24):11500–11509

Snezhko A, Aranson IS (2011) Magnetic manipulation of self-assembled colloidal asters. Nat Mater 10(9):698–703

Snezhko A, Belkin M, Aranson I, Kwok W-K (2009) Self-assembled magnetic surface swimmers. Phys Rev Lett 102(11):118103

Sumpter DJ (2010) Collective animal behavior. Princeton University Press

Sun M, Fan X, Tian C, Yang M, Sun L, Xie H (2021) Swarming microdroplets to a dexterous micromanipulator. Adv Func Mater 31(19):2011193

Tottori S, Zhang L, Peyer KE, Nelson BJ (2013) Assembly, disassembly, and anomalous propulsion of microscopic helices. Nano Lett 13(9):4263–4268

Vicsek T, Zafeiris A (2012) Collective motion. Phys Reports-Rev Sect Phys Lett 517(3–4):71–140

Wang H, Pumera M (2020) Coordinated behaviors of artificial micro/nanomachines: from mutual interactions to interactions with the environment. Chem Soc Rev 49(10):3211–3230

Wang XP, Hu CZ, Schurz L, De Marco C, Chen XZ, Pane S, Nelson BJ (2018) Surface-chemistry-mediated control of individual magnetic helical microswimmers in a swarm. ACS Nano 12(6):6210–6217

Wang Q, Yang L, Wang B, Yu E, Yu J, Zhang L (2019a) Collective behavior of reconfigurable magnetic droplets via dynamic self-assembly. ACS Appl Mater Interfaces 11(1):1630–1637

Wang B, Ji FT, Yu JF, Yang LD, Wang QQ, Zhang L (2019b) Bubble-assisted three-dimensional ensemble of nanomotors for improved catalytic performance. Iscience 19:760–771

Wei YH, Han SB, Kim J, Soh SL, Grzybowski BA (2010) Photoswitchable catalysis mediated by dynamic aggregation of nanoparticles. J Am Chem Soc 132(32):11018–11020

Weinert FM, Braun D (2008) Observation of slip flow in thermophoresis. Phys Rev Lett 101(16):168301

Westrum EF Jr, Grønvold F (1969) Magnetite (Fe_3O_4) heat capacity and thermodynamic properties from 5 to 350 K, low-temperature transition. J Chem Thermodyn 1(6):543–557

Wilhelm C, Browaeys J, Ponton A, Bacri JC (2003) Rotational magnetic particles microrheology: the Maxwellian case. Phys Rev E 67(1):011504

Xie X, Li DY, Tsai TH, Liu J, Braun PV, Cahill DG (2016) Thermal conductivity, heat capacity, and elastic constants of water soluble polymers and polymer blends. Macromolecules 49(3):972–978

Xie H, Sun M, Fan X, Lin Z, Chen W, Wang L, Dong L, He Q (2019) Reconfigurable magnetic microrobot swarm: Multimode transformation, locomotion, and manipulation. Sci Robot 4(28):eaav8006

Yan J, Han M, Zhang J, Xu C, Luijten E, Granick S (2016) Reconfiguring active particles by electrostatic imbalance. Nat Mater 15(10):1095–1099

Yang L, Yu J, Yang S, Wang B, Nelson BJ, Zhang L (2022) A survey on swarm microrobotics. IEEE Trans Rob 38(3):1531–1551

Yigit B, Alapan Y, Sitti M (2019) Programmable collective behavior in dynamically self-Assembled mobile microrobotic swarms. Adv Sci 6(6):1801837

Yu J, Yang L, Zhang L (2018a) Pattern generation and motion control of a vortex-like paramagnetic nanoparticle swarm. Int J Robot Res 37(8):912–930

Yu J, Wang B, Du X, Wang Q, Zhang L (2018b) Ultra-extensible ribbon-like magnetic microswarm. Nat Commun 9:3260

Yu J, Yang L, Du X, Chen H, Xu T, Zhang L (2021) Adaptive pattern and motion control of magnetic microrobotic swarms. IEEE Trans Rob 38(3):1552–1570

Zhou DK, Gao Y, Yang JJ, Li YC, Shao GB, Zhang GY, Li TL, Li LQ (2018) Light-ultrasound driven collective "firework" behavior of nanomotors. Advanced Science 5(7):1800122

Chapter 6
Pattern Transformation Rate Control of Magnetic Microswarms

Abstract Reversible pattern transformation enhances the environmental adaptability of micro/nanorobot swarms, but strategies to control the pattern transformation rate are lacking. In this chapter, pattern transformation rate control strategies of ribbon-like and vortex-like microswarms are presented. First, a theoretical model for characterizing the magnetic and hydrodynamic interactions between nanoparticle chains in the microswarm is established. The relationships between pattern transformation rates of the two types of microswarms and the three field parameters (i.e., field ratio, input field strength, and frequency) are obtained from the calculation results with the model and the experimental results, based on which the pattern transformation rate control strategy is proposed. The control strategy is then validated in viscous Newtonian fluids, non-Newtonian bio-fluids, and flowing fluids, showing wide applicability

Keywords Microswarm · Swarm pattern · Transformation rate · Magnetic control

6.1 Introduction

Collective behaviors in nature enable living organisms to build swarms to complete tasks that are beyond the capability of individuals (Sumpter 2010; Krause et al. 2002; Couzin et al. 2005). Although performing like an entity, the natural swarm generally has a transformable pattern, exhibiting excellent environmental adaptability (Parrish and Edelstein-Keshet 1999; Sumpter 2006). For example, honeybee swarms actively change their morphology to improve pattern stability in the presence of dynamic loading (Peleg et al. 2018), the overall formation shape of birds varies over time during the migratory flight (Portugal et al. 2014), and the shape of a fish school changes under the attack of a predator (Lee et al. 2006). Inspired by fascinating natural swarming behaviors, various microswarms with transformable patterns have been developed (Rubenstein et al. 2014; Li et al. 2019; Zhou et al. 2021; Buttinoni et al. 2013). Active pattern transformation capability endows artificial swarms, especially swarms consisting of micro/nanoscale agents, with environmental adaptability and multiple functions (Wang and Zhang 2018; Wang and Pumera 2020). Patterns of microswarms can be changed to complete various tasks, including adaptive navigation through narrowed areas (Mou et al. 2019), manipulation of objects (Dong and Sitti 2020), overcoming obstacles (Sun et al. 2021), connecting microelectrodes (Jin et al. 2019), and enhancing the contrast of medical imaging (Yu et al. 2019a). These adaptive microswarms hold great potential for applications in diverse fields, especially biomedical applications (Servant et al. 2015; Wang et al. 2021; Yu et al. 2019b; Wu et al. 2020; Yang et al. 2022).

Unpredictable and rapidly changing environments require living organisms to have fast responses to external stimuli and perform activities quickly (Allen et al. 2013; Forterre et al. 2005). Like many individual creatures and biological structures capable of changing their deformable shape rapidly (Kier and Smith 1985; Oliver et al. 2016; Quillin 1999), natural swarms generally complete pattern transformation in a very short timescale, thus having a high transformation efficiency (or transformation rate). For example, the swarm of *Bacillaria paradoxa* (a widely distributed diatom) can perform fast reversible pattern expansion and contraction under light stimulation, and the pattern transformation can be completed in several to tens of seconds (Fig. 6.1a) (Ussing et al. 2005; Schmid 2007; Kapinga and Gordon 1992; Cai et al. 2013; Jahn and Schmid 2007). Although artificial microswarms can achieve multimode, reversible, and large-degree pattern transformation similar to natural swarms, their transformation rates are much slower (Fig. 6.1b). Pattern transformation of artificial microswarms is conducted without controlling transformation rates, and the transformation process can sometimes be extremely long (several minutes or even longer), impeding further applications. Improving pattern transformation rates of microswarms may promise major benefits in shortening the operation time to increase transformation efficiency, maintaining pattern stability, and enhancing environmental adaptability of microswarms, which is of great significance for applications of microswarms in complex environments, especially for in vivo medical applications. However, research on pattern transformation of artificial microswarms mainly concentrates on achieving different forms and large degrees of transformation (Xie et al. 2019; Snezhko and Aranson 2011; Jin et al. 2021; Wang et al. 2012; Yan et al. 2016; Yu et al. 2021, 2018a, b), and methods for controlling pattern transformation rates remain unknown due to the pattern transformation mechanism of microswarms has not yet been fully investigated. Therefore, to realize pattern transformation rate control of microswarms, investigations on the relationship between pattern transformation rates and external inputs are required.

In this chapter, we introduce strategies for controlling pattern transformation rates of the ribbon-like and vortex-like microswarms. The influences of three field parameters, including the field ratio, input field strength, and frequency, on pattern transformation rates of these two microswarms are studied. A mathematical model is constructed to elucidate the underlying mechanism by quantitatively characterizing the magnetic and hydrodynamic interactions between nanoparticle chains inside the microswarm. Experimental and modeling results show that pattern transformation rates of both ribbon-like and vortex-like microswarms can be well controlled by adjusting these three parameters. The ribbon-like microswarm is used

6.2 Modeling of Chain-Chain Interactions

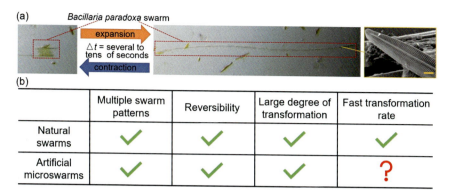

Fig. 6.1 **a** Swarm of *Bacillaria paradoxa* performs reversible pattern transformation between a contracted pattern and a fully stretched pattern (Nance 2012). The SEM image shows a single *Bacillaria paradoxa* (scale bar: 5 μm). Reproduced with permission from reference (Jahn and Schmid 2007). **b** comparison of pattern transformation capabilities between natural swarms and artificial microswarms

to further validate the pattern transformation rate control strategy in viscous fluids (viscosity: 1–4 mPa·s), bio-fluids (i.e., diluted blood, blood plasma, whole blood), and flowing fluid. This result reveals that pattern transformation rates of colloidal microswarms are controllable and provides methods for conducting pattern transformation with fast transformation rates in diverse environments, which is expected to be applied to a variety of swarming systems. This work can also deepen the understanding of collective behaviors and further improve the environmental adaptability of microswarms, which contributes to the application of microswarms in complex practical environments.

6.2 Modeling of Chain-Chain Interactions

6.2.1 Magnetic Interaction

The morphology of the swarm pattern depends on the internal interaction forces. For microswarms formed by paramagnetic nanoparticles, the nanoparticle chains are considered as basic building blocks. These particle chains are subjected to hydrodynamic and magnetic interaction forces in a periodically changing magnetic field. We first quantitatively characterize these two forces by establishing a mathematical model. Assume that two nanoparticle chains are parallel to the direction of the magnetic field in the magnetic field **B**, and they are both composed of $2n + 1$ nanoparticles with a diameter of d, as shown in Fig. 6.2a. Their relative position can be denoted by two parameters: the center-to-center distance R ($R = |\bm{R}_{ab}|$) and the position angle α (the angle between $\bm{R}_{ab}$ and $\bm{l}$). The dipole–dipole interaction force between each pair of particles on these two chains is expressed as (Melle et al. 2003)

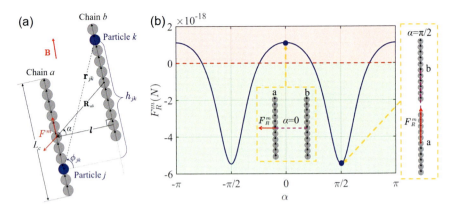

Fig. 6.2 a Schematic illustration of the two magnetic nanoparticle chains. **b** the variation of F_R^m when the position angle α changes from $-\pi$ to π. The insets are schematic diagrams of the relative position of the two chains. R and L_c (the chain length) are set as 150 μm and 80 μm, respectively

$$\mathbf{F}^d = \frac{3\mu_0 m^2}{4\pi}\left(\frac{\left(1-5(\hat{\mathbf{m}}\cdot\hat{\mathbf{r}})^2\right)\hat{\mathbf{r}} + 2(\hat{\mathbf{m}}\cdot\hat{\mathbf{r}})\hat{\mathbf{m}}}{|\mathbf{r}|^4}\right), \quad (6.1)$$

where $\mathbf{r}$ is the vector between the centers of the nanoparticles. $\mathbf{m} = V_p \chi_p \mathbf{B}/\mu_0$ is the induced magnetic dipole moment of the particles. $\mathbf{m} = |\mathbf{m}|$, $\hat{\mathbf{m}}$ and $\hat{\mathbf{r}}$ are the unit vectors of $\mathbf{m}$ and $\mathbf{r}$, respectively. Therefore, the magnetic force exerted on chain a by chain b can be expressed as (Du et al. 2021)

$$\mathbf{F}^m = \sum_{j=1}^{2n+1}\sum_{k=1}^{2n+1}\mathbf{F}_{jk}^d = \frac{3\mu_0 \mathbf{m}^2}{4\pi}\sum_{j=1}^{2n+1}\sum_{k=1}^{2n+1}\left[\frac{\left(1-5\sin^2\phi_{jk}\right)\hat{\mathbf{r}}_{jk} + (2\sin\phi_{jk})\hat{\mathbf{m}}}{|\mathbf{r}_{jk}|^4}\right] \quad (6.2)$$

$$\left(\phi_{jk} = \tan^{-1}(h_{jk}/|l|), \ |\mathbf{r}_{jk}| = \sqrt{h_{jk}^2 + |l|^2}\right),$$

where $\mathbf{F}_{jk}^d$ is the magnetic dipole–dipole interaction force between the kth particle (on chain b) and the jth particle (on chain a). ϕ_{jk} is the position angle of the jth and kth particles (Fig. 6.2a). $h_{jk} = (k-j)d + R\sin\alpha$ and $|l| = R\cos\alpha$. The center-to-center distance R is considered constant in a stable swarm, while the position angle α varies with the magnetic field direction. Figure 6.2b shows the variation of the magnetic force in the direction of $\mathbf{R}_{ab}$ (F_R^m) when α varies from $-\pi$ to π. F_R^m is sometimes positive and sometimes negative as a function of α, indicating that the magnetic interaction force between two chains can be attractive or repulsive, depending on their relative position. In addition, dynamic magnetic fields that actuate the microswarms can have varying field strengths, causing the length of the nanoparticle chains to vary with time. Therefore, the magnetic interaction between the nanoparticle chains in

6.2 Modeling of Chain-Chain Interactions

the microswarm is denoted by the average magnetic force in one actuation cycle (T) as follows:

$$\overline{F}^m = \frac{1}{T}\int_0^T \frac{3\mu_0 m^2}{4\pi} \sum_{j=1}^{2n+1}\sum_{k=1}^{2n+1}\left[\frac{-2\sin\phi_{jk}\sin\varphi - (1 - 5\sin^2\phi_{jk})\cos(\phi_{jk} - \varphi)}{|\mathbf{r}_{jk}|^4}\right]dt. \quad (6.3)$$

It is worth noting that, in order to simplify the analysis, $\overline{F}^m$ here represents the component of the average magnetic force in the horizontal direction (the x-axis). φ is the field angle defined as the angle between the field direction and the y-axis.

We take the ribbon-like microswarm (introduced in Chap. 3) as an example to analyze the magnetic interaction force between nanoparticle chains. The reversible elongation and shrinkage of the ribbon-like swarm are achieved by adjusting the field ratio γ of the oscillating magnetic field. The expression of the oscillating magnetic field is $\mathbf{B} = B_1 \sin(2\pi f t)\hat{\mathbf{x}} + (B_1/\gamma)\hat{\mathbf{y}}$, where B_1 is the input magnetic field strength, and f is the oscillation frequency. Thus, the field angle can be expressed as $\varphi = \tan^{-1}(\gamma \sin(2\pi f t))$. Furthermore, the actual strength of the oscillating magnetic field is a function of time, which is calculated as:

$$B(t) = \sqrt{(B_I\sin(2\pi f t))^2 + (B_I/\gamma)^2}. \quad (6.4)$$

As previously introduced, the length of the nanoparticle chains in the ribbon-like swarm is time-varying, with the chains being the shortest when the magnetic field strength is the smallest. The length of the chain at all time points can be quantitatively calculated by the chain length model (Wang et al. 2020; Yu et al. 2017). Here, we use the average chain length within one actuation cycle to calculate $\overline{F}^m$. In the calculation, the relative position angle α is expressed as $\alpha = \alpha_0 + \varphi$, where α_0 is the relative position angle at the initial moment ($t = 0$, and $\mathbf{B} = (B_I/\gamma)\hat{\mathbf{y}}$). Figure 6.3 shows the variation of the magnetic force (F_m) with time in one oscillating cycle ($T = 0.1$ s). Figure 6.4 shows the variation of the average magnetic force $\overline{F}^m$ with γ, which is calculated using different α_0 and R according to Eq. (6.3). $\overline{F}^m$ decreases with the increase of γ, indicating that the magnetic interaction between particle chains becomes weaker. When γ becomes larger and larger, the effect of the increase of γ on the field strength and chain length becomes smaller, so the curve of $\overline{F}^m$ becomes flatter. Assume that a microswarm consists of $2N + 1$ nanoparticle chains, where the central chain is at the origin of the coordinates and the other chains are labeled from $-N$ to N. For the ath chain, the magnetic force exerted on it by all other nanoparticle chains is given by

$$\overline{F}^m_{sum} = \sum_{b=-N}^{a-1} \overline{F}^m_{ab} + \sum_{b=a+1}^{N} \overline{F}^m_{ab}. \quad (6.5)$$

Fig. 6.3 Variation of F_m with time in one oscillating cycle ($T = 0.1$ s) when $\alpha_0 =$ 0, $\pi/4$, $\pi/3$, and $\pi/2$. The red arrows indicate the direction of the magnetic field. R is 100 μm

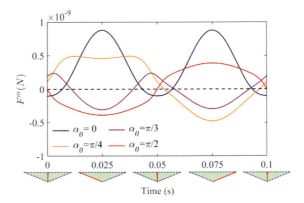

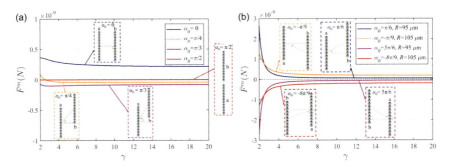

Fig. 6.4 Variation of $\overline{F}^m$ with γ. **a** $\alpha_0 = 0$, $\pi/4$, $\pi/3$, $\pi/2$, and $R = 100$ μm. **b** $\alpha_0 = \pi/6$, $R = 95$ μm; $\alpha_0 = -\pi/9$, $R = 105$ μm; $\alpha_0 = 5\pi/6$, $R = 95$ μm; $\alpha_0 = -8\pi/9$, $R = 105$ μm. The positive value means the direction of the force is to the right

For any nanoparticle chain in the microswarm, $\overline{F}^m_{sum}$ decreases with γ and always points to the center of the pattern. Furthermore, the increase in the chain-chain distance leads to a significant drop in magnetic force. Therefore, we simplify the analysis by considering only the magnetic force from the nearest chains, and the variation of $\overline{F}^m_{sum}$ with the field parameter can be represented by that of $\overline{F}^m$.

6.2.2 Hydrodynamic Interaction

Nanoparticle chains driven by the magnetic field induce fluid flow around them. The chains in the microswarm are affected by the flow induced by the remaining chains, resulting in chain-chain hydrodynamic interactions (Climent et al. 2007; Goto and Tanaka 2015). For the incompressible fluid system, the governing equation is given by the Naiver-Stokes equations:

6.2 Modeling of Chain-Chain Interactions

$$\frac{\partial \mathbf{u}}{\partial t} + \mathbf{u} \cdot \nabla \mathbf{u} = -\frac{1}{\rho}\nabla p + \upsilon \nabla^2 \mathbf{u}, \quad \nabla \cdot \mathbf{u} = 0, \tag{6.6}$$

where $\mathbf{u}$ is the velocity field, p is the pressure, ρ is the fluid density, and υ is the kinematic viscosity. We then analyze the rotation or oscillation of two nanoparticle chains (the ith and jth chains) under the same magnetic field. The flow velocity generated by the jth chain is

$$\mathbf{u}_j(x) = \omega(t) \times \mathbf{r}_x \frac{d_c^2 L_c}{(2r_x)^3}, \tag{6.7}$$

where $\mathbf{r}_x$ is the position vector defined from the center of the jth particle chain. d_c and $\omega(t)$ are the diameter and angular velocity of the nanoparticle chain, respectively. Then, chain-chain hydrodynamic interaction between the two particle chains in the low Reynolds-number environment is estimated. The lift force exerted on the ith chain by the jth chain is directly proportional to the induced flow $u_i = \omega(t)L_c/2$ and the viscosity η. The lift force is caused by fluid inertia, which involves the Reynolds number $\mathrm{Re} = \rho L_c^2 W_j/4\eta$. W_j is the shear rate near the ith chain induced by the jth chain and proportional to $\omega(t)d_c^2 L_c/(2r_{12})^3$. Therefore, the hydrodynamic interaction force exerted by the jth chain on the ith chain is expressed as

$$\mathbf{F}_i^h(t) = c_f \rho \omega(t)^2 d_c^3 \left[\frac{L_c(\gamma, f)}{2}\right]^4 \frac{(\mathbf{r}_i - \mathbf{r}_j)}{|\mathbf{r}_i - \mathbf{r}_j|^4}, \tag{6.8}$$

where c_f is a constant of proportionality. $L_c(\gamma, f)$ is the length of nanoparticle chains that changes with the field parameters. Since $\mathbf{F}_i^h(t)$ points from the center of the jth chain to that of the ith chain, the hydrodynamic interaction force between two nanoparticle chains is repulsive. The angular velocity $\omega(t)$ is time-varying, which is calculated by the derivation of the field angle $\varphi(t)$ with respect to time. According to the expression of the oscillating field, $\omega(t)$ is expressed as

$$\omega(t) = \frac{d}{dt}\varphi(t) = \frac{2\pi\gamma f \cos(2\pi f t)}{1 + \gamma^2 \sin^2(2\pi f t)} \tag{6.9}$$

$\omega(t)$ is maximum when the $\varphi(t)$ is 0° (the magnetic field points toward the y-axis) and decreases to 0 when the angle reaches its maximum. Here, we use the mean angular velocity ω_{mean} over an actuation cycle to calculate the chain-chain hydrodynamic force. ω_{mean} decreases with the increase of γ, while the increase of the frequency makes ω_{mean} become larger, as shown in Fig. 6.5a. Figure 6.5b shows the hydrodynamic force calculated from Eq. (6.8). It can be found that F^h decreases with increasing γ, which is mainly attributed to the decrease of chain length with γ, although ω_{mean} increases with γ. Specifically, according to Eq. (6.8), $F^h \propto \omega_{\mathrm{mean}}^2 L_c^4/R^3$, indicating that changes in L_c have a more significant impact on F^h than ω_{mean}. Similarly, when the oscillating frequency increases, although ω_{mean} increases accordingly, the hydrodynamic force decreases due to the decrease of L_c.

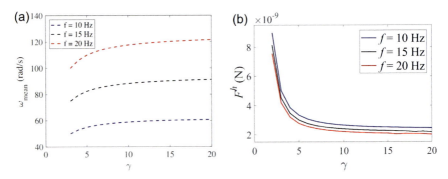

Fig. 6.5 Changes in **a** mean angular velocity and the **b** hydrodynamic force of nanoparticle chains with γ under different oscillating frequencies

Because F^h decreases significantly with increasing chain-chain distances, we also only consider the hydrodynamic interactions between adjacent nanoparticle chains.

6.2.3 Driving Force of Pattern Transformation

For the microswarm in a dynamic-equilibrium state, the chain-chain magnetic and hydrodynamic forces balance each other. The balance of these forces can be disrupted by changing the field ratio, resulting in a resultant force. Then, the nanoparticles reconfigure and move until a new dynamic-equilibrium pattern forms in which these forces are balanced again. In low Reynolds number environments, the inertia of nanoparticle chains is negligible. We consider that the resultant force induced by the broken balance of magnetic and hydrodynamic forces is counterbalanced by the drag force generated by the particle chain motion. The magnitude of this resultant force is proportional to the moving speed of the particle chain and directly affects the time required for pattern transformation and thus is defined as the driving force of pattern transformation. During the pattern transformation, the driving force gradually decreases to 0. Therefore, the driving force at the beginning of the transformation is analyzed and used to predict the transformation rate. In order to calculate the driving force, we denote the chain-chain magnetic and hydrodynamic forces as $\overline{F}^m/\overline{F}_0^m$ and $F^h/F_0^h (F^h \cos\alpha_0/F^h \cos\alpha_0)$, where $\overline{F}_0^m$ and $\overline{F}_0^h$ indicate $\overline{F}^m$ and F^h before tuning the field ratio, respectively. Then, the relative magnitude of the driving force can be expressed as $\Delta F = F^h/F_0^h - \overline{F}^m/\overline{F}_0^m$. ΔF is a dimensionless number, which indicates the state of the microswarm. ΔF is equal to 0 in a stable swarm where magnetic and hydrodynamic forces balance each other. The pattern elongates when $\Delta F > 0$ and shrinks when $\Delta F < 0$. More importantly, the chains move faster when the absolute value of ΔF ($|\Delta F|$) becomes larger, leading to a faster pattern transformation rate. In the following sections, $|\Delta F|$ will be calculated with different field parameters to predict the variation of pattern transformation rates. The calculated results will be compared with experimental results to derive the specific strategy for controlling pattern transformation rates of microswarms.

6.3 Pattern Transformation Rate Control Strategy

6.3.1 Pattern Transformation Rate Control of the Ribbon-Like Microswarm

We define the pattern transformation rate as $|\Delta\delta|/\Delta t$, where $|\Delta\delta|$ is the absolute value of the total change in the aspect ratio of the swarm pattern, and Δt is the spent time. First, the pattern transformation experiment of the ribbon-like microswarm is conducted in deionized water (DI water), and the transformation rates are measured. The effects of three magnetic field parameters, including the field ratio γ, input field strength B_I, and oscillating frequency f, on the pattern transformation rate are independently investigated, and the results are shown in Fig. 6.6.

Figure 6.6a–c shows that a larger γ makes the swarm elongate faster, and a higher pattern shrinkage rate can be obtained by adjusting γ to a smaller value. For example, the elongation rate when γ changes from 2 to 30 is 2.89 ± 0.10 s^{-1}, about 7.5 times that when γ changes from 2 to 5. The shrinkage rate when γ changes from 6 to 1.5 is 1.05 ± 0.21 s^{-1}, about 3.6 times that when γ changes from 6 to 3.5. It is worth noting that the absolute value of the change in γ ($|\Delta\gamma|$) is positively correlated with the pattern transformation rate, which indicates that the pattern transformation can be completed faster by changing the field ratio more drastically. The relationship between pattern transformation rates and the input field strength B_I is shown in Fig. 6.6d–f. The results show that the increase in B_I makes both pattern elongation and shrinkage rates higher. When B_I is increased from 8 to 12 mT, the elongation and shrinkage rates increase by 1.41 and 12.2 times, respectively. The results of the microswarm performing pattern transformation at different oscillating frequencies are shown in Fig. 6.6g–i. The elongation rate of the swarm increases with the frequency in the range of 10–20 Hz and has no obvious change in the range of 20–30 Hz (Fig. 6.6h). This may be due to the fact that the nanoparticle chains are in a step-out state at high frequencies. The elongation process is triggered by increasing γ. In this condition, the actual magnetic field strength weakens, and the angular velocity increases, resulting in the oscillation of the chain being easily out of sync with the field. Therefore, when f increases from 20 to 30 Hz, the magnitude of the driving force has no obvious change. The shrinkage rate of the clusters decreases with increasing frequency (Fig. 6.6i). The shrinkage rate when $f = 30$ Hz is only 0.17 ± 0.01 s^{-1}, which is 8.5 times lower than that when $f = 10$ Hz. The slow rate of shrinkage at high frequencies is due to the hydrodynamic forces that prevent the chains from approaching each other becoming dominant as the chains oscillate faster. The green curves in Fig. 6.6 indicate $|\Delta F|$ calculated by using field parameters the same as that used in the experiments. The

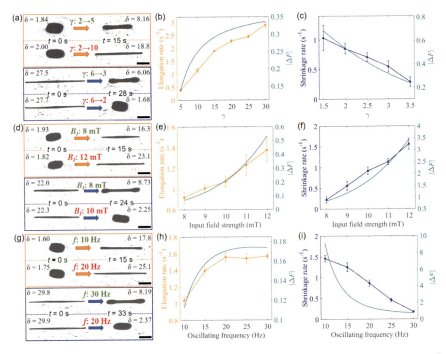

Fig. 6.6 Relationship between pattern transformation rates of the ribbon-like microswarm and different field parameters. **a** pattern transformation process under different field ratios γ. **b, c** variation of elongation rates (γ is increased from 2) and shrinkage rates (γ is decreased from 6) with γ. $f = 10$ Hz. and $B_I = 10$ mT. **d** pattern transformation process under different input field strengths B_I. **e, f** variation of elongation rates (γ is increased from 2 to 9, $f = 15$ Hz) and shrinkage rates (γ is decreased from 6 to 2, $f = 10$ Hz) with B_I. **g** pattern transformation process under different oscillating frequencies f. **h, i** variation of elongation rates (γ is increased from 2 to 9) and shrinkage rates (γ is decreased from 6 to 2) with f. $B_I = 12$ mT. The green lines show $|\Delta F|$ calculated with field parameters used in the experiments. Each error bar denotes the standard deviation from three trials. All of the scale bars are 500 μm

variations of $|\Delta F|$ have good agreement with that of the experimentally measured pattern transformation rates, indicating the accuracy of the theoretical model.

According to the modeling and experimental results, we can derive a strategy to control the pattern transformation rate of the ribbon-like microswarms by adjusting three field parameters: increasing pattern elongation rates by increasing $|\Delta \gamma|$, B_I, and f; improving pattern shrinkage rates by increasing $|\Delta \gamma|$ and B_I and reducing f. This strategy is applied to optimize the pattern transformation of a ribbon-like microswarm navigating in a narrow channel, as shown in Fig. 6.7. The microswarm is in a shrinkage state with an aspect ratio of around 2.7 before entering the narrow channel, and the initial field parameters are $B_I = 8$ mT, $f = 15$ Hz, and $\gamma = 3$. Then the field parameters are tuned to $B_I = 12$ mT, $f = 20$ Hz, and $\gamma = 17$ to elongate the swarm for entering the channel. After the aspect ratio reaches around 20, the elongation process is stopped by turning the field parameters to $B_I = 8$ mT, $f = 15$ Hz, and $\gamma = 6$. It

6.3 Pattern Transformation Rate Control Strategy

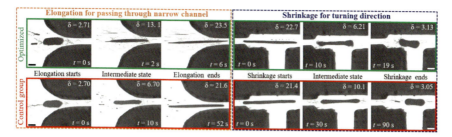

Fig. 6.7 Using pattern transformation rates control strategy during swarm navigation in a confined channel. The scale bars are 500 μm

takes only 6 s (from $t = 0$ s to $t = 6$ s) to complete the elongation process by using the pattern transformation control strategy. If the pattern elongation is conducted by simply increasing γ from 3 to 6 with B_I and f remaining unchanged (the control group in Fig. 6.7), the elongation process will take 52 s, which is about 9.6 times slower than the optimized elongation. After passing through the narrow channel, a contracted swarm pattern is needed for the convenience of turning the moving direction. The field parameters are adjusted to $B_I = 11$ mT, $f = 10$ Hz, and $\gamma = 2$ to trigger the pattern shrinkage. After 19 s, a swarm with an aspect ratio of around 3 is obtained, and the field parameters are turned back to the initial state. The shrinkage rate is approximately 5 times compared with that of simply decreasing γ to 3. This demonstration shows that the pattern transformation of microswarms using general magnetic field tuning methods is relatively slow, limiting the efficiency of performing tasks. The pattern transformation rate control strategy optimizes the time of the entire transformation process from 142 to 25 s, and the pattern transformation rate increases to 568% of the original.

6.3.2 Pattern Transformation Rate Control of the Vortex-Like Microswarm

Pattern transformation rates of other magnetic microswarms can also be realized by changing external inputs to adjust the inner interactions. We further introduce pattern transformation rates control of the vortex-like microswarm here. As introduced in Chap. 2, when changing the field ratio of the rotating field, the vortex-like microswarm is capable of performing reversible transformation between circular and elliptical patterns. Here, we write the expression of the rotating magnetic field as $\mathbf{B} = B_1 \cos(2\pi f t)\hat{\mathbf{y}} - \varepsilon \cdot B_1 \sin(2\pi f t)\hat{\mathbf{x}}$, where ε is used to represent the field ratio of the rotating field (to distinguish from the field ratio γ of the oscillating field). Similarly, the effects of three magnetic field parameters (i.e., field ratio ε, input field strength B_I, and oscillating frequency f) on pattern transformation rates of the vortex-like microswarm are investigated, and the results are presented in Fig. 6.8.

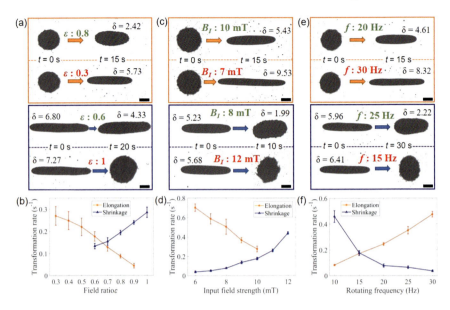

Fig. 6.8 Relationship between pattern transformation rates of the vortex-like microswarm and different field parameters. **a** pattern transformation process under different field ratios ε. **b** variation of elongation rates (ε is decreased from 1, $B_\mathrm{I} = 8$ mT, and $f = 20$ Hz) and shrinkage rates (ε is increased from 0.4, $B_\mathrm{I} = 11$ mT and $f = 20$ Hz) with ε. **c** pattern transformation process under different input field strengths B_I. **d** variation of elongation rates (ε is decreased from 1 to 0.3, $f = 30$ Hz) and shrinkage rates (ε is increased from 0.5 to 1, $f = 20$ Hz) with B_I. **e** pattern transformation process under different oscillating frequencies f. **f** variation of elongation rates (ε is decreased from 1 to 0.3, $B_\mathrm{I} = 8$ mT) and shrinkage rates (ε is increased from 0.5 to 1, $B_\mathrm{I} = 8$ mT) with f. Each error bar denotes the standard deviation from three trials. All of the scale bars are 300 μm

As shown in Fig. 6.8a, b, pattern elongation rates decrease with ε, while pattern shrinkage rates increase with ε, which seems to be the opposite of the effect of γ. If we use the absolute value of the change in ε ($|\Delta\varepsilon|$) to denote the influence of ε on pattern transformation rates of the vortex-like microswarm, a principle similar to that of the ribbon-like microswarm can be obtained; that is, both elongation and shrinkage rates increase with $|\Delta\varepsilon|$. This indicates that the vortex-like microswarm can also obtain a fast pattern transformation rate by changing the field ratio more greatly. The input field strength B_I has opposite effects on the elongation and shrinkage rates of the vortex-like microswarm. The increase in B_I makes the pattern elongation slower and the pattern shrinkage faster (Fig. 6.8c, d). Figure 6.8e and f shows that elongation rates increase with f, and the shrinkage rates decrease with f. Higher frequencies mean larger angular velocities, leading to shorter nanoparticle chains in both vortex-like and ribbon-like microswarms. The repulsive chain-chain hydrodynamic force becomes dominant, which promotes pattern elongation and hinders pattern shrinkage.

The magnetic and hydrodynamic chain-chain interactions in the vortex-like microswarm are also calculated using the theoretical model. The driving forces of pattern transformation for each condition in the experiments are shown in Fig. 6.9.

6.3 Pattern Transformation Rate Control Strategy

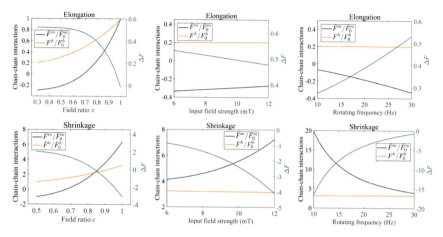

Fig. 6.9 Driving force of pattern transformation of the vortex-like microswarm under different field parameters (green lines). The magnetic field parameters used in the calculations are the same as those used in the experiments

For the rotating magnetic field, the angular velocity is $\omega(t) = 2\pi f$, and the field strength is expressed as

$$B(t) = \sqrt{(B_I \sin(2\pi f t))^2 + (\varepsilon B_I \cos(2\pi f t))^2}. \quad (6.10)$$

The calculation results in Fig. 6.9 show that the variation of the driving force with the three field parameters is in good agreement with the experimental results. It is noted that the inward force extorted by the main vortex promotes pattern contraction and hinders elongation, of which the magnitude is proportional to the rotating speed f_v of the main vortex. The relationship between f_v and the three magnetic field parameters is shown in Fig. 6.10. It can be found that f_v increases with ε and B_I and decreases with f. Therefore, the inward force becomes smaller when increasing f, decreasing ε, or reducing B_I, which makes the pattern elongation faster and the shrinkage slower. The variation of the inward force with the field parameters is consistent with the experimental results and $|\Delta F|$. The increase of inward force with B_I may be the reason for the opposite effect of B_I on the elongation rate of the vortex-like and the ribbon-like microswarms.

Based on the above modeling and experimental results, a pattern transformation rate control strategy for the vortex-like microswarm can be obtained: improve the elongation rate by increasing $|\Delta\varepsilon|$, f, and reducing B_I; improve the shrinkage rate by increasing $|\Delta\varepsilon|$, B_I, and reducing f. Figure 6.11 shows using the proposed control strategy to optimize the pattern transformation process of the vortex-like microswarm. For a vortex-like microswarm with a circular pattern under the rotating field ($B_I = 8$ mT, $f = 20$ Hz, and $\varepsilon = 1$), the general approach to elongate it into an elliptical pattern with an aspect ratio of 6 is directly reducing the field ratio ε to 0.6. By using this method, the elongation process takes 61 s (Fig. 6.11a, control

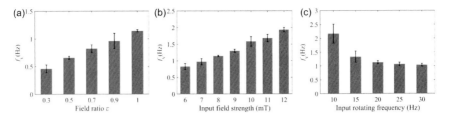

Fig. 6.10 Relationship between the rotating frequency of the main vortex (fv) and different field parameters. **a** changes in fv with the field ratio ε ($B_I = 8$ mT, $f = 20$ Hz). **b** changes in fv with input field strength B_I ($\varepsilon = 1, f = 20$ Hz). **c** changes in fv with input rotating frequency f ($\varepsilon = 1, B_I = 8$ mT). Each error bar denotes the standard deviation from five trials

group). This process is optimized by using the control strategy. The field parameters are adjusted to $B_I = 7$ mT, $f = 30$ Hz, and $\varepsilon = 0.3$, and it takes only 5 s to obtain the same desired pattern (Fig. 6.11a, optimized group). By simply increasing ε from 0.6 to 1 ($B_I = 8$ mT and $f = 20$ Hz), it takes 62 s to shrink an elliptical pattern with an aspect ratio of around 6 into a circular pattern (Fig. 6.11b, control group). According to the control strategy, we optimize the pattern shrinkage process to only 10 s by adjusting the field parameters to $B_I = 9$ mT, $f = 10$ Hz, and $\varepsilon = 1$ (Fig. 6.11b, optimized group). The total time of pattern transformation is reduced from 123 to 15 s, which indicates the pattern transformation efficiency is increased to 820% of the original. Although pattern transformation rate control strategies for ribbon-like and vortex-like microswarms have some differences, the fundamental principle is to coordinate the inner interactions by adjusting external inputs.

6.4 Pattern Transformation Rate Control in Various Environments

Although the proposed control strategy demonstrates an excellent optimization for pattern transformation rates of microswarms in DI water, whether it is effective in other environments is unknown due to the influence of surrounding environments on the behaviors of microswarms can be significant. Therefore, we use the ribbon-like microswarm to further validate the effectiveness of the pattern transformation rate control strategy in viscous Newtonian fluids, non-Newtonian bio-fluids, and flowing fluids. We defined two parameters that integrate the effect of each field parameter on pattern transformation rates: $T_e = B_I \cdot f \cdot |\Delta \gamma|$ for elongation rate control and $T_s = B_I \cdot |\Delta \gamma|/f$ for shrinkage rate control.

Pattern transformation rate control is first conducted in glycerol-water solutions with different concentrations. Figure 6.12 shows the relationship between the measured pattern transformation rates and the two control parameters. As indicated by curves in Fig. 6.12, the elongation and shrinkage rates in both DI water (the viscosity is 1 mPa·s) and glycerol-water solutions (the viscosities are from 2 to

6.4 Pattern Transformation Rate Control in Various Environments

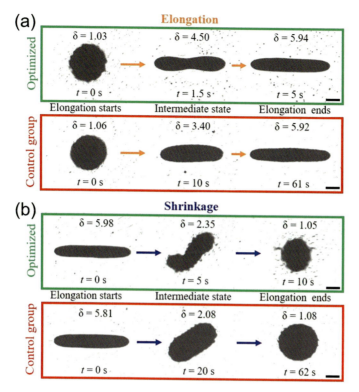

Fig. 6.11 Demonstration of using pattern transformation rates control strategy to optimize the **a** pattern elongation and **b** pattern shrinkage of a vortex-like microswarm. The scale bars are 300 μm

4 mPa·s) increase with these two parameters, respectively. The results indicate that the proposed pattern transformation rate control strategy can be applied in Newtonian fluids with different viscosities. The pattern transformation rate becomes slower with increasing viscosity, which may be attributed to the fact that the movement of nanoparticle chains is more difficult as the Reynolds number becomes smaller. Another reason is that the length of the nanoparticle chains becomes shorter due to the increase in viscosity, which leads to a decrease in the chain-chain interaction force, and the driving force of pattern transformation reduces accordingly. Therefore, a larger T_s or T_e is required to obtain high transformation rates in high-viscosity fluids.

Magnetic microswarms hold great potential in biomedical applications (Chen et al. 2021; Dong et al. 2021; Sun et al. 2022). The complex composition of bio-fluids can have significant effects on microswarms (detailed in Chap. 7). The pattern transformation rate control strategy is further validated in various bio-fluids, including fourfold diluted blood (4× diluted blood), blood plasma, and whole blood. These bio-fluids are all non-Newtonian (shear-thinning) fluids with high viscosity at low

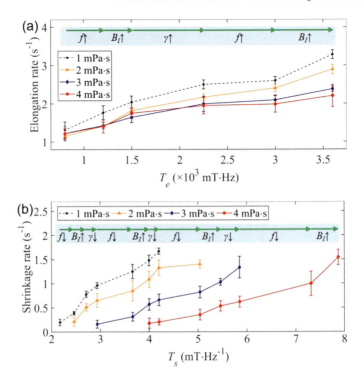

Fig. 6.12 Pattern transformation rate control of the ribbon-like microswarm in viscous Newtonian fluids. **a** The variation of pattern elongation rates with T_e. **b** The variation of pattern shrinkage rates with T_s. The adjustment of the field parameters is marked at the top of the figure. Each error bar denotes the standard deviation from three trials

shear rates. The blood plasma and whole blood have viscosities of around 1.9 and 4–5 mPa·s at a shear rate of 10 s^{-1}, respectively. Since the whole blood contains a large number of red blood cells that block the light, the pattern transformation rates are measured by using ultrasound imaging in the whole blood. Figure 6.13a shows ultrasound images of the ribbon-like microswarm during the pattern transformation process. It can be found that the swarm pattern can be fully observed by ultrasound imaging. Although the imaging contrast sometimes becomes weak, especially during the pattern shrinkage process, the shape of the swarm pattern can still be discerned due to its dynamically changing imaging contrast (Wang et al. 2022). As shown in Fig. 6.13b, c, the elongation rate increases with T_e in the diluted blood, while it increases first and then decreases with T_e in blood plasma and whole blood. The high amount of blood components, such as red blood cells (in whole blood) and proteins, can impede the movement or even cause the step-out state of nanoparticle chains. Therefore, when T_e is large, the elongation rate is inversely proportional to T_e. The pattern shrinkage rates of the microswarm in the three biological fluids all increase with T_s, which is similar to that in glycerol-water solution and DI water. It can also be found from the results that pattern transformation of microswarms

6.4 Pattern Transformation Rate Control in Various Environments

in whole blood is slow or even impossible, which is caused by the high viscosity and blood components. In this case, the proposed pattern transformation rate control strategy is an effective way to enable microswarms to work properly in bio-fluids.

When applied in physiological environments, microswarms may also be affected by flowing fluids (Ahmed et al. 2021). Therefore, the pattern transformation rate control of the ribbon-like microswarm in dynamic flow is studied. A square tube with an inner surface width of 5 mm is used to create a dynamic environment. The applied flow rate is 4 mL/min (average flow velocity: 2.67 mm/s), which is comparable to the flow rate of bio-fluids in some lumens of the human body (such as bile ducts) (Howard et al. 1991). The ribbon-like microswarm can remain stable when drifting with the flow (Fig. 6.14a). At $t = 10$ s, the field parameter is adjusted to different values to trigger the pattern transformation. At the same time, a pitch angle is applied to keep the swarm in place. We define the right side of the swarm as the head and the left side as the tail according to the direction of the fluid flow. During pattern elongation, particles at the tail of the swarm move downstream, while those at the head move upstream, resulting in a swarm pattern with a thinner head and thicker tail (Fig. 6.14a). In contrast, the microswarm has a pattern with a thicker head and thinner tail when performing pattern shrinkage. In the dynamic flow, a higher elongation rate can be obtained by enlarging γ, B_1, and f; reducing γ and f and increasing B_1 make the shrinkage of the swarm faster (Fig. 6.14). These results demonstrate that the proposed pattern transformation rate control strategy is still applicable in flowing fluids.

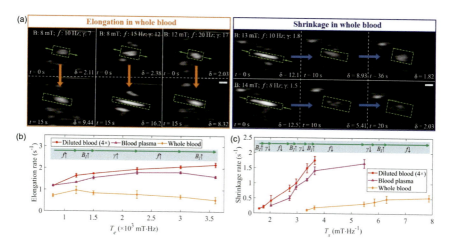

Fig. 6.13 Pattern transformation rate control of the ribbon-like microswarm in non-Newtonian bio-fluids. **a** ultrasound images of the pattern transformation processes in whole blood. The scale bars are 500 μm. **b** the variation of pattern elongation rates with T_e. **c** the variation of pattern shrinkage rates with T_s. The adjustment of the field parameters is marked at the top of the figure. Each error bar denotes the standard deviation from three trials

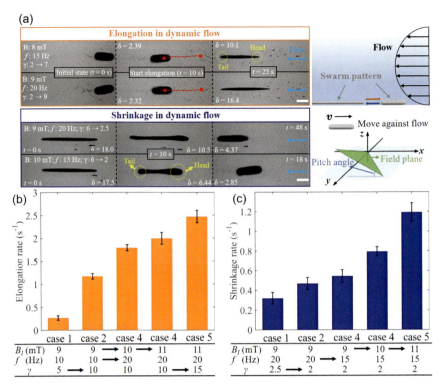

Fig. 6.14 Pattern transformation rate control of the ribbon-like microswarm in flowing fluids. **a** pattern transformation process in flowing fluids. The scale bars are 500 µm. **b** pattern elongation rates in flowing fluids. **c** pattern shrinkage rates in dynamic flow. The average flow velocity is 2.67 mm/s. Each error bar denotes the standard deviation from three trials

6.5 Conclusion

In this chapter, we introduce the pattern transformation control strategy for the ribbon-like and vortex-like microswarms. A mathematical model is constructed to analyze the inner interactions. Owing to the different changes in chain-chain hydrodynamic and magnetic forces when tuning field parameters, the driving force of pattern transformation can be adjusted. The influence of three magnetic field parameters, including the field ratio, the input field strength, and frequency, on pattern transformation rates is investigated. The modeling and experimental results are in good agreement with each other, based on which the pattern transformation rate control strategy is derived, as shown in Fig. 6.15. In consideration of biomedical applications, the effectiveness of the proposed strategy is further validated in various environments, including viscous Newtonian fluids, non-Newtonian bio-fluids, and flowing fluids. This research provides a new perspective for active adjustment of

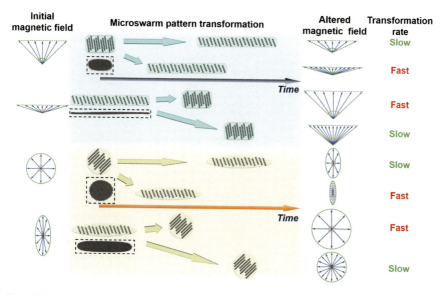

Fig. 6.15 Schematic diagram showing pattern transformation control strategies for ribbon-like and vortex-like microswarms

pattern transformation rate and fundamental understanding of pattern transformation mechanism. The method of analyzing and adjusting the inner interactions of the microswarm can be extended to other adaptive reconfigurable swarming systems.

References

Ahmed D, Sukhov A, Hauri D, Rodrigue D, Maranta G, Harting J, Nelson BJ (2021) Bioinspired acousto-magnetic microswarm robots with upstream motility. Nature Machine Intell 3(2):116–124

Allen JJ, Bell GRR, Kuzirian AM, Hanlon RT (2013) Cuttlefish skin papilla morphology suggests a muscular hydrostatic function for rapid changeability. J Morphol 274(6):645–656

Buttinoni I, Bialké J, Kümmel F, Löwen H, Bechinger C, Speck T (2013) Dynamical clustering and phase separation in suspensions of self-propelled colloidal particles. Phys Rev Lett 110(23):238301

Cai J, Chen M, Wang Y, Pan J, Li A, Zhang D (2013) Culture and motion analysis of diatom bacillaria paradoxa on a microfluidic platform. Curr Microbiol 67(6):652–658

Chen H, Zhang H, Xu T, Yu J (2021) An overview of micronanoswarms for biomedical applications. ACS Nano 15(10):15625–15644

Climent E, Yeo K, Maxey MR, Karniadakis GE (2007) Dynamic self-assembly of spinning particles. J Fluids Eng 129(4):379–387

Couzin ID, Krause J, Franks NR, Levin SA (2005) Effective leadership and decision-making in animal groups on the move. Nature 433(7025):513–516

Dong X, Sitti M (2020) Controlling two-dimensional collective formation and cooperative behavior of magnetic microrobot swarms. Int J Robotics Res 39(5):617–638

Dong Y, Wang L, Yuan K, Ji F, Gao J, Zhang Z, Du X, Tian Y, Wang Q, Zhang L (2021) Magnetic microswarm composed of porous nanocatalysts for targeted elimination of biofilm occlusion. ACS Nano 15(3):5056–5067

Du X, Yu J, Jin D, Chiu PWY, Zhang L (2021) Independent pattern formation of nanorod and nanoparticle swarms under an oscillating field. ACS Nano 15(4):4429–4439

Forterre Y, Skotheim JM, Dumais J, Mahadevan L (2005) How the venus flytrap snaps. Nature 433(7024):421–425

Goto Y, Tanaka H (2015) Purely hydrodynamic ordering of rotating disks at a finite reynolds number. Nat Commun 6:5994

Howard P, Murphy G, Dowling R (1991) Gall bladder emptying patterns in response to a normal meal in healthy subjects and patients with gall stones: Ultrasound study. Gut 32(11):1406–1411

Jahn R, Schmid A-MM (2007) Revision of the brackish-freshwater diatom genus bacillaria gmelin (bacillariophyta) with the description of a new variety and two new species. Eur J Phycol 42(3):295–312

Jin D, Yu J, Yuan K, Zhang L (2019) Mimicking the structure and function of ant bridges in a reconfigurable microswarm for electronic applications. ACS Nano 13(5):5999–6007

Jin D, Yuan K, Du X, Wang Q, Wang S, Zhang L (2021) Domino reaction encoded heterogeneous colloidal microswarm with on-demand morphological adaptability. Adv Mater 33(37):2100070

Kapinga MR, Gordon R (1992) Cell motility rhythms in bacillaria paxillifer. Diatom Res 7(2):221–225

Kier WM, Smith KK (1985) Tongues, tentacles and trunks: The biomechanics of movement in muscular-hydrostats. Zool J Linn Soc 83(4):307–324

Krause J, Ruxton GD, Ruxton G (2002) Living in groups. Oxford University Press

Lee S-H, Pak H, Chon T-S (2006) Dynamics of prey-flock escaping behavior in response to predator's attack. J Theor Biol 240(2):250–259

Li S, Batra R, Brown D, Chang H-D, Ranganathan N, Hoberman C, Rus D, Lipson H (2019) Particle robotics based on statistical mechanics of loosely coupled components. Nature 567(7748):361–365

Melle S, Calderón OG, Rubio MA, Fuller GG (2003) Microstructure evolution in magnetorheological suspensions governed by mason number. Phys Rev E 68(4):041503

Mou F, Zhang J, Wu Z, Du S, Zhang Z, Xu L, Guan J (2019) Phototactic Flocking of Photochemical Micromotors. Iscience 19:415–424

Nance J 2012 Never seen a diatom do that before...(bacillaria paradoxa). Flickr. https://www.flickr.com/photos/nebarnix/6740351653/in/photostream/

Oliver K, Seddon A, Trask RS (2016) Morphing in nature and beyond: a review of natural and synthetic shape-changing materials and mechanisms. J Mater Sci 51(24):10663–10689

Parrish JK, Edelstein- L (1999) Complexity, pattern, and evolutionary trade-offs in animal aggregation. Science 284(5411):99–101

Peleg O, Peters JM, Salcedo MK, Mahadevan L (2018) Collective mechanical adaptation of honeybee swarms. Nat Phys 14(12):1193–1198

Portugal SJ, Hubel TY, Fritz J, Heese S, Trobe D, Voelkl B, Hailes S, Wilson AM, Usherwood JR (2014) Upwash exploitation and downwash avoidance by flap phasing in ibis formation flight. Nature 505(7483):399–402

Quillin KJ (1999) Kinematic scaling of locomotion by hydrostatic animals: Ontogeny of peristaltic crawling by the earthworm lumbricus terrestris. J Exp Biol 202(6):661–674

Rubenstein M, Cornejo A, Nagpal R (2014) Programmable self-assembly in a thousand-robot swarm. Science 345(6198):795–799

Schmid AMM (2007) The "paradox" diatom bacillaria paxillifer (bacillariophyta) revisited 1. J Phycol 43(1):139–155

Servant A, Qiu F, Mazza M, Kostarelos K, Nelson BJ (2015) Controlled in vivo swimming of a swarm of bacteria-like microrobotic flagella. Adv Mater 27(19):2981–2988

Snezhko A, Aranson IS (2011) Magnetic manipulation of self-assembled colloidal asters. Nat Mater 10(9):698–703

References

Sumpter DJ (2006) The principles of collective animal behaviour. Phil Trans Royal Soc b: Biol Sci 361(1465):5–22

Sumpter DJ (2010) Collective animal behavior. Princeton University Press

Sun M, Fan X, Tian C, Yang M, Sun L, Xie H (2021) Swarming microdroplets to a dexterous micromanipulator. Adv Func Mater 31(19):2011193

Sun M, Chan KF, Zhang Z, Wang L, Wang Q, Yang S, Chan SM, Chiu PWY, Sung JJY, Zhang L (2022) Magnetic microswarm and fluoroscopy-guided platform for biofilm eradication in biliary stents. Adv Mater 34(34):2201888

Ussing AP, Gordon R, Ector L, Buczkó K, Van Landingham SL 2005 The colonial diatom" bacillaria paradoxa": chaotic gliding motility, lindenmeyer model of colonial morphogenesis, and bibliography, with translation of of müller (1783)," about a peculiar being in the beach-water", Vol 5. Gantner

Wang H, Pumera M (2020) Coordinated behaviors of artificial micro/nanomachines: from mutual interactions to interactions with the environment. Chem Soc Rev 49(10):3211–3230

Wang Q, Zhang L (2018) External power-driven microrobotic swarm: from fundamental understanding to imaging-guided delivery. ACS Nano 15(1):149–174

Wang W, Castro LA, Hoyos M, Mallouk TE (2012) Autonomous motion of metallic microrods propelled by ultrasound. ACS Nano 6(7):6122–6132

Wang Q, Yu J, Yuan K, Yang L, Jin D, Zhang L (2020) Disassembly and spreading of magnetic nanoparticle clusters on uneven surfaces. Appl Mater Today 18:100489

Wang B, Kostarelos K, Nelson BJ, Zhang L (2021) Trends in micro-/nanorobotics: Materials development, actuation, localization, and system integration for biomedical applications. Adv Mater 33(4):2002047

Wang Q, Yang S, Zhang L (2022) Magnetic actuation of a dynamically reconfigurable microswarm for enhanced ultrasound imaging contrast. IEEE/ASME Trans Mechatron. https://doi.org/10.1109/TMECH.2022.3151983

Wu Z, Chen Y, Mukasa D, Pak OS, Gao W (2020) Medical micro/nanorobots in complex media. Chem Soc Rev 49(22):8088–8112

Xie H, Sun M, Fan X, Lin Z, Chen W, Wang L, Dong L, He Q 2019. Reconfigurable magnetic microrobot swarm: Multimode transformation, locomotion, and manipulation. Science Robotics 4(28):eaav8006

Yan J, Han M, Zhang J, Xu C, Luijten E, Granick S (2016) Reconfiguring active particles by electrostatic imbalance. Nat Mater 15(10):1095–1099

Yang L, Yu J, Yang S, Wang B, Nelson BJ, Zhang L (2022) A survey on swarm microrobotics. IEEE Trans Rob 38(3):1531–1551

Yu J, Xu T, Lu Z, Vong CI, Zhang L (2017) On-demand disassembly of paramagnetic nanoparticle chains for microrobotic cargo delivery. IEEE Trans Rob 33(5):1213–1225

Yu J, Yang L, Zhang L (2018a) Pattern generation and motion control of a vortex-like paramagnetic nanoparticle swarm. Int J Robotics Res 37(8):912–930

Yu J, Wang B, Du X, Wang Q, Zhang L (2018b) Ultra-Extensible Ribbon-like Magnetic Microswarm. Nat Comm 9:3260

Yu J, Wang Q, Li M, Liu C, Wang L, Xu T, Zhang L (2019a) Characterizing nanoparticle swarms with tuneable concentrations for enhanced imaging contrast. IEEE Robotics Auto Lett 4(3):2942–2949

Yu J, Jin D, Chan KF, Wang Q, Yuan K, Zhang L (2019b) Active generation and magnetic actuation of microrobotic swarms in bio-fluids. Nat Commun 10:5631

Yu J, Yang L, Du X, Chen H, Xu T, Zhang L (2021) Adaptive pattern and motion control of magnetic microrobotic swarms. IEEE Trans Rob 38(3):1552–1570

Zhou Z, Hou Z, Pei Y (2021) Reconfigurable particle swarm robotics powered by acoustic vibration tweezer. Soft Rob 8(6):735–743

Chapter 7
Formation and Actuation of Micro/Nanorobot Swarms in Bio-Fluids

Abstract Micro/nanorobot swarms are a promising medical tool for applications such as in vivo targeted drug delivery. Bio-fluids with complex composition have non-negligible effects on formation and locomotion of microswarms. This chapter begins by reviewing three types of magnetic micro/nanorobot swarms, namely medium-induced, magnetic field-induced, and weakly interacted swarms. The effects of bio-fluid properties (including viscosity, ionic strength, and mesh-like structure) on vortex-like microswarms (medium-induced) and ribbon-like microswarms (magnetic field-induced) are investigated, and the selection strategy for optimized swarms in bio-fluids is proposed. The selection strategy is validated in gastric acid, blood plasma, whole blood, fetal bovine serum, hyaluronic acid, and bovine vitreous humor. Finally, the controlled navigation of microswarms in bio-fluids (*e.g.*, whole blood and vitreous humor) and in an ex vivo bovine eyeball is presented.

Keywords Magnetic microswarm · Swarm formation · Bio-fluids · Viscosity · Ionic strength · Mesh-like structure

7.1 Introduction

Thanks to the development of micro/nanomaterials and the application of different energy sources, various micro/nanorobot swarms have been developed (Wang et al. 2015; Wang and Zhang 2018; Jin and Zhang 2022; Solovev et al. 2013). These microswarms can be actuated in many different methods, such as magnetic fields (Sun et al. 2021; Yigit et al. 2019; Xie et al. 2019), acoustic fields (Zhou et al. 2021; Li et al. 2020; Ahmed et al. 2017), electric fields (Chen et al. 2011; Ma et al. 2015; Mao et al. 2013; Yan et al. 2016), light (Ibele et al. 2009; Mou et al. 2019; Ji et al. 2020), and chemical fuel (Kagan et al. 2011). Their building blocks are also diverse, such as colloidal particles (Yu et al. 2018a, 2018b; Massana-Cid et al. 2019), natural pollen and spores (Zhang et al. 2019; Sun et al. 2022), liquid metals (Li et al. 2020). Compared with a single micro/nanorobot, the microswarm has enhanced functionality and environmental adaptability due to its advantages, like reconfigurable morphology. Micro/nanorobot swarms are expected to bring technological innovations to many engineering fields, especially the biomedical field (Chen et al. 2021; Wu et al. 2020; Wang et al. 2021; Yang et al. 2022). Among the biomedical applications of microswarm, the most anticipated is to apply them inside the human body to complete tasks such as targeted drug delivery. Therefore, the practical application environment of microswarms will be in bio-fluids with complex components, such as blood. However, in current studies, the working environments of most microswarms are non-biological fluids, such as pure water or some chemical solutions (Martinez-Pedrero et al. 2016; Mou et al. 2020; Xu et al. 2015), which are quite different from the fluid environments inside living organisms. Studying the behaviors of microswarms in bio-fluids, such as formation, transformation, locomotion, and sensing, is one of the key points for realizing real in vivo applications of microswarms.

In fact, many research works have been carried out to investigate the behavior of individual and collective micro/nanorobots in the real bio-fluid or environment similar to it. For example, magnetically actuated helical artificial bacterial flagella can perform efficient locomotion in viscous fluids (Peyer et al. 2012, 2013) and fibrous environments (Ullrich et al. 2016). Sperm-based microrobots can move in methylcellulose-simulated artificial tubal fluid (Magdanz et al. 2013; Schwarz et al. 2020). The behaviors of bacteria and magnetic particle chains in highly viscous liquids are characterized in detail (Magariyama and Kudo 2002; Belharet et al. 2014). Magnetic fluorescent spore microrobots can be used as a tool for rapid real-time detection of bacterial toxins in patient feces (Zhang et al. 2019). Surface functionalized magnetic micropropellers can actively penetrate gastric acid and mucin gels (Walker et al. 2015; Ávila et al. 2017). A swarm of slippery helical micropropellers can be navigated through the vitreous to the retina under the magnetic field (Wu et al. 2018). The swarm of nanoparticle-shelled microbubbles is used for thrombolysis in the blood environment (Wang et al. 2020). These studies show that micro/nanorobots and their swarms have great application prospects in biological environments, but the complex components in bio-fluids can significantly affect their behavior, especially the agent-agent interactions. Therefore, it is necessary to systematically study the effects of various physical and chemical properties of bio-fluids on the function of micro/nanorobot swarms.

Due to differences in behaviors of different types of micro/nanorobot swarms in bio-fluids, this chapter first classifies the existing microswarms. Active magnetic micro/nanorobot swarms are classified into three categories, namely medium-induced, magnetic field (MF)-induced, and weakly interacted swarms, according to the differences in the dominant inner interactions of the swarm. The vortex-like swarm (introduced in Chap. 2) and the ribbon-like swarm (introduced in Chap. 3) are chosen as representatives of the medium-induced and MF-induced swarms, respectively. The influences of bio-fluid properties, including viscosity, ionic strength, and mesh-like structure, on the behavior of these two types of swarms are analyzed individually, and the strategy to obtain efficient swarms in specific bio-fluids is proposed accordingly. The optimized strategy is then validated in several different types of fluids, *i.e.*, gastric acid, blood plasma, whole blood, fetal bovine serum (FBS), hyaluronic acid (HA), and bovine vitreous humor. Subsequently, the formation and

7.2 The Library of Magnetic Active Micro/Nanorobot Swarms

Active magnetic microswarms with flexible morphology and good controllability are able to be navigated to hard-to-reach regions inside the human body and therefore have great potential for in vivo applications like targeted drug delivery. Therefore, we have divided the magnetic microswarms into three types according to the dominant inner interactions, as introduced in detail in Chap. 1. When navigated in bio-fluids, the inner agent-agent interactions of the microswarms can be significantly affected, which may prevent the microswarm from completing the specified tasks. Here, we first briefly review these three types of microswarms, *i.e.*, the medium-induced swarm, magnetic field-induced swarm, and weakly interacted swarm.

The medium of microswarms has an important influence on swarm behaviors, and most micro/nanorobot swarms are in liquid environments. Therefore, in many swarms, the hydrodynamic interaction among building blocks is more powerful than other interactions and is the basis for swarm formation. A swarm with this characteristic is called the medium-induced swarm, and several representative ones are shown in Fig. 7.1. For instance, the vortex-like swarm introduced in Chap. 2 is a typical medium-induced swarm (Fig. 7.1a) (Yu et al. 2018a). The hydrodynamic interaction between particle chains and the inward force exerted by the main vortex on the chains are based on the medium, and they act as the main force maintaining the vortex-like swarm pattern. Similarly, the three-dimensional tornado-like swarm introduced in Chap. 5 is constructed based on the light-induced fluid flow (Ji et al. 2020). The active magnetic colloids confined at the interface between oil and water can self-assemble into aster arrays due to the circular wave generated from the deformation of the interface (Fig. 7.1b) (Snezhko and Aranson 2011). A self-assembled snake-like swarm pattern at the liquid–air interface performs propulsion by breaking the symmetry of the surface flow (Fig. 7.1c) (Snezhko et al. 2009). The helical microrobots in water can also form a circular swarm pattern under a rotating magnetic field (Fig. 7.1d) (Vach et al. 2017). Overall, the formation of these swarms and their collective behaviors primarily depend on medium-based interactions, such as fluid flow and deformation of interfaces.

In the magnetic field-induced swarm (Fig. 7.2), the magnetic agent-agent interaction induced by the input magnetic field is dominant. The magnetic interaction force is the main factor for the formation of MF-induced swarms, while the hydrodynamic interaction force only plays an auxiliary role. Representative MF-induced swarms include the ribbon-like swarm described in Chaps. 3 and 5 (Fig. 7.2a). Whether the ribbon-like swarm is composed of Fe_3O_4 nanoparticles (Yu et al. 2018b) or nickel

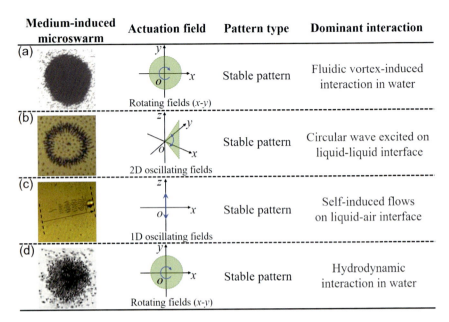

Fig. 7.1 Representative medium-induced swarms. **a** the vortex-like swarm formed by paramagnetic Fe$_3$O$_4$ nanoparticles, in which the hydrodynamic force induced by the vortices is dominant. Reproduced with permission from reference (Yu et al. 2018a). **b** Self-assembled colloidal asters formed at the oil–water interface based on the circular wave generated from the deformation of the interface. Reproduced with permission from reference (Snezhko and Aranson 2011). **c** Magnetic particles form a snake-like swarm pattern at the liquid–air interface. Reproduced with permission from reference (Snezhko et al. 2009). **d** Helical microrobots actuated by the rotating magnetic field form a circular swarm pattern in the water. Reproduced with permission from reference (Vach et al. 2017)

nanorods (Du et al. 2021), the pattern formation and transformation processes mainly depend on the magnetic interaction between particle chains (or nanorod chains). This is also evidenced by the different pattern transformation behaviors of nanoparticle swarms versus nanorod swarms due to differences in the magnetic properties of the building blocks. The magnetic particles form a tightly bound circular pattern under the rotating magnetic field (Fig. 7.2b), which can perform a neutrophil-like rolling motion along the wall after applying the acoustic field (Ahmed et al. 2017). Due to strong field-induced forces, this swarm is also capable of moving against the flow (Ahmed et al. 2021). Figure 7.2c shows magnetic particles driven by a three-dimensional oscillating field self-assemble into a closely arranged carpet-like swarm pattern (Martinez-Pedrero and Tierno 2015). Peanut-shaped hematite particles can form a chain-like swarm pattern under the rotating magnetic field (Fig. 7.2d) (Xie et al. 2019). Magnetic droplet swarm turns into a solid-like state when the applied rotating magnetic field has a high frequency and can roll like a wheel as an entity in the vertical plane (Fig. 7.2e) (Sun et al. 2021). Due to dominant field-induced

7.2 The Library of Magnetic Active Micro/Nanorobot Swarms

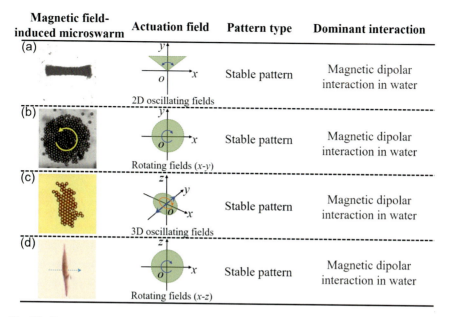

Fig. 7.2 Representative magnetic field-induced swarms. **a** the ribbon-like swarm consisting of paramagnetic Fe$_3$O$_4$ nanoparticles, in which the chain-chain magnetic force is dominant. Reproduced with permission from reference (Yu et al. 2018b). **b** magnetic particles form a circular rolling swarm actuated by a rotating field. Reproduced with permission from reference (Ahmed et al. 2017). **c** magnetic colloidal particles self-assemble into a carpet-like swarm under oscillating fields. Reproduced with permission from reference (Martinez-Pedrero and Tierno 2015). **d** the magnetic droplet swarm exhibits a solid-like state and can roll like a wheel under a high-frequency rotating magnetic field. Reproduced with permission from reference (Sun et al. 2021)

interactions among agents, MF-induced swarms are generally more stable and less susceptible to fluid flow than medium-induced swarms.

Except for medium-induced swarms dominated by hydrodynamic interactions and MF-induced swarms dominated by magnetic interactions, the agent-agent interactions in some swarms are very weak. Each individual in the swarm behaves relatively independently and is hardly affected by other individuals around it, so this type of swarm is defined as the weakly interacted swarm (Fig. 7.3). For example, there is no obvious interaction between the individuals in a swarm of nanoparticle chains performing the tumbling motion (Fig. 7.3a) (Yu et al. 2017). The swarm of artificial bacterial flagella with near-infrared probes can navigate in the body of mice, but the swarm has no fixed pattern (Fig. 7.3b) (Servant et al. 2015). The weakly interacted swarm is characterized by the lack of inner interactions that exert constraints on the building blocks, resulting in a loose and unstable swarm pattern with constantly changing boundaries (*i.e.*, free pattern).

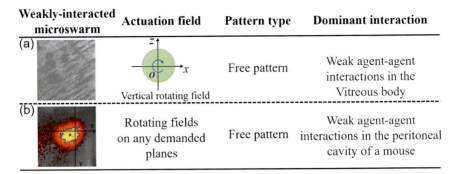

Fig. 7.3 Representative weakly-interacted swarms. **a** A swarm of nanoparticle chains performing the tumbling motion. Reproduced with permission from reference (Yu et al. 2017). **b** The swarm of artificial bacterial flagella is navigated in the peritoneal cavity of mice. Reproduced with permission from reference (Servant et al. 2015). The rotating plane of the magnetic field varies in three-dimensional space, which enables the microswimmers to move in any direction

7.3 Influence of Fluid Properties on Microswarms

7.3.1 Main Physical Properties of Bio-Fluids

For biomedical applications in the human body, confining all building blocks inside the swarm pattern is crucial for microswarms to realize targeted delivery with high efficiency and access rate. In this regard, medium-induced and MF-induced swarms are considered more suitable than weakly interacted swarms for applications such as in vivo drug delivery. Furthermore, since inner agent-agent interactions of weakly-interacted swarms are negligible, their swarming behaviors in bio-fluids can be inferred from the properties of individual building blocks. For the medium-induced and MF-induced swarms where the inner interactions are indispensable, the influence of bio-fluids needs to be systematically investigated. Here, the vortex-like swarm (Fig. 7.4a) is chosen as a representative of the medium-induced swarm, and the ribbon-like swarm (Fig. 7.4b) is chosen as a representative of the MF-induced swarm to study how the properties of bio-fluids affect the behavior of microswarms. The formation mechanisms and characteristics of these two swarms are introduced in detail in Chaps. 2 and 3.

Bio-fluids with diverse compositions have different physical properties, resulting in very different effects on microswarms. We believe that the influence of bio-fluids on microswarm mainly depends on three properties of bio-fluids, *i.e.*, fluid viscosity, ionic strength, and polymeric mesh-like structure. Therefore, the effects of these three properties on swarm formation and locomotion are analyzed separately. Glycerol-water solutions with different concentrations are used to obtain fluids with different viscosities, phosphate-buffered saline (PBS) is used to adjust the ionic strength, and hyaluronic acid (HA) is used to simulate the fibrous mesh-like structure in bio-fluids.

7.3 Influence of Fluid Properties on Microswarms

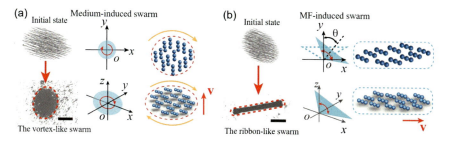

Fig. 7.4 Schematic diagrams and experimental images of the vortex-like and ribbon-like swarms. **a** the rotating field-actuated vortex-like paramagnetic nanoparticle swarm is a typical medium-induced swarm. **b** the oscillating field-actuated ribbon-like paramagnetic nanoparticle swarm is a typical MF-induced swarm. The magnetic fields are represented by the blue areas. The scale bar is 500 μm

7.3.2 Swarm Formation in Fluids with Controlled Physical Conditions

The generation process of these two swarms is first characterized. The effective region of the swarm is indicated by the red dashed line in Fig. 7.4. The ratio of the area of the effective region to the total distributed area of nanoparticles is defined as the relative effective area of the swarm, which can represent the gathering effect of microswarms in fluids of different ionic strengths and viscosities, and the results are shown in Fig. 7.5. The ionic strength of PBS solution is defined as 1.0 × , which serves as a standard to represent the ionic strength of other liquids. The results in Fig. 7.5a show that an increase in ionic strength leads to a weaker gathering effect of the vortex-like swarm. The relative effective area decreases by 12.5% when the ionic strength increases from 0.2× to 0.4×. When the ionic strength continues to increase above 0.4×, the vortex-like swarm cannot even be successfully formed. In contrast, the gathering effect of the ribbon-like clusters is less affected by the ionic strength. Although increasing the ionic strength from 0.2× to 0.4× decreases the relative effective area from 92 to 75%, continuing to increase the ionic strength to 1.0× does not change the relative effective area significantly. The influence of viscosity on swarm formation is presented in Fig. 7.5b. The increase in fluid viscosity results in a decrease in the relative effective area of the vortex-like swarm. The relative effective area decreases from 82 to 54% when the viscosity of the fluid increases from 2 to 40 cp. The ribbon-like swarm has larger relative effective areas in solutions with viscosities of 2 cp and 5 cp, but it cannot be successfully formed in fluids with a viscosity higher than 10 cp. The above results indicate that the ionic strength has a greater effect on the medium-induced swarm, while the MF-induced swarm is more sensitive to the viscosity of the fluid.

Since bio-fluids with fibrous mesh-like structures generally have high viscosity, only the effect of fibrous meshes on the formation of medium-induced swarms is investigated. In the experiments, the fibrous mesh of the fluid is obtained by using HA

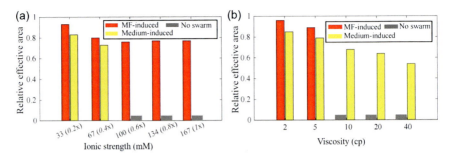

Fig. 7.5 Swarm formation in fluids with different ionic strengths and viscosities. **a** the relative effective area of medium-induced and MF-induced swarms in fluids with different ionic strengths. **b** the relative effective area of medium-induced and MF-induced swarms in fluids with different viscosities

solutions at concentrations of 3 mg/mL and 4.6 mg/mL, and their dynamic viscosities are 60 cp and 200 cp at a shear rate of 5 s^{-1}, respectively. Meanwhile, glycerol-water solutions with concentrations of 80% and 89% are prepared as controls. Figure 7.6 shows a comparison of the swarm formation in viscous fluids with and without fibrous meshes. The results show that in the glycerol-water solution, despite the absence of fibrous meshes, the excessively high fluid viscosity leads to a poor gathering effect of the swarm. In the HA solution, most of the nanoparticles could be aggregated into the swarm pattern, indicating that the fibrous mesh is beneficial to the formation of medium-induced swarms. The possible reason is that the nanoparticle chains and fibrous meshes are entangled with each other, causing them to have a larger influence area in a rotating magnetic field and, thus, being more likely to gather together. The specific relative effective area is shown in Fig. 7.7. Although both the density of the fibrous mesh and the fluid viscosity increase with the concentration of HA, medium-induced swarms are always successfully generated in fluids with fibrous meshes.

Fig. 7.6 Formation of the vortex-like swarm in viscous fluids with and without fibrous meshes. The viscosity (60 cp and 200 cp) here means the dynamic viscosity at a shear rate of 5 s^{-1}. The scale bar is 500 μm

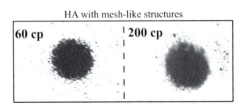

7.3 Influence of Fluid Properties on Microswarms

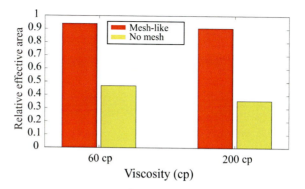

Fig. 7.7 Relative effective area of vortex-like swarms in viscous fluids with and without fibrous meshes

7.3.3 Swarm Locomotion and Transformation in Ionic and Viscous Fluids

After the formation in bio-fluids, microswarms need to be navigated to the targeted area to perform specific tasks. The locomotion ability of medium-induced and MF-induced swarms is investigated in fluids with different ionic strengths and viscosities. Although the increase in fluid viscosity slows the movement of the vortex-like swarm, the swarm is still able to move efficiently in viscous fluids (Fig. 7.8a). The vortex-like swarm can move at a velocity of 100 μm/s even when the viscosity reaches 80 cp. Since MF-induced swarms can only form in less viscous fluids, only their locomotion velocities within the viscosity range of successful formation are measured. The ribbon-like swarm moves much faster in fluid with a viscosity of 2 cp than in DI water (Fig. 7.8b). And a further increase in viscosity leads to a decrease in the swarm velocity and even the collapse of the swarm pattern at large pitch angles. Moving velocities of the vortex-like and ribbon-like swarms in fluids with different ionic strengths are presented in Fig. 7.9. The ionic strength has no significant effect on the moving speed of either swarm.

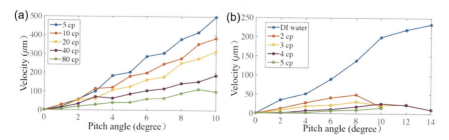

Fig. 7.8 Moving velocity of medium-induced and MF-induced swarms in fluids with different viscosities. **a** the relationship between moving velocity of the vortex-like swarm and the pitch angle in five fluids with different viscosities. **b** the relationship between moving velocity of the ribbon-like swarm and the pitch angle in five fluids with different viscosities

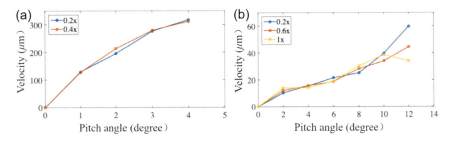

Fig. 7.9 Moving velocity of medium-induced and MF-induced swarms in fluids with different ionic strengths. **a** changes in moving velocity of the vortex-like swarm with the pitch angle in fluids with different ionic strengths. **B** changes in moving velocity of the ribbon-like swarm and the pitch angle in fluids with different ionic strengths

Pattern transformation ability is crucial for microswarms to conduct tasks such as passing through narrow channels and manipulating tiny objects. The influence of fluid viscosity and ionic strength on swarm morphology and pattern transformation behavior of the ribbon-like swarm is presented in Fig. 7.10. At the same field ratio γ, the higher ionic strength leads to a smaller aspect ratio of the swarms (Fig. 7.10a). In addition, the ionic strength also has an effect on the minimum field ratio required for swarm formation. For example, when the ionic strengths are 0.2×, 0.4×, and 0.6×, the swarm can be formed at $\gamma = 3$. When the ionic strengths are 0.8×, 1×, and 2×, the minimum field ratios are 4, 5, and 6, respectively. Ions in the solution are able to enhance interactions between particles inside the swarm, and therefore, nanoparticle chains in fluids with high ionic strength are longer and more difficult to be broken and reassembled under the same magnetic field. The reconfiguration process of the ribbon-like swarm may be hindered, and the chain-chain attraction is also enhanced, resulting in the ribbon-like cluster having smaller aspect ratios in high ionic strength fluids. In contrast, the fluid viscosity is proportional to the aspect ratio of the ribbon-like swarm, which is due to the shortening of particle chain lengths as the viscosity increases. Although the fluid properties have an effect on the morphology of the swarm pattern, pattern transformation of the ribbon-like swarm can still be successfully conducted in ionic and viscous fluids.

7.4 Selection Strategy for Optimized Swarms in Bio-Fluids

After the independent effects of each fluid parameter on the microswarm are obtained, we further investigate the formation of swarms in artificial fluids with different ionic strengths and viscosities. The selected artificial fluids have ionic strengths ranging from 83.5 to 417.5 mM (*i.e.*, 0.5 × to 2.5 × ionic strength of PBS) and viscosities ranging from 2 to 8 cp. Phase diagrams for the formation of medium-induced and MF-induced swarms are shown in Figs. 7.11a and Fig. 7.11b, respectively. As shown by illustrations at the bottom of Fig. 7.11a, the green circle (○) indicates

7.4 Selection Strategy for Optimized Swarms in Bio-Fluids

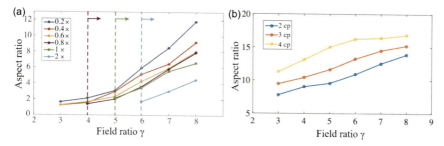

Fig. 7.10 Influence of ionic strength and viscosity on swarm transformation. **a** the relationship between the aspect ratio of the ribbon-like swarm and field ratio in fluids with different ionic strengths. **b** the relationship between the aspect ratio of the ribbon-like swarm and field ratio in fluids with different viscosities

that the swarm can be successfully formed, the blue cross (×) indicates that the generated swarm has a loose structure and certain particles (about 40%) that are not gathered, and the black cruciform symbol (+) indicates that the swarm cannot be formed. For medium-induced swarms, increased ionic strength makes the formation difficult, but swarms can still be successfully formed in fluids with both high ionic strength and high viscosity. This indicates that increasing the viscosity can reduce the adverse effect of high ionic strength on the formation of medium-induced swarms to a certain extent. The region above the red line in Fig. 7.11a represents the fluid parameters applicable for the formation of medium-induced swarms. Similarly, the red line in Fig. 7.11b represents the dividing line between the successful and unsuccessful formation regions of MF-induced swarms in these fluids. The green circle (○) and blue cross (×) indicate that the swarm is able to perform reversible pattern transformation with large aspect ratios (≥ 8) and small aspect ratios (≤ 3), respectively. The black cruciform symbol (+) indicates that big clusters are formed. The successful formation region is below the horizontal red line, indicating that MF-induced swarms are only sensitive to viscosity and that changes in ionic strength have no significant effect on swarm formation.

From the above results, fluid parameters suitable for the formation of medium-induced and MF-induced swarms are summarized, as shown in the red and blue regions in the integrated phase diagram (Fig. 7.12), respectively. The inset on the right in Fig. 7.12 is the enlarged image of the black dashed rectangle on the left. In the overlapping part of the two regions, both types of swarms can be successfully formed. The physical properties of several different actual bio-fluids, including whole blood, blood plasma, FBS, gastric acid, and vitreous humor, are listed in Table 7.1, according to which they are marked in the corresponding positions in Fig. 7.12. Based on the region in which each bio-fluid is located in the phase diagram, it is possible to predict the formation of different types of swarms in it. For example, both types of swarms can be successfully generated in whole blood and plasma. Only MF-induced swarms can be generated in FBS and gastric acid, while only medium-induced swarms are applicable in the vitreous humor. These results provide a detailed reference for the application of microswarms in bio-fluids; that is, according to the properties of a

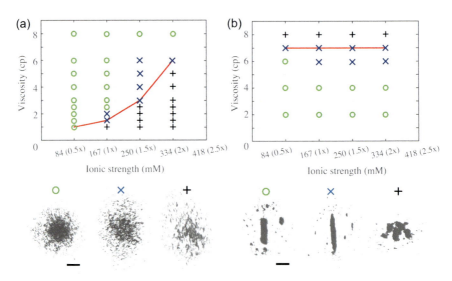

Fig. 7.11 Phase diagrams for the formation of microswarms in fluids with different ionic strengths and viscosities. **a** phase diagram for the formation of medium-induced swarms. The green circle (○) represents successful formation, the blue cross (×) represents successful formation with particle loss (about 40%), and the black cruciform symbol (+) represents failed formation. **b** phase diagram for the formation of MF-induced swarms. The green circle (○) and blue cross (×) indicate that the swarm can perform reversible pattern transformation with large aspect ratios (≥ 8) and small aspect ratios (≤ 3), respectively. The black cruciform symbol (+) represents failed formation with big clusters formed. The inserts show images of different formation states of the microswarms. The scale bars are 500 μm

particular bio-fluid, the optimal swarm type can be selected for applications in such environments.

7.5 Swarm Formation and Navigation in Bio-Fluids

7.5.1 Swarm Formation and Pattern Transformation in Bio-Fluids

Experiments on the formation of medium-induced and MF-induced swarms in different bio-fluids are conducted to verify the selection strategy. It is worth noting that the red blood cells severely block the light for optical microscopes, and ultrasound imaging is employed in the experiments to observe the swarm states in whole blood. As shown in Fig. 7.13, the active generation of both swarms is successfully realized in plasma and whole blood. In HA and vitreous humor, only medium-induced swarms are effective. In gastric acid and FBS, only MF-induced swarms can be formed. These results agree well with the predictions in Fig. 7.12, indicating that

7.5 Swarm Formation and Navigation in Bio-Fluids

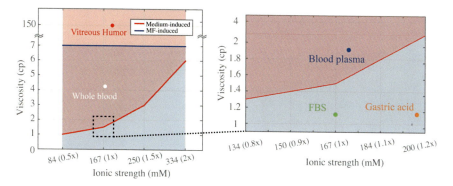

Fig. 7.12 Integrated phase diagram for the formation of medium-induced and MF-induced swarms. The red and blue regions represent the successful formation region of the medium-induced and MF-induced swarms, respectively. The inset on the right is an enlarged image of the black dashed rectangle on the left. Different types of bio-fluids, *i.e.*, whole blood, blood plasma, FBS, gastric acid, and vitreous humor, are labeled in the figure according to their physical parameters with white, blue, green, orange, and red circles, respectively. The viscosities are obtained at a shear rate of 5 s^{-1}

Table 7.1 Physical parameters of different fluids

Fluid type	Viscosity (cp)	Ionic strength (mM)	Polymetric mesh
DI water	~0.85	0	No
Plasma	~1.9	~180 (1.1 × PBS)	No
Whole blood	4–5	~180 (1.1 × PBS)	No
FBS	~1.1	~170 (1.0 × PBS)	No
Gastric acid	~1.12	~200 (1.2 × PBS)	No
Vitreous humor	~150	~170 (1.0 × PBS)	Yes
HA (5 mg/mL)	~200	0	Yes

the selection strategy is applicable for microswarms in real bio-fluids. The formation conditions of medium-induced and MF-induced swarms in different bio-fluids are presented in Figs. 7.14. The inset on the right side of Fig. 7.14 shows experimental images of swarm patterns under different field parameters. As shown in Fig. 7.14a, for the MF-induced swarm, zig-zag patterns appear when the field ratio γ is small (black cross). Straight patterns that tend to elongate continuously are indicated by blue asterisks. When increasing γ, ribbon-like patterns are formed with continuous pattern reconfiguration (orange diamond). But if γ becomes too large, multiple slender patterns parallel to each other will form (red cross). Green circles

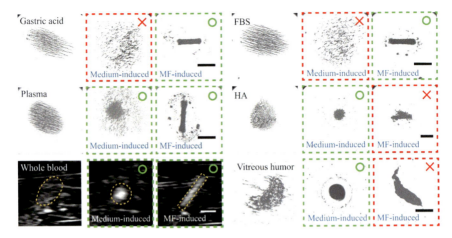

Fig. 7.13 Formation states of medium-induced and MF-induced swarms in gastric acid, FBS, plasma, HA, whole blood, and vitreous humor. The green dashed squares with the symbol "○" indicate successful formation, and red dashed squares with the symbol " × " indicate failed formation. All scale bars are 800 μm

(green regions) indicate the successful formation of stable MF-induced swarms. For the medium-induced swarm (Fig. 7.14b), when the strength of the magnetic field is low, the chain cannot rotate due to the strong drag of the medium (black cross). By increasing the field strength, vortex-like swarm patterns with small and loose cores are formed (blue asterisks). Similarly, the green region with green circles means the medium-induced swarm can be successfully formed under corresponding field parameters.

Quantitative analysis of the gathering effect of microswarms in different bio-fluids is provided in Fig. 7.15. The assembly rate, which is defined as the ratio of the core region to the total area of the swarm, is used to denote the gathering effect. When both medium-induced and MF-induced swarms can be formed, the latter usually exhibits a better gathering effect compared to the former swarm. The medium-induced swarm has a significantly higher assembly rate in HA and vitreous humor but does not work well in plasma. In contrast, the MF-induced swarm is more adaptive to ionic bio-fluids, including gastric acid and FBS. The statistical result in whole blood is not presented because of the interference from ultrasound imaging noise. The pattern transformation ability of the ribbon-like swarm in bio-fluids is also investigated. The aspect ratio of the ribbon-like swarm increases with the field ratio γ in FBS, gastric acid, and blood plasma (Fig. 7.16). The variation trend of the aspect ratio is similar to that in DI water, indicating that the high ionic strength of bio-fluids has little effect on the pattern transformation ability of ribbon-like swarms.

7.5 Swarm Formation and Navigation in Bio-Fluids

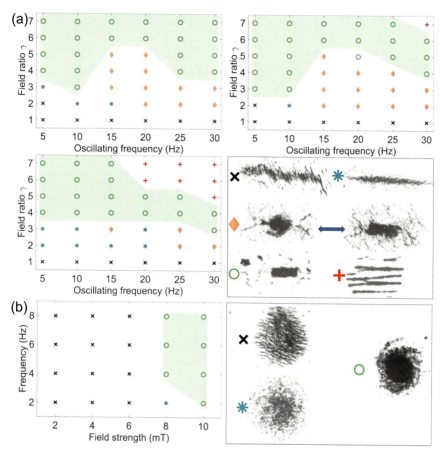

Fig. 7.14 Generation conditions of medium-induced and MF-induced swarms in bio-fluids. **a** generation conditions of MF-induced swarms in gastric acid, plasma, and FBS, respectively. **b** generation conditions of medium-induced swarms in HA. Green regions indicate successful formation. The inset on the right side shows states of the two swarms under different field parameters

Fig. 7.15 Assembly rate of medium-induced and MF-induced swarms in different bio-fluids. The assembly rates in DI water are also provided for comparison

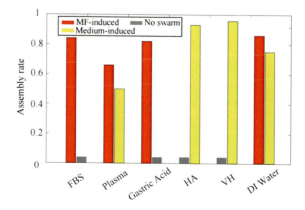

200   7 Formation and Actuation of Micro/Nanorobot Swarms in Bio-Fluids

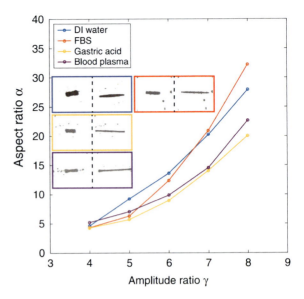

Fig. 7.16 Change in the aspect ratio of the ribbon-like swarm in different bio-fluids. The insets in blue, yellow, purple, and red rectangles show pattern elongation processes in DI water, gastric acid, blood plasma, and FBS, respectively

7.5.2 Swarm Navigation in Bio-Fluids

After the formation of stable microswarms in various bio-fluids, their locomotion capabilities are further evaluated. It is found that the motion trajectories of these microswarms can be well controlled by elaborately tuning the pitch angle and direction angle of the programmed magnetic fields, as shown in Fig. 7.17. For example, the ribbon-like swarm is able to navigate along the letters "C" and "U" in FBS and gastric acid, respectively, while the vortex-like swarm displays the trajectories of "H" and "K" in HA and plasma, respectively.

Figure 7.18 shows the effect of pitch angles on the translational velocity of medium-induced and MF-induced swarms in different bio-fluids. The velocity of the vortex-like swarm is set to zero in FBS and gastric acid as it cannot be formed in these two bio-fluids. Similarly, the velocity of the ribbon-like swarm in HA is always

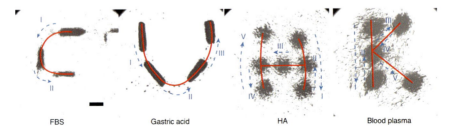

Fig. 7.17 Locomotion trajectories of "C", "U", "H" and "K" achieved by ribbon-like and vortex-like swarms in various bio-fluids. The numbers I, II…V indicate the motion sequence of swarms. The scale bar is 500 μm

7.5 Swarm Formation and Navigation in Bio-Fluids

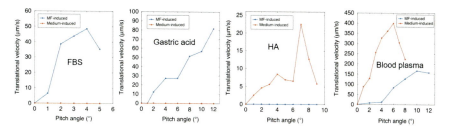

Fig. 7.18 Effect of the pitch angle of external magnetic fields on the translational velocity of swarms in various bio-fluids. A speed equal to zero at all means that the swarm cannot be generated

equal to zero. A maximum moving speed of ~50 μm/s is obtained by the ribbon-like swarm in FBS at a pitch angle of 4°, and further increasing pitch angle makes the swarm unstable and thus deteriorates its motion capability. In HA with a concentration of 5 mg/mL, the vortex-like swarm can only reach a maximum velocity of 23 μm/s due to the influence of mesh-like structure and the high viscosity. In addition, both swarms exhibit good locomotion performance in blood plasma, and their translational velocities are significantly higher than those in the other three bio-fluids.

7.5.3 Swarm in Whole Blood

Vascular system is one of the most common fluidic transport networks inside the human body for microswarms to realize the in vivo applications. So, the fluidic conditions of whole blood should be cautiously analyzed. The dynamic blood flow in arteries may impose a significant effect on the formation and locomotion of microswarms. Although the flow velocity is relatively low near the surface of blood vessels where microswarms navigate, it is still very challenging to maintain the stable patterns and perform reversible transformation during locomotion. A potential solution is to use the recent clinic catheter techniques. Building blocks are firstly delivered to a spacious region near the target by catheter, and then the equipped balloon is inflated to occlude blood vessel. In this manner, the blood flow is effectively restrained, facilitating the navigation of microswarms in blood vessels.

The formation processes of vortex-like and ribbon-like swarms in whole blood have already been demonstrated in the above section, and here it comes to the navigation of microswarms in such environments, as shown in Fig. 7.19. The green and red dots represent the initial and real-time positions of microswarms, respectively, while the red dashed shapes indicate the outlines of microswarms. It is obvious that effective navigation of both swarms is achieved under the guidance of ultrasound imaging. Figure 7.20 presents maximum velocities of swarms in whole blood and glycerol-water solution with similar viscosity (4–5 cp). The results show that the ribbon-like swarms reach maximum velocities of about 30 μm/s (pitch angle: 10°) in both whole blood and glycerol-water solution. However, the maximum velocity

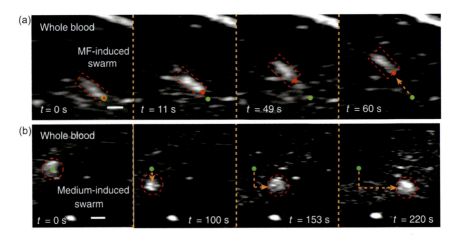

Fig. 7.19 Controlled navigation of **a** the ribbon-like swarm and **b** the vortex-like swarm in whole blood under the guidance of ultrasound imaging. The swarms are in the area enclosed by the red dashed line. The green and red dots represent the initial and real-time positions of the swarm, respectively. The scale bar is 1 mm

(pitch angle: 6°) of the vortex-like swarm in whole blood (~23 μm/s) is significantly smaller than that in glycerol-water solution (~280 μm/s). According to the analysis above (Fig. 7.9), the ionic strength has no obvious effect on the moving velocity of both types of swarms. Therefore, the velocity decay of vortex-like swarms in whole blood may be caused by red blood cells (RBCs); that is, the colloidal jamming effects of RBCs are significantly different for the two types of swarms. For the vortex-like swarm, the strong inward force of the main vortex traps nanoparticle chains inside the swarm pattern. Meanwhile, RBCs around the swarm will also be attracted to the swarm center and then influence the rotation of nanoparticle chains, which results in a decrease of moving velocity.

In order to study the interaction between the ribbon-like swarm and red blood cells, the whole blood is diluted by four times so that blood cells and swarms can be clearly observed by using an optical microscope. Figure 7.21a shows that the red blood cells are pushed away during the formation of the ribbon-like swarm. The blank regions represent areas where the RBC concentration is low, as highlighted by red arrows. After the formation of the ribbon-like swarm, adjacent RBCs are continuously pushed away by the induced flow. Therefore, these blank regions still existed along the outline of the swarm during the direction-turning process, which actively reduces the resistance from surrounding environments (Fig. 7.21b). In other words, the ribbon-like swarm always navigates in fluids with lower RBC concentration, which is similar to the situation in the blood plasma. As a result, it is much easier for the ribbon-like swarm to perform locomotion in whole blood than the vortex-like swarm. The navigated locomotion of the ribbon-like swarm is shown in Fig. 7.21c. Besides, the influence of protein corona in blood on the ribbon-like swarm is investigated using corona-coated nanoparticles. The preparation process is performed by

7.5 Swarm Formation and Navigation in Bio-Fluids

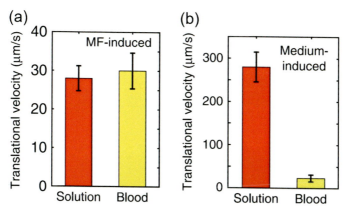

Fig. 7.20 Maximum velocities of the ribbon-like and vortex-like swarms in whole blood and glycerol-water solution with similar viscosity (4–5 cp). **a** the maximum velocity of the ribbon-like swarm. The pitch angle is 10°. **b** the maximum velocity of the vortex-like swarm. The pitch angle is 6°. Each error bar denotes the standard deviation from five trials

adding Fe_3O_4 nanoparticles into blood plasma and then incubating them for 10 min at room temperature (Hadjidemetriou et al. 2016). Subsequently, the obtained magnetic nanoparticles with well-coated protein corona are separated from the plasma using a magnet and washed with HEPES buffer to remove weakly bound proteins. The fabricated corona-coated nanoparticles are used to form the ribbon-like swarm in plasma, and the generated swarm is capable of performing reversible pattern transformation and navigated locomotion, as shown in Fig. 7.22. This suggests that the magnetic interaction within the ribbon-like swarm consisting of corona-coated nanoparticles still dominates, despite the influence of protein corona and extra charges.

7.5.4 Swarm in Vitreous Humor and Bovine Eyeballs

The effectiveness of vortex-like microswarms in animal eyeballs is also investigated as eyeball therapies are one of the most attractive applications of micro/nanorobots. The ribbon-like microswarm is not studied here because the highly viscous fluidic environment inside eyeballs severely restricted the formation of ribbon-like swarm as described above. Besides, Fe_3O_4 nanoparticles are surface functionalized with hydrophobic property to serve as the building blocks in order to improve the locomotion capability of generated microswarms. Initial experiments are carried out in bovine vitreous humor as shown in Fig. 7.23. Actuated by a rotating magnetic field (the field strength is 7 mT, and the frequency is 18 Hz), the vortex-like swarm moves along a rectangular trajectory with excellent controllability. Besides, almost all the building blocks are gathered into the dynamically stable pattern even during the locomotion, which is beneficial to minimize any side effects after swarming process.

204   7 Formation and Actuation of Micro/Nanorobot Swarms in Bio-Fluids

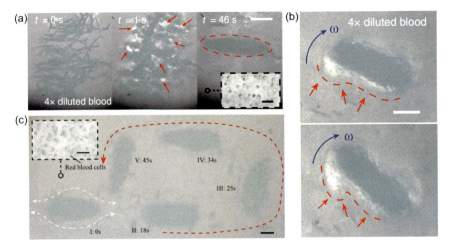

Fig. 7.21 Pattern formation, turning direction, and navigated locomotion of the ribbon-like swarm in 4× diluted blood. **a** the generation process of the ribbon-like swarm in 4× diluted blood. **b** the ribbon-like swarm turns direction in 4× diluted blood. **c** navigated locomotion of the ribbon-like swarm in 4× diluted blood. The red arrows and dashed lines in **a** and **b** highlight blank regions caused by the lack of RBCs. The white and red dashed lines in **c** represent trajectories of RBCs and the swarm pattern, respectively. The insets in **a** and **c** are enlarged optical microscope images of red blood cells with the scale bar of 20 μm. The scale bars are 500 μm in **a** and **b** and 200 μm in **c**

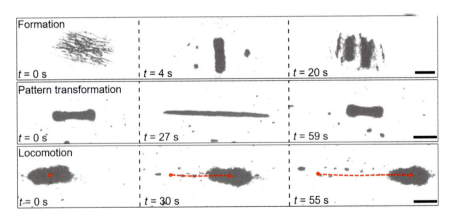

Fig. 7.22 Pattern formation, transformation, and locomotion of the ribbon-like swarm consisting of corona-coated Fe_3O_4 nanoparticles in blood plasma. The red dashed lines indicate the trajectory of the swarm. The scale bars are 500 μm

This is also a prominent advantage in ophthalmic treatment, as even a small amount of nanoparticle residue in the eye can affect vision.

It is worth noting that the locomotion behavior of vortex-like swarms in vitreous humor is different from that in other bio-fluids. During the locomotion process, the swarm pattern attracts other particles and impurities at the micro/nanoscale in

7.5 Swarm Formation and Navigation in Bio-Fluids

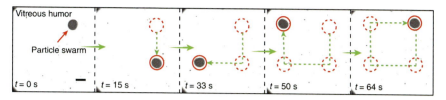

Fig. 7.23 Vortex-like swarm performs navigated locomotion along a rectangular trajectory in bovine vitreous humor. The scale bar is 600 μm

vitreous humor (indicated by black dots in Fig. 7.23) through fluidic interactions. Impurities entering the vortex have a significant effect on the movement of the swarm in the presence of the dense mesh-like structure; that is, the swarm vibrates slightly during locomotion. Nonetheless, the moving direction of the swarm is not affected by vibration, and high navigation accuracy is still ensured. The previously reported swarm of slippery helical micropropellers passes through the vitreous body with a relatively loose pattern (Wu et al. 2018). These helical microswimmers (300 nm in diameter) move mainly through the gaps of the mesh-like structure (the mesh size is about 500 nm), and their propulsion force is only at the pico-newton level (Wang et al. 2019, 2017). Although the particle chains in the swarm are also at the micro/nanoscale, the swarm is moving as an entirety with its size at the millimeter/submillimeter scale. By observing the motion state of particles around the swarm, the propulsion force of the swarm is estimated to be about 1–8 μN according to Stokes' law and Weissenberg number, which is comparable to a magnetic object with a similar size (Pokki et al. 2015; Fabris et al. 1999). Therefore, forces generated by the swarm pattern can distort and deform the mesh of the vitreous humor, allowing the vortex-like swarm to perform effective locomotion in the vitreous humor.

The addition of magnetic nanoparticles into the vitreous humor is conducted by injecting the nanoparticle suspension, which may lead to local dilution of the vitreous humor at the site of swarm formation. To investigate the effect of local dilution on locomotion of the vortex-like swarm, the nanoparticle suspension is stained with blue dye before injection. The blue areas in Fig. 7.24 indicate dilution regions due to the injected solution. Actuated by the rotating magnetic field, the dispersed nanoparticles rapidly gather into a vortex-like swarm (Fig. 7.24a, b) and can move in the diluted region (Fig. 7.24c). Subsequently, the swarm moves outside the blue area and is still able to perform effective locomotion (Fig. 7.24d–f). This suggests that the locomotion of the vortex-like swarm in the vitreous humor is not dependent on the injection-induced local dilution. In addition, it can also be found that the spontaneous diffusion of the dye is limited by the high viscosity and mesh-like structure of the vitreous humor, and the change of the contour of the blue area is affected by the swarm-induced flow, as shown by the yellow dashed arrow in Fig. 7.24d. The rotation of the swarm pattern (counterclockwise, indicated by the red arrows) causes the dye above it to be pushed back to the left and the dye below it to be pulled to the right.

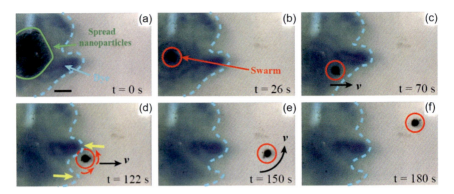

Fig. 7.24 Influence of local dilution of the vitreous humor on locomotion of the vortex-like swarm. **a** the blue nanoparticle suspension diffuses in the vitreous humor after injection. **b** formation of the vortex-like swarm after applying a rotating magnetic field. The swarm pattern is highlighted by red circles. **c** The swarm moves within the blue diluted area. **d–f** The swarm moves outside the blue area. Black, red, and yellow arrows indicate the swarm's moving direction, rotating direction, and the contour changes of the stained area, respectively. The scale bar is 2 mm

In addition, the influence of overall dilution of the vitreous humor on swarm locomotion is investigated. The PBS solution with dispersed nanoparticles (200 μL) is injected into 5 mL of vitreous humor. The viscoelastic property of the vitreous humor is characterized after two minutes of swarm formation and locomotion. In another set of experiments, the same amount of PBS solution is injected into the vitreous humor, and the measurements are conducted after the same time. The control group is set to the original vitreous humor, and the results are shown in Fig. 7.25. The viscoelasticity of the three solutions has no significant difference, indicating that the injection does not result in obvious overall dilution of the vitreous humor. Furthermore, the movement of the swarm does not significantly alter the viscoelasticity of the vitreous humor. Therefore, the injection of small amounts of particle suspension is not critical for swarm locomotion in such environments.

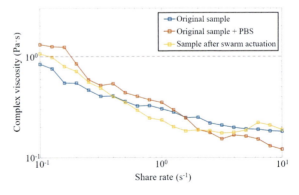

Fig. 7.25 Viscoelasticity of fresh vitreous humor, vitreous humor injected with PBS solution, and vitreous humor after two minutes of swarm locomotion

7.5 Swarm Formation and Navigation in Bio-Fluids

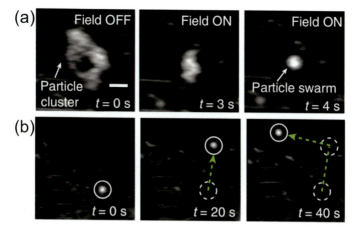

Fig. 7.26 Ultrasound images of the vortex-like swarm in vitreous humor. **a** ultrasound imaging feedback of microswarm during the generation process. **b** ultrasound imaging feedback of microswarm during locomotion. The scale bar is 600 μm

Ultrasound imaging is a widely adopted method in ophthalmic diagnosis to identify whether there is any lesion inside eyeballs, such as retinal detachment, vitreous opacity, and vitreous hemorrhage (Silverman 2009). Ultrasound images of the vortex-like swarm in vitreous humor are shown in Fig. 7.26. Initially, the nanoparticles are dispersive, as indicated by the irregular white pattern (Fig. 7.26a). Once the programmed rotating magnetic field is applied, the white signal shrinks quickly and eventually forms a bright circular pattern, whose contrast is stronger than that at the initial stage, thus facilitating the observation and navigation of microswarms. For instance, under the guidance of ultrasound imaging, the vortex-like swarm is controlled to move along a right-angled path, as shown in Fig. 7.26b.

Finally, ex vivo experiments are conducted by performing the active generation and locomotion of vortex-like swarms in a bovine eyeball. The magnetic control setup integrated with an ultrasound imaging system is shown in Fig. 7.27a. An eyeball is placed in the center of the integrated setup, with the pupil pointing toward the side edge, and the ultrasound transducer is slightly attached to the eyeball surface. The nanoparticle solution is then injected into the eyeball with a syringe, and meanwhile, the rotating magnetic field is applied (Fig. 7.27b). The monitored generation process of vortex microswarm inside the eyeball is presented in Fig. 7.27c. The signal of syringe needle is very bright so that the injection point can be located easily. Similar to the ultrasound imaging results in vitreous humor, the injected building blocks initially generate a signal with a white irregular pattern, as shown by the red dotted area. After applying the magnetic field, the pattern of signal becomes an ellipse after ~15 s, indicating the successful generation of the vortex-like microswarm. Figure 7.28a, b shows the controlled locomotion of microswarm inside the eyeball with different injection dosages. The purple and green dots represent the injection points and swarm formation points, respectively, while the red dots indicate the final position after navigation. The size of microswarm changes with different injection dosages, and it can be found that effective locomotion inside eyeball is achieved by the vortex-like swarm, demonstrating its promising applications in eyeball therapies.

In addition, the application of microswarms in the gastrointestinal tract is also demonstrated by formatting and actuating the vortex-like swarm on a porcine intestinal tract. The mucous membranes of the intestinal tract are kept in the experiment, and the nanoparticles are directly added onto the sample surface. However, due to the strong adhesion of the intestinal surface, the nanoparticles are stuck in place and cannot be actuated. To solve this problem, the nanoparticles are first added to a tank filled with mucus. The rotating field is applied to form a vortex-like swarm. Because the mucus also contains fibrous meshes, nanoparticle chains inside the swarm are

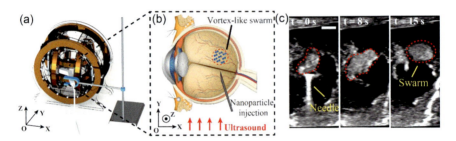

Fig. 7.27 Generation of the vortex-like swarm in a bovine eyeball. **a** schematic of the experimental setup. **b** schematic of nanoparticle injection. **c** generation process of the vortex microswarm inside the eyeball with ultrasound imaging feedback. The red dashed lines represent the outline of the swarm. The scale bar is 1 mm

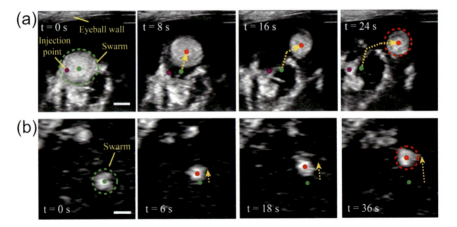

Fig. 7.28 Navigated locomotion of the vortex-like swarm with **a** low and **b** high injection dosages in a bovine eyeball under the guidance of ultrasound imaging. The purple and green dots represent the injection points and swarm formation points, respectively. The red dots indicate the final position after navigation. Motion trajectories are indicated by the yellow dashed arrows. The scale bars are 800 μm

7.5 Swarm Formation and Navigation in Bio-Fluids

tangled with fibers, which significantly increases the magnetic torque and force. Then, the nanoparticles are retracted from the mucus and injected onto the surface of the porcine intestinal tract. Readded particles can form a vortex-like swarm on the sample surface with mucus (Fig. 7.29). After applying a pitch angle, effective locomotion of the swarm can also be realized.

To further validate the potential of microswarms for biomedical applications, the cytotoxicity of nanoparticles that form the swarm is tested. Cell viabilities of normal cells (3T3 cells) and tumor cells (HeLa cells and HepG2 cells) are tested after being cultured with nanoparticles in different concentrations (0.01, 0.1, 1, and 10 mg/mL) for 24 h. The results show that nanoparticles are not significantly toxic to both normal cells and tumor cells at low concentrations and only have slight toxicity to normal cells at high concentrations (Fig. 7.30).

Fig. 7.29 Formation and actuation of the vortex-like swarm on the porcine intestinal tract. The red dotted circles and blue arrows indicate the location and moving direction of the swarm, respectively. The scale bar is 2 mm

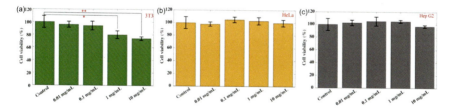

Fig. 7.30 Cytotoxicity test of the nanoparticle. The cell viability of **a** 3T3 cells, **b** HeLa cells, and **c** HepG2 cells after being cocultured with nanoparticles with concentrations of 0.01, 0.1, 1, and 10 mg/mL for 24 h. The error bars denote the standard deviation from three to five trials. The symbol "*" denotes $p < 0.05$, and "**" denotes $p < 0.01$

7.6 Conclusion

This chapter discusses the role of bio-fluids in the formation and actuation of micro/nanorobot swarms. Based on dominant inner interactions, we first categorize general microswarm systems into three categories, *i.e.*, medium-induced, MF-induced, and weakly interacted swarms. Here, the vortex-like and ribbon-like swarms are selected as representatives of the medium-induced and MF-induced swarms, respectively. The active formation and controlled locomotion of the vortex-like and ribbon-like swarms in bio-fluids are systematically investigated. Artificial fluids with controlled physical parameters, including ionic strength, viscosity, and polymeric mesh structure, are first prepared to study the collective behaviors of swarms. It is found that vortex-like swarms are more sensitive to the change of ions in fluids, and the formation of ribbon-like swarms would be hindered by viscous fluids, while the mesh-like structure facilitates the formation of vortex-like swarms. Through analyzing and summarizing the performance of microswarms, a selection strategy for predicting the optimized type of microswarms in a bio-fluid with specific physical properties is proposed. Then six types of fluids, *i.e.*, gastric acid, FBS, blood plasma, HA, whole blood, and vitreous humor, are employed to validate the strategy, and the experimental results show excellent agreement with predictions based on the proposed strategy. Furthermore, as important application scenarios, the formation and navigation of microswarms in whole blood and eyeballs are successfully demonstrated using the feedback of ultrasound imaging. Corresponding mechanisms regarding the colloidal jamming effects by blood cells and swarm locomotion in meshed environments are also properly explained. We believe the research results presented here will advance the fundamental understanding of magnetic colloidal microswarms, which is an important step further toward the in vivo targeted delivery applications of micro/nanorobot swarms.

References

Ahmed D, Baasch T, Blondel N, Laubli N, Dual J, Nelson BJ (2017) Neutrophil-inspired propulsion in a combined acoustic and magnetic field. Nat Commun 8:770

Ahmed D, Sukhov A, Hauri D, Rodrigue D, Maranta G, Harting J, Nelson BJ (2021) Bioinspired acousto-magnetic microswarm robots with upstream motility. Nature Machine Intelligence 3(2):116–124

Belharet K, Folio D, Ferreira A (2014) Study on rotational and unclogging motions of magnetic chain-like microrobot. In: 2014 IEEE/RSJ international conference on intelligent robots and systems. IEEE, 834–839

Chen Q, Bae SC, Granick S (2011) Directed self-assembly of a colloidal kagome lattice. Nature 469(7330):381–384

Chen H, Zhang H, Xu T, Yu J (2021) An overview of micronanoswarms for biomedical applications. ACS Nano 15(10):15625–15644

de Ávila BE-F, Angsantikul P, Li J, Angel Lopez-Ramirez M, Ramírez-Herrera DE, Thamphiwatana S, Chen C, Delezuk J, Samakapiruk R, Ramez V (2017) Micromotor-enabled active drug delivery for in vivo treatment of stomach infection. Nat Commun 8:272

References

Du X, Yu J, Jin D, Chiu PWY, Zhang L (2021) Independent pattern formation of nanorod and nanoparticle swarms under an oscillating field. ACS Nano 15(3):4429–4439

Fabris D, Muller SJ, Liepmann D (1999) Wake measurements for flow around a sphere in a viscoelastic fluid. Phys Fluids 11(12):3599–3612

Hadjidemetriou M, Al-Ahmady Z, Kostarelos K (2016) Time-evolution of in vivo protein corona onto blood-circulating PEGylated liposomal doxorubicin (DOXIL) nanoparticles. Nanoscale 8(13):6948–6957

Ibele M, Mallouk TE, Sen A (2009) Schooling behavior of light-powered autonomous micromotors in water. Angew Chem 121(18):3358–3362

Ji F, Jin D, Wang B, Zhang L (2020) Light-driven hovering of a magnetic microswarm in fluid. ACS Nano 14(6):6990–6998

Jin D, Zhang L (2022) Collective behaviors of magnetic active matter: recent progress toward reconfigurable, adaptive, and multifunctional swarming micro/nanorobots. Acc Chem Res 55(1):98–109

Kagan D, Balasubramanian S, Wang J (2011) Chemically triggered swarming of gold microparticles. Angew Chem 123(2):523–526

Li Z, Zhang H, Wang D, Gao C, Sun M, Wu Z, He Q (2020) Reconfigurable assembly of active liquid metal colloidal cluster. Angew Chem Int Ed 59(45):19884–19888

Ma F, Wang S, Wu DT, Wu N (2015) Electric-field-induced assembly and propulsion of chiral colloidal clusters. Proc Natl Acad Sci 112(20):6307–6312

Magariyama Y, Kudo S (2002) A mathematical explanation of an increase in bacterial swimming speed with viscosity in linear-polymer solutions. Biophys J 83(2):733–739

Magdanz V, Sanchez S, Schmidt OG (2013) Development of a sperm-flagella driven micro-bio-robot. Adv Mater 25(45):6581–6588

Mao X, Chen Q, Granick S (2013) Entropy favours open colloidal lattices. Nat Mater 12(3):217–222

Martinez-Pedrero F, Tierno P (2015) Magnetic propulsion of self-assembled colloidal carpets: efficient cargo transport via a conveyor-belt effect. Phys Rev Appl 3(5):051003

Martinez-Pedrero F, Cebers A, Tierno P (2016) Dipolar rings of microscopic ellipsoids: magnetic manipulation and cell entrapment. Phys Rev Appl 6(3):034002

Massana-Cid H, Meng F, Matsunaga D, Golestanian R, Tierno P (2019) Tunable self-healing of magnetically propelling colloidal carpets. Nat Commun 10:2444

Mou F, Zhang J, Wu Z, Du S, Zhang Z, Xu L, Guan J (2019) Phototactic Flocking of Photochemical Micromotors. Iscience 19:415–424

Mou F, Li X, Xie Q, Zhang J, Xiong K, Xu L, Guan J (2020) Active micromotor systems built from passive particles with biomimetic predator-prey interactions. ACS Nano 14(1):406–414

Peyer KE, Zhang L, Nelson BJ (2013) Bio-inspired magnetic swimming microrobots for biomedical applications. Nanoscale 5(4):1259–1272

Peyer KE, Qiu F, Zhang L, Nelson BJ (2012) Movement of artificial bacterial flagella in heterogeneous viscous environments at the microscale. In: 2012 IEEE/RSJ International Conference on Intelligent Robots and Systems. IEEE, 2553–2558

Pokki J, Ergeneman O, Sevim S, Enzmann V, Torun H, Nelson BJ (2015) Measuring localized viscoelasticity of the vitreous body using intraocular microprobes. Biomed Microdevice 17(5):1–9

Schwarz L, Karnaushenko DD, Hebenstreit F, Naumann R, Schmidt OG, Medina-Sánchez M (2020) A rotating spiral micromotor for noninvasive zygote transfer. Advanced Science 7:2000843

Servant A, Qiu F, Mazza M, Kostarelos K, Nelson BJ (2015) Controlled in vivo swimming of a swarm of bacteria-like microrobotic flagella. Adv Mater 27(19):2981–2988

Silverman RH (2009) High-resolution ultrasound imaging of the eye—a review. Clin Experiment Ophthalmol 37(1):54–67

Snezhko A, Aranson IS (2011) Magnetic manipulation of self-assembled colloidal asters. Nat Mater 10(9):698–703

Snezhko A, Belkin M, Aranson I, Kwok W-K (2009) Self-assembled magnetic surface swimmers. Phys Rev Lett 102(11):118103

Solovev AA, Sanchez S, Schmidt OG (2013) Collective behaviour of self-propelled catalytic micromotors. Nanoscale 5(4):1284–1293

Sun M, Fan X, Tian C, Yang M, Sun L, Xie H (2021) Swarming microdroplets to a dexterous micromanipulator. Adv Func Mater 31(19):2011193

Sun M, Chan KF, Zhang Z, Wang L, Wang Q, Yang S, Chan SM, Chiu PWY, Sung JJY, Zhang L (2022) Magnetic microswarm and fluoroscopy-guided platform for biofilm eradication in biliary stents. Adv Mater 34(34):2201888

Ullrich F, Qiu F, Pokki J, Huang T, Pané S, Nelson BJ (2016) Swimming characteristics of helical microrobots in fibrous environments. In 2016 6th IEEE International Conference on Biomedical Robotics and Biomechatronics (BioRob). IEEE, 470–475

Vach PJ, Walker D, Fischer P, Fratzl P, Faivre D 2017. Pattern formation and collective effects in populations of magnetic microswimmers. J Phys D: Appl Phys 50(11):11LT03

Walker D, Käsdorf BT, Jeong H-H, Lieleg O, Fischer P (2015) Enzymatically active biomimetic micropropellers for the penetration of mucin gels. Sci Adv 1(11):e1500501

Wang Q, Zhang L (2018) External power-driven microrobotic swarm: from fundamental understanding to imaging-guided delivery. ACS Nano 15(1):149–174

Wang W, Duan W, Ahmed S, Sen A, Mallouk TE (2015) From one to many: dynamic assembly and collective behavior of self-propelled colloidal motors. Acc Chem Res 48(7):1938–1946

Wang X, Luo M, Wu H, Zhang Z, Liu J, Xu Z, Johnson W, Sun Y (2017) A three-dimensional magnetic tweezer system for intraembryonic navigation and measurement. IEEE Trans Rob 34(1):240–247

Wang B, Kostarelos K, Nelson BJ, Zhang L (2021) Trends in micro-/nanorobotics: Materials development, actuation, localization, and system integration for biomedical applications. Adv Mater 33(4):2002047

Wang X, Ho C, Tsatskis Y, Law J, Zhang Z, Zhu M, Dai C, Wang F, Tan M, Hopyan S (2019) Intracellular manipulation and measurement with multipole magnetic tweezers. Sci Robot 4(28):eaav6180

Wang S, Guo X, Xiu W, Liu Y, Ren L, Xiao H, Yang F, Gao Y, Xu C, Wang L (2020) Accelerating thrombolysis using a precision and clot-penetrating drug delivery strategy by nanoparticle-shelled microbubbles. Sci Adv 6(31):eaaz8204

Wu Z, Chen Y, Mukasa D, Pak OS, Gao W (2020) Medical micro/nanorobots in complex media. Chem Soc Rev 49(22):8088–8112

Wu Z, Troll J, Jeong H-H, Wei Q, Stang M, Ziemssen F, Wang Z, Dong M, Schnichels S, Qiu T, Fischer P (2018) A swarm of slippery micropropellers penetrates the vitreous body of the eye. Science Adv 4(11):eaat4388

Xie H, Sun M, Fan X, Lin Z, Chen W, Wang L, Dong L, He Q 2019 Reconfigurable magnetic microrobot swarm: Multimode transformation, locomotion, and manipulation. Sci Robot 4(28):eaav8006

Xu T, Soto F, Gao W, Dong R, Garcia-Gradilla V, Magana E, Zhang X, Wang J (2015) Reversible swarming and separation of self-propelled chemically powered nanomotors under acoustic fields. J Am Chem Soc 137(6):2163–2166

Yan J, Han M, Zhang J, Xu C, Luijten E, Granick S (2016) Reconfiguring active particles by electrostatic imbalance. Nat Mater 15(10):1095–1099

Yang L, Yu J, Yang S, Wang B, Nelson BJ, Zhang L (2022) A survey on swarm microrobotics. IEEE Trans Rob 38(3):1531–1551

Yigit B, Alapan Y, Sitti M (2019) Programmable collective behavior in dynamically self-assembled mobile microrobotic swarms. Advanced Science 6(6):1801837

Yu J, Xu T, Lu Z, Vong CI, Zhang L (2017) On-demand disassembly of paramagnetic nanoparticle chains for microrobotic cargo delivery. IEEE Trans Rob 33(5):1213–1225

Yu J, Yang L, Zhang L (2018a) Pattern generation and motion control of a vortex-like paramagnetic nanoparticle swarm. Int J Robotics Res 37(8):912–930

Yu J, Wang B, Du X, Wang Q, Zhang L (2018b) Ultra-Extensible Ribbon-like Magnetic Microswarm. Nat Commun 9:3260

References

Zhang Y, Zhang L, Yang L, Vong CI, Chan KF, Wu WK, Kwong TN, Lo NW, Ip M, Wong SH (2019). Real-time tracking of fluorescent magnetic spore–based microrobots for remote detection of C. diff toxins. Sci Adv 5(1):eaau9650

Zhou Z, Hou Z, Pei Y (2021) Reconfigurable particle swarm robotics powered by acoustic vibration tweezer. Soft Rob 8(6):735–743

Chapter 8
Localization of Microswarms Using Various Imaging Methods

Abstract Fast and accurate localization is crucial for in vivo applications of microswarms. This chapter begins with using B-mode ultrasound imaging for real-time localization and navigation of magnetic vortex-like microswarms. Since the length and direction of the nanoparticle chains change with the rotating magnetic field, the ultrasound image of the vortex-like microswarm also varies periodically. Based on theoretical analysis and experimental results, strategies for enhancing the ultrasound imaging contrast of the vortex-like microswarm and the optimal driving frequency for locating the swarm in different conditions are obtained. In addition, fluorescence imaging and photoacoustic imaging of microswarms are introduced.

Keywords Microswarm · Localization · Imaging-guide navigation · Ultrasound imaging · Fluorescence imaging · Photoacoustic imaging

8.1 Introduction

Micro/nanorobots are promising for applications in various fields, especially the biomedical field, such as targeted drug delivery (Lenaghan et al. 2013; Sitti et al. 2015), biosensing (Yang et al. 2020), micromanipulation (Zhang, et al. 2009; Jing et al. 2019; Wang et al. 2018), and minimally invasive surgery (Bergeles and Yang 2014; Bergeles et al. 2012). Versatile micro/nanorobots actuated various strategies (e.g., magnetic fields (Xie et al. 2019; Shahrokhi et al. 2019; Wright et al. 2017; Erin et al. 2019; Yang et al. 2018), chemical fuels (Khalil et al. 2014), acoustic field (Youssefi and E. Diller 2019), and biohybrid methods (Medina-Sanchez et al. 2015; Zhang et al. 2016)) have been developed. Real-time localization of micro/nanorobots is crucial for in vivo biomedical applications, and closed-loop control systems can be constructed based on the tracked positions (Medina-Sánchez and Schmidt 2017). For example, magnetic resonance imaging (MRI) is applied for localizing and navigating micro/nanorobots along predesigned trajectories (Martel et al. 2009; Lastname et al. 2019). The helical microswimmers inside rodent stomachs can be imaged by MRI (Yan et al. 2017). Positron emission tomography (PET) is applied to track catalytic tubular micromotors in a circular phantom (Vilela et al. 2018). The artificial bacterial flagella functionalized with fluorophores can be observed in the subcutaneous tissue of mice by infrared fluorescence imaging (Servant et al. 2015). Among these medical imaging modalities, ultrasound imaging stands out as one of the most promising imaging methods for micro/nanorobots and their swarms. Compared with MRI, PET/CT, and X-ray imaging, ultrasound imaging has advantages of high spatial and temporal resolution (fast imaging speed), low cost, and no ionizing radiation, which enables motion control and path planning of millimeter-scale robots in real time (Pané et al. 2019; Guo et al. 2018; Wang and Zhang 2018, 2020; Zhou and Zheng 2015; Scheggi et al. 2017). The ultrasound feedback has also been used for navigating miniature helical robots to rub blood clots in vitro (Khalil et al. 2018).

Ultrasound imaging depends on gradients of acoustic impedance, and only micro/nanorobots larger than the sonographic detection limit are able to be localized inside a living body by ultrasound imaging. The real-time localization of micro/nanorobots is challenged by the low signal-to-noise ratio (imaging contrast). A simple way to solve this problem is to increase the size of the imaged object, such as using millimeter-scale robots (Scheggi et al. 2017; Khalil et al. 2018; Chen et al. 2019a; Hu et al. 2018). However, the large size limits further applications in confined environments, especially in bifurcated microvasculature. Another approach is to use microbubbles to scatter ultrasound to enhance the imaging contrast (Ackermann and Schmitz 2016; Lastname et al. 2018a). Some micro/nanorobots generate bubbles through chemical catalytic reactions to achieve self-propulsion and thus can be indirectly located by ultrasound imaging (Lastname et al. 2014). However, the moving direction and velocity of these microrobots are difficult to control, and they can only be applied in specific environments, such as the stomach (Olson et al. 2013). Micro/nanorobot swarms are beneficial to effective localization and control in vivo due to the enhanced imaging contrast of various medical imaging methods, including MRI (Martel et al. 2009), PET (Vilela et al. 2018), and fluorescence imaging (Servant et al. 2015). The advantage of microswarms over millimeter-scale robots lies in their reconfigurability, i.e., reversible disassembly and reassembly (Yu et al. 2018). Tiny objects in confined environments can form microswarms that can be observed by medical imaging and perform specified tasks in an assembled or disassembled manner after being navigated to the targeted area (Yu et al. 2019; Wu et al. 2018; Lastname et al. 2018b; Wang et al. 2020a). Magnetite nanoparticles, which can be functionalized in a wide size range (from a few nanometers to micrometers) and with different magnetism, are considered as promising candidates in the fields of medical imaging-guided therapy and biosensing (Tay et al. 2018; Min et al. 2015; Hao et al. 2010; Cai et al. 2013).

In this chapter, we first introduce the localization and real-time navigation of a magnetic nanoparticle swarm under the guidance of B-mode ultrasound imaging (Fig. 8.1). A theoretical model is constructed to study the variation of nanoparticle chains within the swarm pattern under rotating magnetic fields, including the changes in length and orientation. The contrast of ultrasound images changes periodically, and an image processing method is used to improve the signal-to-noise ratio. The relationship between imaging contrast and the area density of nanoparticles is analyzed and presented. Then, an optimized actuation strategy for enhancing

8.1 Introduction

the ultrasound imaging contrast of a magnetic microswarm is introduced. Modeling and simulations are performed to reveal the fundamental mechanism. The optimal driving frequency for localizing rotating microswarms at different temporal resolutions and imaging depths is obtained by theoretical analysis and experiments. The appropriate pattern transformation behavior of microswarms can further enhance the ultrasound imaging contrast, and ex vivo swarm localization at different depths is studied. Microswarms can adapt to surrounding environments with enhanced ultrasound contrast through pattern transformation while navigating in confined environments. Finally, fluorescence imaging and photoacoustic imaging of microswarms are introduced.

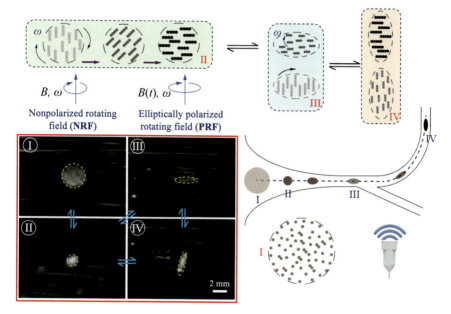

Fig. 8.1 Actuation, deformation, and dynamic ultrasound imaging contrast of the rotating colloidal swarm. Magnetic nanoparticle swarm (stage II) is formed using a non-polarized rotating magnetic field (NRF), and it exhibits deformation by tuning the rotating field to an elliptically polarized rotating field (PRF). Ultrasound contrast is affected by the orientation of nanoparticle chains inside the swarm (stages III and IV)

8.2 Localization of a Rotating Colloidal Microswarm Under Ultrasound Guidance

8.2.1 Mathematical Modeling and Simulations

The vortex-like microswarm is formed under an in-plane rotating magnetic field (Fig. 8.2a), and the locomotion is realized by adding a small pitch angle (γ) to the rotating plane (Fig. 8.2b). The locomotion relies on friction asymmetry caused by the boundary, i.e., the drag force coefficient increases near the substrate (Sing et al. 2010). Figure 8.2c schematically illustrates formation process of the microswarm. As described in Chap. 2, nanoparticles form chains due to the induced dipolar attractive forces. Meanwhile, the hydrodynamic interaction among nanoparticle chains contributes to the gathering effects. A region with a high area density of nanoparticles is observed and keeps expanding after trapping more nanoparticles. Finally, a dynamic-equilibrium vortex-like microswarm is formed.

The contrast of ultrasound imaging is highly dependent on the orientation and lengths of nanoparticle chains within the swarm pattern; i.e., the ultrasound contrast

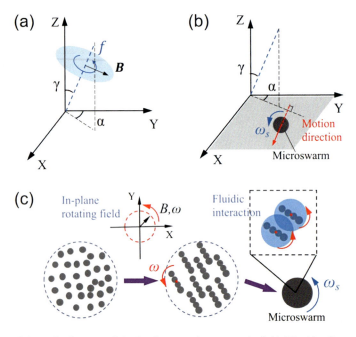

Fig. 8.2 **a** Schematic diagram of the in-plane rotating magnetic field. The blue line and arrow indicate the normal line and rotating direction of the field, respectively. B is the field strength, f is the rotating frequency, α is the yaw angel, and γ is the pitch angle. **b** Schematic diagram of the locomotion of the microswarm. ω_s indicates the angular velocity of the microswarm. **c** Schematic diagram of the formation of the vortex-like microswarm

8.2 Localization of a Rotating Colloidal Microswarm Under Ultrasound ...

is time-varying. A mathematical model is constructed to estimate the variation of chain lengths based on the balance of torques. Paramagnetic Fe$_3$O$_4$ nanoparticles are treated as spheres with a radius of a. The induced dipole moment of a nanoparticle under an external magnetic field (strength B) is expressed as $\mathbf{m} = 4/3\pi a^3 \mu_0 \chi \mathbf{B}$, where χ and μ_o are the magnetic susceptibility and permeability of vacuum, respectively. The induced magnetic forces between nanoparticles are (Biswal and Gast 2004)

$$\mathbf{F} = \frac{3m^2}{4\pi \mu_0 r^4}(3\cos^2\beta - 1)\hat{r} + \frac{3m^2}{4\pi \mu_0 r^4}\sin(2\varphi)\hat{\theta} \qquad (8.1)$$

where r indicated the distance between centers of nanoparticles, and φ indicates the phase lag. The in-plane rotating field can be considered as the superposition of two sinusoidal fields with a phase lag of $\pi/2$, and therefore, we give its expression as. $\mathbf{B}(t) = B\sin(2\pi ft)\hat{x} + B\cos(2\pi ft)\hat{y}$.

Considering a nanoparticle chain consisting of N nanoparticles with a chain length of $L = 2Na$. The Reynolds number is estimated as 1×10^{-3}, which is a typical low Reynolds number regime. Thus, the counterbalance relationship between the driving magnetic torque and viscous drag torque exerted on a particle chain governs the chain's formation. The driving magnetic torque can be calculated by summing up the magnetic interactions among neighboring nanoparticles (Singh and Hatton 2005; Petousis et al. 2007)

$$\Gamma_m = \frac{3\mu_0 m^2 (N-1)}{4\pi (2a)^3}\sin(2\varphi) \qquad (8.2)$$

The viscous drag torque of a nanoparticle chain with the consideration of shape factor is calculated as (Lastname et al. 2003).

$$\Gamma_d = \frac{8\pi a^3}{3}\frac{N^3}{\ln\left(\frac{N}{2}\right) + \frac{2.4}{N}}\eta\omega \qquad (8.3)$$

where η is the viscosity of the surrounding fluid, and ω is the angular velocity. Based on the counterbalance relationship between the two torques, we can have $\Gamma_d = \Gamma_m$. The length of nanoparticle chains is estimated using Mason number, a ratio defined as the viscous torque divided by the magnetic torque (Melle et al. 2003). The modified Mason number in this case is expressed as

$$R_T = 16\frac{\eta\omega\mu_0}{\chi^2 B}\frac{N^3}{(N-1)\left(\ln\left(\frac{N}{2}\right) + \frac{2.4}{N}\right)} \qquad (8.4)$$

$R_T > 1$ means the viscous torque is larger than the driving magnetic torque, and fragmentation of the nanoparticle chain will occur, which is able to be induced by suddenly decreasing B or increasing ω. $R_T = 1$ represents a critical case, i.e., the two torques balance each other, and nanoparticle chains remain stable correspondingly.

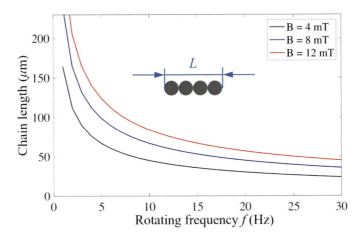

Fig. 8.3 Variation of the chain length with input rotating frequency at different field strengths

Therefore, the relationship between the chain length and input rotating frequency can be calculated by setting $R_T = 1$, as shown in Fig. 8.3. The length decreases when increasing frequency since the drag becomes larger. Increasing the magnetic field strength while keeping the input frequency constant results in longer chain lengths.

A finite element simulation is used to investigate the flow induced by the rotation of nanoparticle chains and the chain-chain hydrodynamic interaction. The nanoparticle chains are modeled as cylinders with lengths of 60–100 μm, and they perform rotation around their own geometric center at frequencies of 6–8 Hz (Fig. 8.4). The simulation results show that after ten complete rotations, the chain-chain hydrodynamic interaction is enhanced by increasing the frequency while keeping the chain length constant (Fig. 8.4a, b). However, the chain length becomes shorter with an increased input frequency, leading to a decrease in the hydrodynamic interaction (Fig. 8.4b–d). The decreased hydrodynamic interaction yields a swarm pattern with a relatively lower area density of nanoparticles. In addition, the input field strength cannot exceed a critical value because an over-high strength will cause clustering of nanoparticles. Thus, there is a trade-off between the driving frequency and nanoparticle chain length. Figure 8.4e presents the distribution of induced fluid flow at the central area of the rotating chains, quantitatively showing the chain-chain hydrodynamic interaction. Based on the analysis in Figs. 8.3 and 8.4, the driving frequency and field strength are set at 6 Hz and 7 mT, respectively, in order to obtain a swarm pattern with a relatively high area density.

8.2 Localization of a Rotating Colloidal Microswarm Under Ultrasound ...

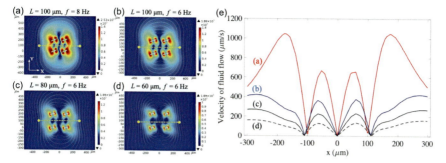

Fig. 8.4 **a–d** Simulation results of the chain-chain hydrodynamic interaction. The contour indicates the magnitude of the tangential velocity (μm/s), marked by the color legend on the right-hand side. White lines represent streamlines of the induced flow. **e** The distribution of induced fluid flow along the central lines (yellow dashed lines in (**a–d**)) at different L and f

8.2.2 Localization of a Microswarm Using Ultrasound Feedback

Because the chain-chain hydrodynamic interaction dominates the swarm formation process, we first investigate the ultrasound feedback of static nanoparticle chains. Nanoparticle chains are assembled along the field direction of an in-plane static magnetic field (Fig. 8.5a). The orientation of chains is able to be changed by adjusting the yaw angle (α) of the static field. After the nanoparticle chains reach static states, the imaging contrast is quantitatively investigated. Ultrasound images are acquired at yaw angles from 0° to 180° with an interval of 15°. The optimal imaging contrast is obtained when $\alpha = 0°$ or 180°, in which the orientation of nanoparticle chains is perpendicular to the wave propagation direction. The contrast decreases when the yaw angle increases from 0° to 90°, and then it gradually increases when α changes from 90° to 180°. To investigate whether the ultrasound wave influences the state of chains, ultrasound images are taken under an in-plane static field for continuous 10 min (Fig. 8.5b). No obvious disturbances on nanoparticle chains are observed during experiments, indicating that the influence from ultrasound waves is negligible. The mean pixel intensity (MPI) of the chain area is quantitatively investigated to reveal the relationship between imaging contrast and chain orientation. The region of interests (ROIs) is defined as a rectangle area that includes all the nanoparticle chains. The MPI of a ROI is calculated and plotted in Fig. 8.6. The intensity reaches the maximum value when $\alpha = 0°$ or 180°, and the curves are approximately symmetry with respect to $\alpha = 90°$. This result indicates that the scattered ultrasound waves reach their strongest when chains are perpendicular to the propagation direction of waves. However, only little ultrasound waves are scattered when the chains are parallel to the wave propagation direction (i.e., $\alpha = 90°$). In addition, the imaging contrast has no obvious change when adding small pitch angles of 2°–6° on the field, showing that a small tilt of the chain has a weak effect on the scattered wave (Fig. 8.6).

222   8 Localization of Microswarms Using Various Imaging Methods

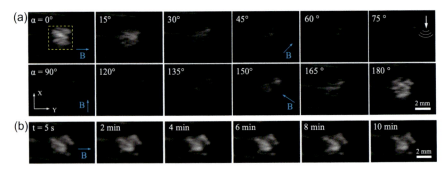

Fig. 8.5 **a** Ultrasound images of nanoparticle chains under an in-plane static field. The yaw angles are adjusted from 0° to 180° with an interval of 15°. The yellow rectangle indicates the region of interest (ROI). The white arrow and curves show the propagation direction of ultrasound waves. All the images are marked with the same ROI. The input field strength is 7 mT. **b** Ultrasound images of nanoparticle chains at $t = 0$–10 min when $\alpha = 0°$

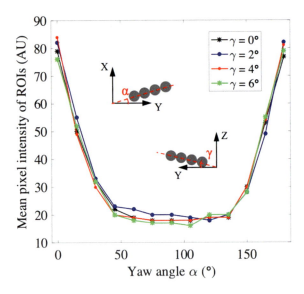

Fig. 8.6 MPI of ROIs under static magnetic fields at $\alpha = 0°$–180° and $\gamma = 0$–6°

Localization of the microswarm is then investigated under a rotating magnetic field. Figure 8.7a shows the image contrast of the initial area and the swarm, which are defined as the area of nanoparticles before (0–5 s) and after (40–45 s) the formation of a dynamic-equilibrium swarm pattern, respectively. The rotating field is applied at $t = 0$ s, and two areas are chosen as the ROIs to investigate the changes in ultrasound contrast. The intensity data are acquired frame by frame. The microswarm is actuated by a rotating field with a frequency of 6 Hz, and the ultrasound system has an output frame rate of fps. Therefore, each data point indicates the case that the orientation of chains is parallel (lowest contrast) or perpendicular (highest contrast) to the wave propagation direction. Periodic changes in ultrasound contrast are observed.

8.2 Localization of a Rotating Colloidal Microswarm Under Ultrasound ...

No obvious disturbances caused by the ultrasound waves are observed due to the relatively low power of ultrasound waves and the friction between microswarms and gelatin substrates. Furthermore, the high area density of nanoparticles (~5 μg/mm^2) contributes to the imaging contrast. By using a dynamic field to trigger the disassembly of the microswarm, nanoparticles with different area densities can be obtained (Fig. 8.7b) (Wang et al. 2020b). Then a rotating magnetic field is applied, and imaging contrast is recorded at the same time. The MPI is measured from continuous 48 frames (i.e., 2 s duration), and the mean values of intensity represent the mean of MPI of the 48 frames. The results show that the imaging intensity of the microswarm nonlinearly increases with the particle area density. An average intensity higher than 70 is able to be obtained when the swarm areas have an area density in the range of 4–5 μg/mm^2. Interestingly, even nanoparticles with a very low area density (e.g., 1 μg/mm^2) contribute to the ultrasound contrast.

The maximum value of the swarm area intensity is 90, as shown in image II of Fig. 8.7a. In Fig. 8.5a, the maximum value is 80 ($\alpha = 0°$), even though the area density of nanoparticles is relatively low (1.45 μg/mm^2). This high contrast is mainly attributed to the length of nanoparticle chains. Under a static field, the millimeter-scale nanoparticle chains are able to increase the number of scattered ultrasound

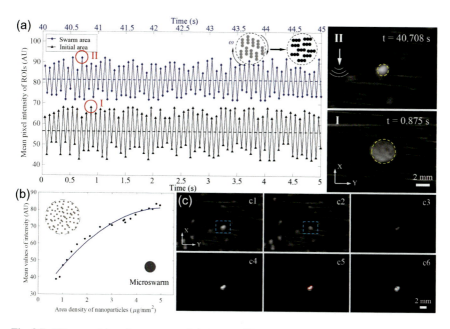

Fig. 8.7 Ultrasound imaging contrast of the vortex-like microswarm. **a** Changes in MPI of two ROIs (dashed circles) under the rotating field. The rotating frequency is 6 Hz, and the ultrasound imaging system has a frame rate of 24 fps. The dotted line represents the mean of the intensity, which is 81.5 and 56.4 for the swarm area and initial area, respectively. **b** Mean values of intensity versus the particle area density. The dots refer to experimental data, and the curve represents the fitting line of the data. **c** Image processing

waves. Previous analysis shows that the microswarm has nanoparticle chains with a length of 80 μm in it (Fig. 8.3). However, the maximum intensity is higher than that in the static-field case because the gathering effect of the microswarm results in a higher area density of nanoparticles. The results above show that the ultrasound contrast is affected by both the area density of nanoparticles and the length of chains. The average pixel intensity of the microswarm is 81.5 (Fig. 8.7a), while that in the case of static fields obtained using the same calculation method (i.e., calculating only the maximum and minimum pixel values over one actuation period) is about 50. The maximum and minimum contrast reaches around 80 and 20, respectively. The area density of nanoparticles has a more significant effect on ultrasonic contrast than the chain length.

Image differencing-based processing is utilized for swarm localization due to the dynamic and enhanced imaging contrast. The implemented steps of image processing are shown in Fig. 8.7c. Since the ultrasound contrast of microswarms is time-varying (Fig. 8.7a), the highest and lowest contrast can be obtained continuously by adjusting the input rotating frequency and the display frame rate of the ultrasound system. For instance, the display frame rate is adjusted to 20 fps when the imaging depth is increased from 3 to 4 cm, and continuous images changing between the highest and lowest contrast are obtained by decreasing the frequency from 6 to 5 Hz. Notably, the microswarm still remains in a dynamic-equilibrium state, and the signal-to-noise ratio is improved through image processing (Fig. 8.7c3). After the pixel intensity amplification, the microswarm is fitted as a circle by adding a threshold, and the swarm position is represented by the circle center (Fig. 8.7c6).

8.3 Magnetic Actuation of a Dynamically Reconfigurable Microswarm for Enhanced Ultrasound Imaging Contrast

8.3.1 Estimation of the Imaging Contrast of a Rotating Microswarm

The above results demonstrate that the imaging contrast of nanoparticle chains relies on the angle between chain orientation and ultrasound propagation direction (Wang et al. 2020). To simplify the description of the angle, we define the angle between the long axis of chains and the fixed wave propagation direction as θ_c with the range of 0°–90°. As schematically illustrated, the optimal ultrasound imaging contrast of the nanoparticle chains is obtained when $\theta_c = 90°$, i.e., the field angle of an in-plane non-polarized rotating field (NRF), is 0° or 180°, in which the backscattered ultrasound waves reach the maximum value (Fig. 8.8a). The imaging contrast decreases with a decreasing θ_c and reaches the minimum value when $\theta_c = 0°$; i.e., nanoparticle chains become parallel to the propagation direction. An in-plane static field is applied to quantitatively investigate the relationship between ultrasound contrast and θ_c, and the

8.3 Magnetic Actuation of a Dynamically Reconfigurable Microswarm … 225

orientation of particle chains is adjusted by changing the angle $\alpha(t)$ (Fig. 8.8b). Here the MPI$_{max}$ is defined as the maximum value of MPI in one rotating cycle; that is, $\alpha(t)$ is adjusted between 0° and 180° with an interval of 15°. Particle chains with different lengths are formed by adjusting the field strength from 5 to 8 mT. The MPI$_{max}$ are $95.2 \pm 4.7, 99.2 \pm 2.9, 102.7 \pm 3.2$, and 105.8 ± 3.5 under field strengths of 5 mT, 6 mT, 7 mT, and 8 mT, respectively. The results show that the chain length can slightly influence the imaging contrast, and only 11.1% of contrast difference can be observed between the 5 mT and 8 mT cases. The ratio MPI/MPI$_{max}$ is used to quantitatively investigate the influence of θ_c on the imaging contrast, which eliminates the effect of chain length under different input field strengths. The relationship between MPI/MPI$_{max}$ and θ_c proves that the imaging contrast of particle chains is highly dependent on θ_c, and the MPI$_{max}$ exists when $\theta_c = 90°$.

A dynamic θ_c results in a periodically changing ultrasound contrast when forming a dynamic swarm using a rotating field. The states of chains captured by the ultrasound imaging system at different input rotating frequencies are shown in Fig. 8.9a. Therefore, the changing ultrasound contrast of a microswarm is able to be estimated by calculating θ_c in real time. We take the 6 Hz case as an example. The five continuous ultrasound images represent one cycle of the imaging contrast because the temporal resolution of the ultrasound system is set as 30 fps. The mean imaging contrast within 1 s is calculated to be 49.48% of MPImax. By increasing the rotating frequency to 8.5 Hz, the mean value of MPI becomes 62.05% of MPImax, while 46.08% of MPImax is calculated with an input frequency of 8.0 Hz. Therefore, there is an optimal input frequency for optimal ultrasound contrast, i.e., the highest mean

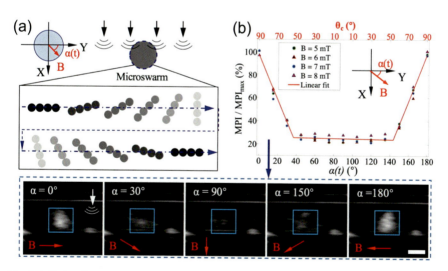

Fig. 8.8 Estimation of the ultrasound imaging contrast of a rotating microswarm. **a** Schematic illustration of the change in orientation of a nanoparticle chain inside the microswarm. **b** Relationship between the ultrasound imaging contrast and θ_c under an in-plane static field. Each data point represents the mean value of three trials. The scale bar is 2 mm

value of MPI over a period of duration. The mean value of MPI (the time duration is 1 s) at rotating frequencies of 1.0–10.0 Hz with an interval of 0.1 Hz is estimated. The results show that the maximum contrast is obtained when $f = 7.5$ Hz, in which the mean value of MPI becomes 62.05% of MPImax (Fig. 8.9b). However, the temporal resolution of the ultrasound system changes with the imaging depth, and the optimal-driven frequencies need to be calculated for different situations.

The optimal input frequency (f_{op}) for obtaining the best imaging contrast with the imaging depths of 2–7 cm is summarized in Table 8.1. An over-low rotating frequency (e.g., <4.0 Hz) weakens the chain-chain hydrodynamic interactions, while an overhigh rotating frequency (e.g., >10.0 Hz) shortens the chain length and weakens

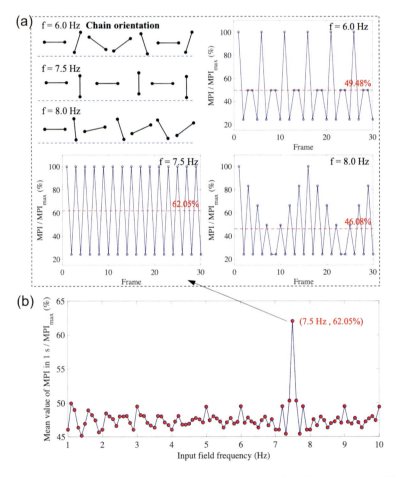

Fig. 8.9 a Estimation of the changes in the ultrasound contrast of a microswarm with different input rotating frequencies. The red dashed lines indicate the mean values. The temporal resolution is set as 30 ft/s. **b** Estimation of the change in the ultrasound contrast ratio (mean MPI in 1 s/ MPI_{max}) of an NRF-actuated microswarm with different input frequencies

8.3 Magnetic Actuation of a Dynamically Reconfigurable Microswarm ...

Table 8.1 Optimal-driven frequency (f_{op}) for localizing an NRF-driven rotating microswarm at different imaging depths

Imaging depth (cm)	Temporal resolution (fps)	f_{op} (Hz)	Estimated mean MPI in 1 s/ MPI_{max} (%)
2	49	9.8	50.00
3	30	7.5	62.05
4	25	6.2, 6.3	57.28
5	20	5.0 (10.0)	62.05 (100)
6	16	4.0 (8.0)	62.05 (100)
7	14	9.7, 9.8 (7.0)	58.71 (100)

the gathering effect, which may lead to an unstable microswarm. Therefore, we set the frequency as 4.0–10.0 Hz and use an interval of 0.1 Hz to estimate the f_{op}. Each f_{op} inside the bracket in Table 8.1 indicates that the ultrasound system keeps acquiring images with $\theta_c = 90°$; i.e., all images have the optimal ultrasound contrast (MPI_{max}). However, this is difficult to achieve in practice. First, in order to obtain continuous images with $\theta_c = 90°$, the temporal resolution of the ultrasound system and the field frequency should be coordinated with each other. But the temporal resolution disturbance may result in misalignment of the frames, and even one frame of disturbance causes the misalignment of $\theta_c = 90°$. Second, the loss of dynamic contrast leads to a decrease in the success rate of microswarm localization, especially in noisy environments. Therefore, we defined alternative optimal frequencies (i.e., the second-highest mean MPI) to localize microswarms at different imaging depths.

8.3.2 Experiments of Ultrasound Contrast Under the Actuation of NRF

Experiments are conducted to validate the optimization actuation parameters of f_{op}. The NRFs with different field strengths and frequencies are applied to form microswarms. The influence of chain length should be excluded when studying the effect of rotating frequencies on ultrasound contrast of the microswarm. For example, the chain length is estimated to be 59.164 μm under an NRF with a strength of 10 mT and a frequency of 10.0 Hz. The chain length is kept consistent by simultaneously adjusting the field strength and rotating frequency (Table 8.2). The swarm region is defined as the ROI for measuring the changing ultrasound contrast. The changes in ultrasound contrast under different driven frequencies are experimentally investigated (Fig. 8.10a). The average MPI over 3 s at different rotating frequencies shows that f_{op} at an imaging depth of 3 cm is 7.5 Hz, which is in good agreement with the estimates (Fig. 8.10b). Then, different field strengths are applied to investigate the effect of chain length on the ultrasound contrast ($f = f_{op} = 7.5$ Hz, Fig. 8.10c). Results show that the imaging contrast increases when the field strength is adjusted from 6

Table 8.2 Magnetic field parameters to form microswarms with adjustable chain lengths

f (Hz)	B (mT)	Chain length (μm)	Area density (μg/mm², three trials)
10.0	10.000	59.164	4.63 ± 0.19
8.0	8.054		4.71 ± 0.24
7.5	7.551		4.52 ± 0.30
6.0	6.041		4.50 ± 0.28
5.0	5.034		4.38 ± 0.41
7.5	6	52.153	4.38 ± 0.30
	7	56.904	4.51 ± 0.29
	8	61.357	4.43 ± 0.32
	9	65.565	4.37 ± 0.36
	10	69.697	4.12 ± 0.15
	12	77.061	3.88 ± 0.47

to 9 mT. Interestingly, the contrast decreases when the field strength is increased to over 10 mT. According to Table 8.2, the chain length becomes longer by increasing the strength from 7 to 9 mT; meanwhile, the area density changes from 4.51 μg/mm² to 4.37 μg/mm². Although the area density of the 7-mT case is higher than that of the 7-mT case, a better ultrasound contrast is able to be observed with a field strength of 9 mT due to the longer chain length. However, under a field with a strength of 12 mT, even though the chain length increases to 77.061 μm, the significant decrease in area density leads to a smaller MPI. This indicates that the contribution of chain length to ultrasonic contrast can only be clearly reflected when the area density is greater than ~4 μg/mm². In summary, magnetic field parameters are closely related to f_{op}, and relatively high area density and chain length should be maintained simultaneously to obtain high imaging contrast of microswarms.

8.3.3 Swarm Transformation and Ultrasound Contrast Under the Actuation of the PRF

The influence of the pattern shape on the imaging contrast is further investigated. The vortex-like microswarm is able to perform pattern transformation under the actuation of a polarized rotating field (PRF). The ultrasound images show that a microswarm with an elliptical pattern is formed by changing the field ratio ξ from 1 to 0.2 (Fig. 8.11a1), and the swarm transforms back to a circular shape by increasing ξ to 1 (Fig. 8.11a2). Two approaches can be applied for forming an elliptic swarm with the long-axis parallels to the wave propagation direction: directly adjust the direction angle of the elliptical field (β) or adjust β followed by decreasing the field ratio. The latter approach is adopted to investigate the contrast change during pattern transformation. The relationship between the imaging contrast and ξ is investigated

8.3 Magnetic Actuation of a Dynamically Reconfigurable Microswarm …   229

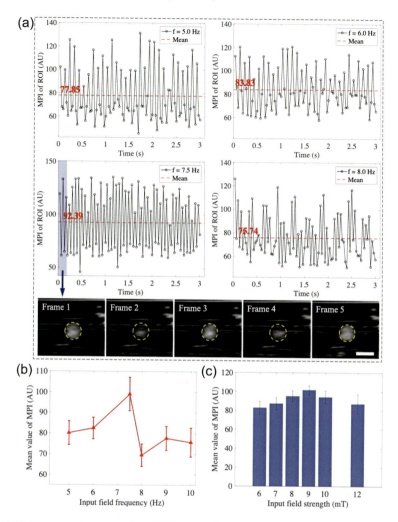

Fig. 8.10 Ultrasound contrast of an NRF-actuated microswarm. **a** Changes in the ultrasound contrast within 3 s (30 fps) under different rotating frequencies. Red dashed lines indicate the mean values. The scale bar is 2 mm. **b** Relationship between the mean value of MPI within 3 s under different rotating frequencies. The corresponding field strengths in (**a**) and (**b**) are listed in Table 8.2. **c** Relationship between the mean value of MPI within 3 s under different input field strengths. The rotating frequency is 7.5 Hz. Each error bar denotes the standard deviation from three trials

with $\beta = 0°$ or $90°$ (Fig. 8.11b). The contrast increases with ξ when the swarm's long axis is perpendicular to the wave propagation ($\beta = 90°$). Conversely, the contrast decreases by increasing the ξ from 0.2 to 1 when $\beta = 0°$.

We analyze the contrast change from two aspects: the change in the chain length and the adjustment of pattern shape. Results in Fig. 8.10 prove that the length and

Fig. 8.11 a Ultrasound imaging of microswarms actuated by PRF. The circular pattern is obtained by increasing ξ from 0.2 to 1 with $\beta = 90°$ (a1 to a2). β is adjusted from 90° to 0°, followed by decreasing ξ from 1 to 0.2 (a2 to a3). An elliptic swarm pattern with its long-axis parallel to the direction of wave propagation is obtained. All scale bars are 3 mm. **b** The change in the mean value of MPI within 2 s with different ξ. The red arrow indicates the pattern transformation process. Each error bar denotes the standard deviation from three trials

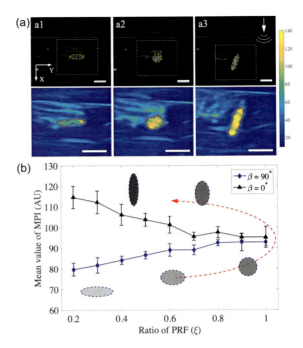

orientation of nanoparticle chains both influence the imaging contrast. In the case of $\beta = 0°$, the direction of the magnetic field strength at its maximum is perpendicular to the direction of wave propagation, i.e., $B(t) = B$ when $\theta_c = 90°$. Therefore, nanoparticle chains reach the maximum length and enhance the imaging contrast. In addition, since the long axis of the elliptical pattern is parallel to the direction of wave propagation, more nanoparticle chains capable of scattering ultrasound waves exist in the elliptical pattern compared to a circular pattern, further enhancing the imaging contrast. In contrast, nanoparticle chains have the shortest length when they are perpendicular to the wave propagation direction in the $\beta = 90°$ case ($B(t) = \xi \cdot B$, $\theta_c = 90°$). The chains become shorter when decreasing ξ, and the swarm pattern continuously shrinks along the wave propagation direction. Thus, a lower imaging contrast is obtained when decreasing the field ratio from 1 to 0.2. In addition, during the pattern transformation process, the change in the area density of the nanoparticles is negligible. These results indicate that the ultrasound contrast of the microswarm can be enhanced through active pattern transformation.

8.3.4 Swarm Navigation and Transformation in a Confined Environment Ex Vivo

Swarm navigation in a narrow channel is performed to demonstrate the active environmental adaptability of ultrasound-guided microswarms (Fig. 8.12a). The channel is wrapped in chicken tissue to simulate an ex vivo environment. NRF is first applied to form the microswarm, and then the swarm is navigated to the entrance of the narrow channel based on real-time ultrasonic localization ($t = 25$ s). A pitch angle of $1°$–$4°$ is applied during the navigation. The swarm pattern transforms to elliptical by reducing ξ from 1 to 0.3, avoiding collisions with the sidewalls ($t = 63$ s). Finally, the microswarm passes through the narrow channel and reaches the targeted position ($t = 90$ s).

Figure 8.12b presents the change in imaging contrast during the experiment (the imaging depth is 3 cm). According to results in Table 8.1 and Fig. 8.10, the magnetic field strength and rotating frequency are set to 9 mT and 7.5 Hz (f_{op}) for optimal ultrasonic contrast. The navigation process consists of four steps, i.e., navigation

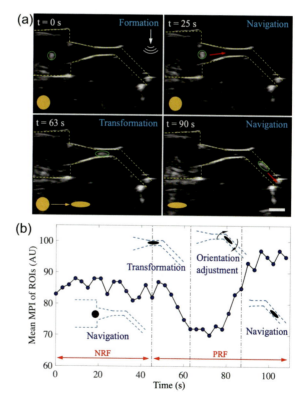

Fig. 8.12 Ultrasound-guided navigation of a microswarm in a confined environment. **a** Ultrasound images of the microswarm. The pattern elongation is achieved by decreasing ξ from 1 to 0.3. The imaging depth is 3 cm, and the scale bar is 3 mm. **b** The change in ultrasound contrast of the microswarm. The process is divided into four steps, as schematically illustrated

using NRF, transformation using PRF, swarm orientation adjustment, and navigation using PRF. The ultrasound contrast of the microswarm has a negligible disturbance and decreases during pattern transition, which is consistent with the results in Fig. 8.11 ($\beta = 90\circ$). A significant increase in imaging contrast is observed when adjusting the swarm orientation to align with the channel ($\beta = 45°$), which is caused by the enhanced swarm-scattered ultrasound waves. The active pattern transformation not only enhances the ultrasound imaging contrast, but also improves the ability of the microswarm to adapt to the surrounding environment, which is essential for imaging-guided biomedical applications, such as targeted drug delivery.

The experimental results show that the imaging contrast of microswarms at different depths is affected by the input rotating frequency and the temporal resolution of ultrasound imaging, and the relationship between them is in good agreement with the theoretical analysis. The optimal input frequency f_{op} can be derived from the relationship between θ_c and the imaging contrast. The area density of the nanoparticles and the length of the nanoparticle chains are two important factors influencing ultrasonic contrast. To obtain better imaging contrast, the area density should maintain a relatively high value, and the chain length should reach its maximum state when $\theta_c = 90°$. In addition, the imaging contrast is further enhanced by adjusting the field ratio and direction of the rotating field to form an elliptical microswarm with its long-axis parallel to the direction of ultrasound propagation. Utilizing microswarms to obtain enhanced imaging contrast is able to reduce the minimum dose of nanoparticles, avoiding the potential toxicity of excess particles. This strategy also provides a reference for the location selection of imaging probes. It is worth noting that the direction of ultrasound propagation should be kept at a small angle to the long axis of the microswarm, which can be achieved by placing multiple imaging probes or using robotic ultrasound imaging systems (Chen et al. 2019b; Wang et al. 2020; Heunis et al. 2020).

8.4 Fluorescence Imaging and Photoacoustic Imaging (PAI) of the Microswarm

Fluorescent magnetic nanoparticles are used to form vortex-like microswarms, and their synthesis process is shown in Fig. 8.13a. Fluorescence imaging of the swarm formation and spreading of the nanoparticles is presented in Fig. 8.13b, d, respectively. Nanoparticles are suspended in DI water, and the fluorescent imaging has weak imaging feedback due to the low area density. The nanoparticles are gathered into a relatively smaller region ($t = 6$ s) under a magnetic field gradient, and a rotating magnetic field is then applied for swarm formation (10–16 s). Finally, a microswarm with multiple subswarms is formed. The pattern is not an ideal circle because the average diameter of the fluorescent nanoparticles is only 20 nm, resulting in a weak response to the applied magnetic fields. A circular pattern is able to be formed faster using magnetic fields with higher strength.

8.4 Fluorescence Imaging and Photoacoustic Imaging (PAI) ...

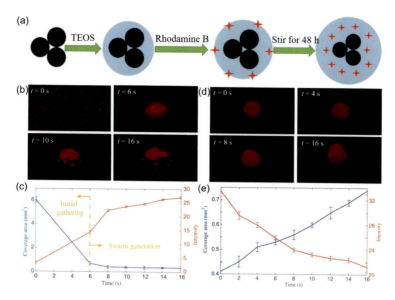

Fig. 8.13 Fluorescence imaging of the microswarm. **a** The synthesis process of fluorescent magnetic nanoparticles. **b** The real-time imaging of the formation process of a vortex-like microswarm. **c** The changes in swarm area and light intensity in (**b**). **d** The real-time imaging during the spreading process. **e** The changes in swarm area and imaging intensity in (**d**). The error bars indicate the standard deviation from three trials. The imaging intensities are calculated from the coverage regions of the swarms

To accurately estimate the coverage area of nanoparticles, imaging binarization, noise removal, and connected-body recognition are conducted. The largest area of all the connected bodies is regarded as the coverage area of the major swarm. After obtaining the coverage areas at several time points, the data are summarized, and the change in imaging intensity is presented (Fig. 8.13c, e). Figure 8.13c shows weak intensity of the suspended nanoparticles, and the curve of intensity rapidly increases to approximately 14 after the initial gathering. Swarm formation causes a significant effect on the imaging contrast. From 6 to 8 s, the swarm area shrinks from 0.7 mm^2 to 0.42 mm^2; i.e., the area density of the particles increases by 67%; meanwhile, the fluorescence intensity increases approximately by 56%. The swarm area continues to decrease when $t = 6$–16 s and finally reaches 0.35 mm^2, whereas the intensity gradually increases to 30. Then the dynamic magnetic field is applied to spread the concentrated nanoparticles (fluorescence images in Fig. 8.13d). The swarm is enlarged during the process, while the fluorescence intensity of the nanoparticle swarm becomes lower (Fig. 8.13e). From $t = 0$–16 s, the swarm area spreads from 0.42 mm^2 to 0.74 mm^2, showing that the area density of the nanoparticles and the intensity decrease by 43.2% and 36% during the process, respectively. During this short period, the photobleaching effect is negligible. These results show that the particle area density can significantly influence the fluorescence intensity, and

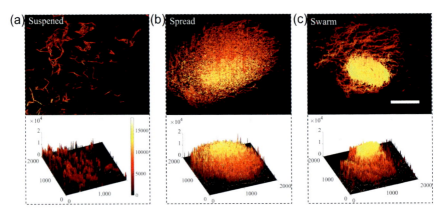

Fig. 8.14 Photoacoustic imaging of the microswarms with different area densities. **a** Suspended state. The nanoparticle area density is 0.7 μg/mm². **b** Spreading state. The nanoparticle area density is 4.3 μg/mm². **c** Swarming state. The nanoparticle area density is 40.5 μg/mm². The color bar indicates the intensity of the imaging feedback. The scale bar is 1 mm

the swarm formation-caused high density of nanoparticles enhances the fluorescent imaging feedback.

Photoacoustic imaging is also capable of showing nanoparticle swarms with different area densities. As shown in Fig. 8.14a, the particle suspension is dropped into the tank filled with DI water, and the coverage area of the particles reaches 100 mm², leading to a low particle area density (0.7 μg/ mm²). In this case, the imaging feedback is very weak, and most of the nanoparticles are not sufficiently clear. Then most of the particles are attracted into a cluster using a magnetic field gradient, and then the spreading process is conducted under the guidance of photoacoustic imaging (PAI, Fig. 8.14b). Compared with Fig. 8.14a, the signal is stronger, and the contour of the swarm becomes clear. Finally, a concentrated elliptical swarm pattern is formed using rotating magnetic fields. Much concentrated and stronger signals are emitted and shown by the bright yellow ellipsoidal region in Fig. 8.14c.

The statistical data of the feedback intensity using PAI are shown in Fig. 8.15. The blue, green, and yellow bars indicate the results of the suspended state, spread state, and swarm state, respectively. When the nanoparticles are in a suspended state, the percentage of signal points in the low intensity reaches 95%. The image of the suspended particles has very few signal points in the middle and high ranges of intensity. When the particles are gathered into the spread state, the percentage of signal points gradually decreases with the raising of the intensity range, as shown by the green curve. In this case, the percentage in the middle- and high-intensity ranges reaches approximately 45%, which is significantly higher compared with the results of the suspended state. After the nanoparticle swarm is formed, almost half of the signal points are in high-intensity range with another 30% points in the middle range. Therefore, the PAI imaging feedback strength of nanoparticles can be significantly enhanced by inducing swarm formation.

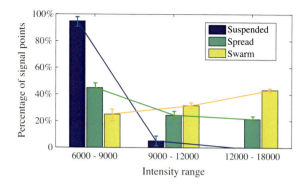

Fig. 8.15 Statistical data of PAI imaging feedback intensity. The entire images are statistically analyzed by calculating the number of pixels with the corresponding intensity range. The percentages of the signal points on the photoacoustic images in different intensity ranges are presented. The bars show the relationship between the results with different particle area densities in the same intensity range, while the curves indicate the percentage difference among the intensity ranges with the same area density. The error bars indicate the standard deviation from three trials

References

Ackermann, Schmitz G (2016) Detection and tracking of multiple microbubbles in ultrasound b-mode images. IEEE Trans Ultrason Ferroelect Freq Control 63(1):72–82

Azizi et al (2019) Using the fringe field of a clinical MRI scanner enables robotic navigation of tethered instruments in deeper vascular regions. Sci Robot 4(36). Art. no. eaax7342

Bergeles C, Yang G-Z (2014) From passive tool holders to microsurgeons: safer, smaller, smarter surgical robots. IEEE Trans Biomed Eng 61(5):1565–1576

Bergeles C, Kratochvil BE, Nelson BJ (2012) Visually servoing magnetic intraocular microdevices. IEEE Trans Robot 28(4):798–809

Biswal SL, Gast AP (2004) Rotational dynamics of semiflexible param- agnetic particle chains. Phys Rev E 69(4)

Cai H et al (2013) Facile hydrothermal synthesis and surface functionalization of polyethyleneimine-coated iron oxide nanoparticles for biomedical applications. ACS Appl Mater Interfaces 5(5):1722–1731

Chen Q et al (2019a) Ultrasound tracking of the acoustically actuated microswimmer. IEEE Trans Biomed Eng 66(11):3231–3237

Chen Q, Liu F-W, Xiao Z, Sharma N, Cho SK, Kim K (2019b) Ultra- sound tracking of the acoustically actuated microswimmer. IEEE Trans Biomed Eng 66(11):3231–3237

Erin et al (2019) Elevation and azimuth rotational actuation of an untethered millirobot by MRI gradient coils. IEEE Trans Robot 35(6):1323–1337

Guo J, Shi C, Ren H (2018) Ultrasound-assisted guidance with force cues for intravascular interventions. IEEE Trans Autom Sci Eng 16(1):253–260

Hao R et al (2010) Synthesis, functionalization, and biomedical applications of multifunctional magnetic nanoparticles. Adv Mater 22(25):2729–2742

Heunis CM, Wotte YP, Sikorski J, Furtado GP, Misra S (2020) The ARMM system-autonomous steering of magnetically-actuated catheters: towards endovascular applications. IEEE Robot Automat Lett 5(2):705–712

Hu W et al (2018) Small-scale soft-bodied robot with multimodal locomotion. Nature 554(7690):81–85

Jing W et al (2019) A microforce-sensing mobile microrobot for automated micromanipulation tasks. IEEE Trans Autom Sci Eng 16(2):518–530

Khalil S et al (2014) The control of self-propelled microjets inside a microchannel with time-varying flow rates. IEEE Trans Robot 30(1):49–58

Khalil S et al (2018) Mechanical rubbing of blood clots using helical robots under ultrasound guidance. IEEE Robot Autom Lett 3(2):1112–1119

Lenaghan SC et al (2013) Grand challenges in bioengineered nanorobotics for cancer therapy. IEEE Trans Biomed Eng 60(3):667–673

Li et al (2018a) Light-responsive biodegradable nanorattles for cancer theranostics. Adv Mater 30(8). Art. no. 1706150

Martel S et al (2009) MRI-based medical nanorobotic platform for the control of magnetic nanoparticles and flagellated bacteria for target interventions in human capillaries. Int J Robot Res 28(9):1169–1182

Medina-Sanchez M et al (2015) Cellular cargo delivery: toward assisted fertilization by sperm-carrying micromotors. Nano Let 16(1):555–561

Medina-Sánchez M, Schmidt OG (2017) Medical microbots need better imaging and control. Nature 545(7655):406–408

Melle S et al (2003) Microstructure evolution in magnetorheological suspensions governed by mason number. Phys Rev E 68(4). Art. no. 041503

Min KH et al (2015) ph-controlled gas-generating mineralized nanoparticles: a theranostic agent for ultrasound imaging and therapy of cancers. ACS Nano 9(1):134–145

Olson S et al (2013) Toward in vivo detection of hydrogen peroxide with ultrasound molecular imaging. Biomaterials 34(35):8918–8924

Pané S et al (2019) Imaging technologies for biomedical micro-and nanoswimmers. Adv Mater Technol 4(4):1–16. Art. no. 1800575

Petousis et al (2007) Transient behaviour of magnetic micro-bead chains rotating in a fluid by external fields. Lab Chip 7(12):1746–1751

Sánchez et al (2014) Magnetic control of self-propelled microjets under ultrasound image guidance. In: Proceedings IEEE RAS and EMBS international conference on biomedical robotics and biomechatronics, 169–174

Scheggi S et al (2017) Magnetic motion control and planning of untethered soft grippers using ultrasound image feedback. In: Proceedings of Conference on Robotics and Automation, 6156–6161

Servant et al (2015) Controlled in vivo swimming of a swarm of bacteria- like microrobotic flagella. Adv Mater 27(19):2981–2988

Shahrokhi S et al (2019) Exploiting nonslip wall contacts to position two particles using the same control input. IEEE Trans Robot 35(3):577–588

Sing E et al (2010) Controlled surface-induced flows from the motion of self-assembled colloidal walkers. Proc Nat Acad Sci 107(2):535–540

Singh PELH, Hatton TA (2005) Rigid, superparamagnetic chains of permanently linked beads coated with magnetic nanoparticles. synthesis and rotational dynamics under applied magnetic fields. Langmuir 21(24):11 500–11 509

Sitti M et al (2015) Biomedical applications of untethered mobile milli/ micro/nanorobots. Proc IEEE 103(2):205–224

Tay ZW et al (2018) Magnetic particle imaging-guided heating in vivo using gradient fields for arbitrary localization of magnetic hyperthermia therapy. ACS Nano 12(4):3699–3713

Vilela D et al (2018) Medical imaging for the tracking of micromotors. ACS Nano 12(2):1220–1227

Wang Q, Zhang L (2020) Ultrasound imaging and tracking of micro/nanorobots: from individual to collectives. IEEE Open J. Nanotechno. 1:6–17

Wang Q et al (2018) Collective behavior of reconfigurable magnetic droplets via dynamic self-assembly. ACS Appl Mater Interfaces 11(1):1630–1637

Wang T, Hu W, Ren Z, Sitti M (2020) Ultrasound-guided wireless tubular robotic anchoring system. IEEE Robot Automat Lett 5(3):4859–4866

References

Wang Q, Yang L, Yu J, Chiu PWY, Zheng Y-P, Zhang L (2020) Real-time magnetic navigation of a rotating colloidal microswarm under ultrasound guidance. IEEE Trans Biomed Eng 67(12):3403–3412

Wang, Zhang Y, Zhang L (2018) Recent progress on micro-and nano- robots: towards in vivo tracking and localization. Quantitative Imag Med Surg 8(5):461–479

Wang Q et al (2020a) Reconfigurable magnetic microswarm for thrombolysis under ultrasound imaging. In: International conference on robotics and automation, accepted

Wang Q et al (2020b) Disassembly and spreading of magnetic nanoparticle clusters on uneven surfaces. Appl Mater Today 18:1–10. Art. no. 100489

Wang et al (2018b) Reconfigurable swarms of ferromagnetic colloids for enhanced local hyperthermia. Adv Functional Mater 28(25):1–12. Art. no. 1705701

Wilhelm et al (2003) Rotational magnetic particles microrheology: the Maxwellian case. Phys Rev E 67(1). Art. no. 011504

Wright SE et al (2017) The spherical-actuator-magnet manipulator: a permanent-magnet robotic end-effector. IEEE Trans Robot 33(5):1013–1024

Wu Z et al (2018) A swarm of slippery micropropellers penetrates the vitreous body of the eye. Sci Adv 4(11). Art. no. eaat4388

Xie H et al (2019) Programmable formation and motion control of a snakelike magnetic microrobot swarm. IEEE/ASME Trans Mechatronics 24(3):902–912

Yan X et al (2017) Multifunctional biohybrid magnetite micro/nanorobots for imaging-guided therapy. Sci Robotics 2(12). Art. no. eaaq1155

Yang L, Wang Q, Zhang L (2018) Model-free trajectory tracking control of two-particle magnetic microrobot. IEEE Trans Nanotechno 17(4):697–700

Yang L et al (2020) An automated microrobotic platform for rapid detection of C. diff toxins. IEEE Trans Biomed Eng 67(5):1517–1527

Youssefi, Diller E (2019) Contactless robotic micromanipulation in air using a magneto-acoustic system. IEEE Robot Autom Lett 4(2):1580–1586

Yu J et al (2018) Ultra-extensible ribbon-like magnetic microswarm. Nat Commun 9. Art. no. 3260

Yu J et al (2019) Active formation and magnetic actuation of microrobotic swarms in bio-fluids. Nat Commun 10. Art. no. 5631

Zhang L et al (2009) Artificial bacterial flagella: fabrication and magnetic control. Appl Phys Lett 94(6):1–3. Art. no. 064107

Zhang C et al (2016) Modeling and analysis of bio-syncretic micro-swimmers for cardiomyocyte-based actuation. Bioinspiration Biomimetics 11(5). Art. no. 056006

Zhou G-Q, Zheng Y-P (2015) Automatic fascicle length estimation on muscle ultrasound images with an orientation-sensitive segmentation. IEEE Trans Biomed Eng 62(12):2828–2836

Chapter 9
Formation and Navigation of Microswarms in Dynamic Environments

Abstract This chapter presents the color Doppler ultrasound imaging-guided formation and navigation of rotating magnetic microswarms in dynamic environments. Rotating magnetic fields with gradients are generated by a rotating permanent magnet system and used to gather magnetic nanoparticles into a dynamical-equilibrium microswarm near the boundary. The fluid drag is reduced and the inner interactions among building blocks are enhanced, which enable the microswarm to navigate upstream and downstream. Doppler ultrasound imaging is used to locate microswarms in dynamic environments. The real-time imaging and targeted delivery strategies are validated by generating and navigating microswarms in ex vivo porcine coronary arteries with flowing blood.

Keywords Magnetic microswarm · Real-time navigation · Doppler ultrasound imaging · Blood flow · Endovascular delivery

9.1 Introduction

Delivering functionalized nanoparticles into the blood circulation system is an effective strategy for treating vascular-related diseases, such as vascular occlusion, atherosclerosis, and tumorigenesis. Intravenously injected nanoparticles can reach most organs and tissues through passive diffusion in the blood circulation system to achieve disease diagnosis and treatment. However, the first-pass metabolism in the liver may lead to side effects such as potential hepatotoxicity and systemic toxicity. The small size of nanoparticles makes their distribution and location hard to monitor in real time, increasing difficulties of targeted delivery. Micro/nanorobots provide a promising approach to overcoming these challenges by active and targeted material delivery (Nelson et al. 2010; Sitti 2017; Li et al. 2017). Their controllability and multimodal actuation enable navigation in hard-to-reach and complex environments (Ceylan et al. 2019; Schwarz et al. 2017), and related studies have been conducted in cells (Venugopalan et al. 2020), fluid-filled cavities (Wu et al. 2019; Yan et al. 2017), and blood (Cheng et al. 2014; Li et al. 2020; Alapan et al. 2020; Wang et al.

2020a). However, the small volume or surface of individual micro/nanorobots limits the drug-loading capability and challenges real-time imaging and control in vivo, especially in dynamic environments.

Micro/nanorobot swarms are considered to be an effective way to achieve delivery in various complex environments. The navigation of microswarms in relatively stagnant environments (e.g., eyes and stomach) has been studied, showing superior performance of using individual micro/nanorobots (Servant et al. 2015; Felfoul et al. 2016; Wu et al. 2018; Yu et al. 2019). Although microswarms have demonstrated potential in places with negligible flow, controllable navigation with high access rates in dynamic environments still remains a challenge. For instance, the pattern of microswarm navigated in blood vessels may be disrupted by strong impacts from the bloodstream or complex blood components (Qiu et al. 2019). When using microswarms for delivery tasks in dynamic environments, the swarm pattern can be stabilized in two ways. First, the interactions between building blocks should be enhanced by inducing agent-agent attractive forces. Meanwhile, the attractive force should not be too large; otherwise, it would lead to the formation of non-dispersible aggregates instead of reconfigurable microswarms. The second method is to reduce the impact of blood flow. The flow velocity near the boundary of blood vessels is much lower than the average. Natural organisms or cells (e.g., sperm, bacteria, and plankton) have the tendency to remain close to boundaries when performing upstream motion. Sperm-hybrid micromotors have been demonstrated to perform active locomotion near a channel surface to overcome the bloodstream and deliver multiple cargo or drugs (Xu et al. 2020). Therefore, the hydrodynamic drag can be significantly reduced by navigating microswarms near the boundary of blood vessels.

As we introduced in Chap. 8, real-time tracking of micro/nanorobots is crucial for targeted delivery and therapy in the human body (Medina-Sánchez and Schmidt 2017; Zhao and Kim 2019). Ultrasound imaging is a widely used clinical imaging method that can provide high temporal resolution (i.e., fast imaging speed) with minimum adverse effects (Aziz et al. 2020). The typical B-mode ultrasound depends on the gradients of acoustic impedance and has fast feedback, which enables real-time motion control and path planning of micro/nanorobots (Hu et al. 2018; Wang et al. 2020b). However, the low signal-to-noise ratio limits the application of B-mode ultrasound in dynamic environments. Unlike B-mode ultrasound imaging, Doppler ultrasound (DUS) imaging is based on the Doppler effect. The imaging process is realized by measuring the frequency shifts of reflected ultrasonic waves that induced by the motion of objects. Generally, it estimates blood flow by bouncing high-frequency sound waves from circulating red blood cells (RBCs) (Oglat et al. 2018), and the DUS images are overlaid on B-mode images to display the flow data. It is reasonable to hypothesize that a rotating colloidal microswarm (RCMS) can disturb the surrounding normal bloodstream and the motion of RBCs, and the disturbances generate DUS signals in situ. Both the RCMS itself and the affected region can be imaged by using the locally induced DUS signals. Therefore, the microswarm can be indirectly localized, and the strategy that integrates microswarm control and DUS imaging is of great potential for realizing active delivery in dynamic environments.

In this chapter, we introduce the formation and navigation of a rotating colloidal microswarm in dynamic environments under the real-time guidance of color DUS imaging for active endovascular delivery (Fig. 9.1). Actuated by rotating magnetic fields with gradients, paramagnetic Fe$_3$O$_4$ nanoparticles gather into a dynamical-equilibrium microswarm near the boundary. The reduced fluid drag and enhanced interactions between nanoparticles enable the upstream and downstream navigation of microswarms in flowing flows with an average velocity of up to 40.8 mm/s and an access rate of over 90%. The 3D bloodstream induced by the RCMS affects the motion of RBCs and disrupts the bloodstream. The Doppler effect is induced by the RCMS when emitting ultrasound waves to the microswarm, which can be detected by DUS imaging using different viewing configurations. Therefore, the 2D pattern of RCMS can be easily tracked. The fast feedback of DUS is beneficial to real-time tracking, navigation, and targeted delivery of microswarms in different flow conditions, including stagnant, flowing, and pulsatile flowing blood. The swarm-induced DUS signals enable the detection of fluid flow at a mean velocity of up to 50.24 mm/s. The delivery strategy is validated by generating and navigating microswarms in the porcine coronary artery. Besides, the switchable viewing configurations facilitate the formation and tracking of RCMS during delivery. Overall, medical imaging-guided navigation and targeted delivery of microswarms in the bloodstream are realized, demonstrating the great potential of integrating swarm control and medical imaging techniques in biomedical applications, especially active delivery in dynamic environments.

9.2 Formation of a Magnetic Nanoparticle Microswarm in Whole Blood

Formation and navigation of the nanoparticle swarm are conducted by using a spherical permanent magnet-based control system. Figure 9.2a shows the simulation results of magnetic field distribution when the north and south poles of the permanent magnet are placed horizontally. The magnetic field is considered parallel to the x–y plane near the top space of the permanent magnet. The swarm plane (s-plane) is defined as the surface where swarm formation and navigation are magnetically controlled, and d_{ms} denotes the distance between the s-plane and the top surface of the magnet. The field distribution on the x–y plane at $d_{ms} = 20$ mm is simulated, and results show that the field strength gradually decreases from the centerline of the magnet (Fig. 9.2b). The field strengths along the x-axis with different d_{ms} are plotted to quantitatively analyze the field distribution. The field strength decreases with both the d_{ms} and in-plane distance to the centerline, as shown in Fig. 9.2c. A rotating magnetic field can be generated on the s-plane by rotating the permanent magnet. During the swarm formation process, nanoparticle chains are first formed due to the induced attractive interaction among nanoparticles and rotate with the field. In this low Reynolds number regime (Re is estimated to be 0.03), the formation

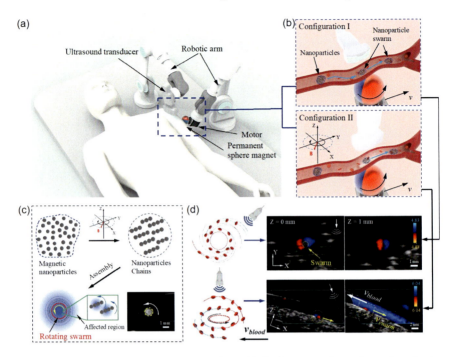

Fig. 9.1 Schematic illustration of DUS imaging-guided microswarm formation and navigation in blood vessels. **a** Schematic diagram of the swarm navigation in blood vessels. The RCMS is formed, navigated, and tracked in blood vessels with different viewing configurations. **b** The system configurations. **c** The formation process of a RCMS. **d** DUS signal around a RCMS in blood

of nanoparticle chains depends on the counterbalance between magnetic and fluidic drag torques (Singh et al. 2005). According to the chain length model, longer chains are formed when increasing the field strength, whereas the chains disassemble when increasing the rotating frequency due to the enlarged fluidic drag (Fig. 9.3). Here, we also treat the nanoparticle chains as the basic units and analyze the swarm formation with the consideration of chain-chain magnetic and hydrodynamic interactions. Under the actuation of the rotating field, the induced magnetic interaction between chains is induced and changes periodically. The average magnetic force with an actuation cycle is attractive (Giovanazzi et al. 2002). Meanwhile, the magnetic gradient leads to an enhanced gathering effect of the microswarm, that is, the nanoparticles gradually gather into a more concentrated region. The locally induced rotational flow also plays a crucial role in the swarm formation process. The hydrodynamic force between the two flows is attractive over the long range, which shortens the distance between nanoparticle chains. Then, the two flows will merge when the chain-chain distance reaches a critical value (Fig. 9.4a) (Josserand and Rossi 2007). After the chains reach a stable state, a dynamical-equilibrium RCMS is formed.

The strength and rotating frequency of the magnetic field significantly influence the formation process of the microswarm in the blood (Fig. 9.4b). The phase diagram

9.2 Formation of a Magnetic Nanoparticle Microswarm in Whole Blood

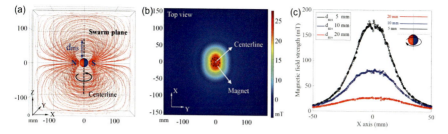

Fig. 9.2 **a** Magnetic field distribution of a spherical permanent magnet with a diameter of 25 mm. The swarm plane represents the position where the RCMS is formed and actuated. **b** A horizontal slice ($z = 32.5$ mm) of field strength distribution at $d_{ms} = 20$ mm. **c** Field strength along the x-axis at d_{ms} of 5 to 20 mm. Dots and lines indicate simulated data and fitted curves, respectively

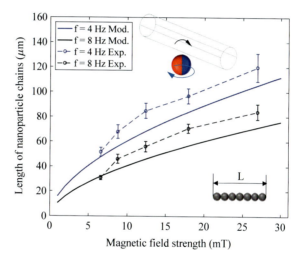

Fig. 9.3 Length of nanoparticle chains under different field parameters. Mod. and Exp. denote the calculated results using the mathematical model and experimental results in the glycerol-water solution with a viscosity of 4 cp, respectively. Each error bar denotes the standard deviation from three trials

shows that relatively long nanoparticle chains are formed in region I, and the chain-chain hydrodynamic repulsion makes them repel each other (Snezhko and Aranson 2011; Han et al. 2020). Longer chains disassemble into shorter ones by adjusting d_{ms} and frequency f, and dynamic-equilibrium RCMSs are formed (region II). Two criteria are defined to evaluate whether swarm formation is completed. First, the area density of nanoparticles in a dynamic pattern should be larger than 4.0 µg/mm². High area densities of the microswarm are beneficial to increase the contrast of the B-mode ultrasound imaging (Wang et al. 2020b). Second, the microswarm should be able to perform locomotion as an entity, while loosely interacted particle clusters or short nanoparticle chains cannot be categorized as a stable RCMS. Short chains are formed in region III, where loosely coupled rotating chains with an area density of 1.2–3.8 µg/mm² limit the gathering effect because the magnetic and hydrodynamic interactions are weak. RCMSs with different sizes can be obtained by changing doses of nanoparticles, as shown in Fig. 9.5. Swarm formations on the flat surface and in

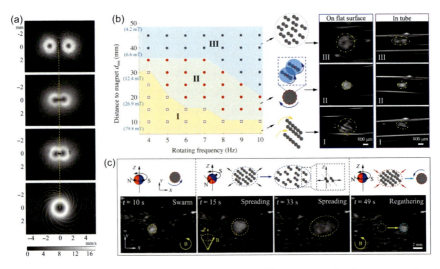

Fig. 9.4 Swarm formation in blood on a flat substrate. **a** Simulation results show two rotational flows merge with each other. The grayscale and white lines indicate the flow velocity and streamlines, respectively. **b** The phase diagram shows the gathering state of nanoparticles under different strengths and frequencies of the magnetic field. The insets show representative images of particles in the three regions. **c** Experimental results of the reversible spreading-regathering process of nanoparticles

confined spaces (circular tubes) are conducted, where the phase diagrams in both cases maintain the same parameters. RCMS has an elliptical pattern with the major axis along the tube due to the constraints of the curved surface (Fig. 9.6).

The RCMS is capable of spreading to a larger coverage area by adjusting the permanent magnet (Fig. 9.4c). A precessing field is generated by tilting the orientation of N-S poles, which induces repulsion between nanoparticle chains and enlarges the

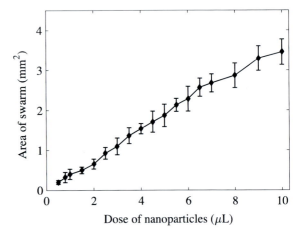

Fig. 9.5 Area of RCMSs in the blood formed by different doses of nanoparticles. The area is measured according to B-mode ultrasound images. Each error bar denotes the standard deviation from three trials

9.2 Formation of a Magnetic Nanoparticle Microswarm in Whole Blood

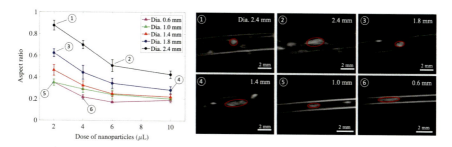

Fig. 9.6 Swarm formation in circular tubes filled with stagnant blood. The inner diameters are 0.6 to 2.4 mm. The aspect ratio is defined as the length ratio between the short and long axis of the swarm pattern. Each error bar denotes the standard deviation from three trials

coverage area of the nanoparticles ($t = 15$–33 s) (Wang et al. 2020c). After the spreading process, the nanoparticles can be regathered into a RCMS by applying a rotating field ($t = 49$ s), which demonstrates the spreading-regathering ability of the microswarm. The reformed microswarm has the same size and position as the original one. If the position of the magnet is moved after the spreading process, the RCMS will be reformed at the new position above the permanent magnet (Fig. 9.7).

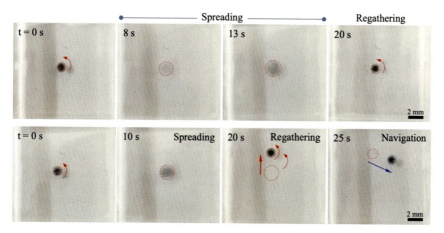

Fig. 9.7 Experimental results of the reversible spreading-regathering process of the microswarm in glycerol-water solution (viscosity: 4 cp). The magnet stays in the same position during the whole process (top) or is moved to a new position during the regathering process (bottom)

9.3 Rotating Microswarm Under Doppler Ultrasound Imaging

The influence of RCMSs on surrounding environments is investigated. Simulation results show that the RCMS locally generates rotational flow around it and above the s-plane (Fig. 9.8a, b). The induced flow velocities at different vertical distances are presented in Fig. 9.8c, which quantitatively shows the flow distribution around the RCMS. The flow velocity reaches up to 2.3 mm/s at a 2-mm vertical distance above the RCMS (RCMS radius: 800 μm), which indicates that the rotation of the swarm can affect the surrounding blood and movement of RBCs in the three-dimensional space. The movement of RBCs generates detectable Doppler shifts to the source (ultrasound transducer) when emitting ultrasound waves to the swarm region. Microswarms are first formed in stagnant blood and observed using color DUS mode to investigate the swarm-induced DUS signal (Fig. 9.8d). Here, configuration I is used, in which the direction of ultrasound propagation is parallel to the s-plane. The ultrasound images are composed of the DUS signals overlaid on the B-mode images, which display both the DUS signal (red and blue colors) and the RCMS (grayscale image). The red and blue colors appear simultaneously because the RBCs on the left and right sides of the RCMS have opposite moving directions (i.e., moving toward or away from the ultrasound source) under the influence of the rotational flow. This indicates that although the RCMS cannot be directly localized using the B-mode signal when the ultrasound waves propagate above the swarm, an indirect localization can be achieved based on the swarm-induced DUS signal. It is noted that the DUS signals are influenced by the input frequency. The RCMS has a larger affected region and generates stronger DUS signals when the input frequency is increased from 4 to 8 Hz. To quantitatively investigate the relationship between the Doppler color area and field parameters, we define the area ratio as the sum of Doppler color area divided by the area of RCMS. The changes in area ratio with different input frequencies are presented in Fig. 9.8e. The area ratio increases when the input frequency increases from 4 to 10 Hz. A larger swarm is observed by using a higher dose of nanoparticles, resulting in an enlarged Doppler color area (Fig. 9.9).

It is essential to generate and detect continuous Doppler signals when conducting DUS-guided navigation of microswarms. The influence of a moving microswarm on surrounding stagnant blood is first investigated, as shown by the simulation results in Fig. 9.10a. The effect on RBCs under different frequencies and locomotion velocities of the microswarm is also investigated. The swarm-induced flow exerts two main forces on the RBCs, i.e., hydrodynamic drag force and trapping force. The trapping force is generated by the attractive interaction in the flow regions of high vorticity (Chong et al. 2013). Therefore, in case I, RBCs are trapped and orbiting the moving RCMS. In case II, when increasing the locomotion velocity of the RCMS, RBCs are gradually released from the rotational flow due to the increased drag force. The surrounding RBCs are affected to exhibit rotational motion as well. In case III, the trapping force is enhanced by increasing the input rotating frequency, which drives more RBCs to rotate and induce the Doppler effect. In addition to the field parameters,

9.3 Rotating Microswarm Under Doppler Ultrasound Imaging

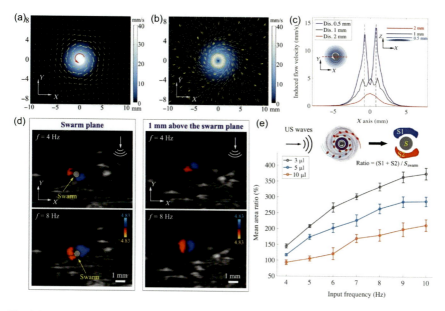

Fig. 9.8 Swarm formation and navigation in stagnant blood under DUS imaging. Simulation results of induced flow **a** on the s-plane and **b** 1 mm above the RCMS. The input frequency is 8 Hz. **c** Distribution of induced flow velocity above the s-plane. **d** DUS signals on and above the s-plane. The ultrasound propagation direction is marked on the top right corner. **e** Each error bar denotes the standard deviation from three trials

Fig. 9.9 Mean area of Doppler colors in stagnant blood in 3 s (14 fps) under different input frequencies and doses of nanoparticles. Each error bar denotes the standard deviation from three trials

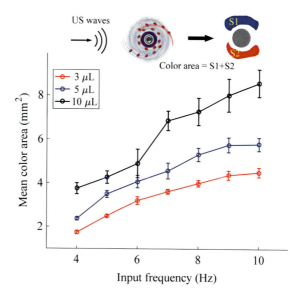

the DUS signal is influenced by two ultrasound parameters: the insonation angle (α) and pulse repetition frequency (PRF). The insonation angle indicates the angle between the moving direction of the object and ultrasound propagation. The Doppler shift is expressed as $f_d = 2(v/c)f_0\cos(\alpha)$, where c is the propagation speed of waves in a specific medium (e.g., tissues and blood), f_0 is the emitted frequency, and v is the velocity of the reflector (RBC). An insonation angle near 90° leads to a weak signal and the detected velocity close to zero (Cassin and Quinton 2019). PRF refers to the number of ultrasound pulses emitted by the transducer within a specified time period. When two pulses (pulse 1 and 2) are emitted with a time interval of 1/PRF, the small distance that the reflector (RBC) travels between the two pulses causes the returned pulse 2 to have a different phase from pulse 1. Multiple pulses are emitted to obtain a new Doppler curve with a frequency equal to f_d. Thus, the moving velocity of the reflector is detected, and the moving direction is marked in red or blue. When observed with configuration II (fixed PRF, Fig. 9.10b1), the DUS signal increases with input frequency, which agrees well with the simulation results (Fig. 9.10a). It is worth noting that increasing the PRF reduces the sensitivity of the ultrasound imaging system to blood with low flow rates (Fig. 9.10b2).

Configuration II has two advantages over configuration I. First, it is hard to keep tracking the s-plane due to the thickness of the swarm (<100 μm). Microswarms are easier to localize in configuration II, even if the imaging plane is off the center of the swarm. Second, the DUS signal above the s-plane can be detected along the x-axis and z-axis directions. Therefore, configuration II is adopted to navigate the microswarm, which is achieved by steering the magnet (Fig. 9.10c). Since the friction and drag forces from the boundary and blood components hinder the movement of the RCMS, high moving velocities may cause the disruption of the swarm pattern. The swarm navigation is performed at an average velocity of 1 mm/s to avoid disruption.

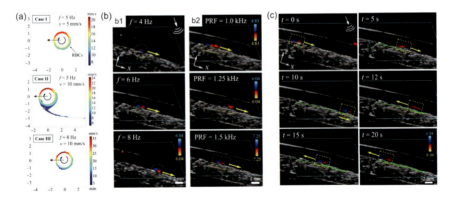

Fig. 9.10 **a** Simulation results of trajectories of simulated RBCs (6-μm-diameter microparticles) influenced by a RCMS. The arrows indicate the moving direction of the RCMS. **b** The DUS signal under different (b1) input frequencies and (b2) PRF values. PRF is 1.25 kHz in (b1) and f is 6 Hz in (b2). **c** Navigation of the RCMS under the guidance of DUS imaging. PRF is 1.25 kHz and f is 6 Hz

9.4 Swarm Formation and Navigation in Flowing Blood

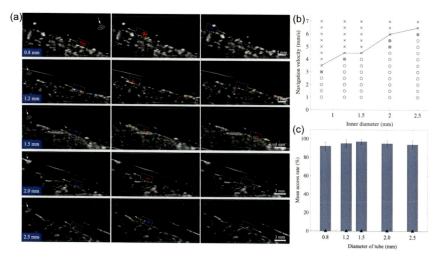

Fig. 9.11 a Ultrasound images of microswarms navigating in tubes with different diameters (filled with stagnant porcine whole blood). Yellow arrows indicate the moving directions. **b** Navigation velocities of the microswarm. Blue circles and black crosses indicate successful navigation and swarm disruption, respectively. The overlap between circles and crosses indicates that both conditions occurred in different trials. **c** Mean access rates of nanoparticles after the navigation process. The navigation distance is 40 mm. Each error bar denotes the standard deviation from five experiments

The navigation velocity is also influenced by the confined environment (Fig. 9.11). The maximum velocity in tubes with a diameter of 0.8 to 2.5 mm is 3 to 5.5 mm/s. Besides, continuous DUS signals are not detected when no rotation of the nanoparticles is induced. In this case, the nanoparticles stick to the inner wall and remain stationary, thus cannot form a dynamic pattern. Therefore, interactions inside the rotating microswarm play a crucial role in generating the DUS signal.

9.4 Swarm Formation and Navigation in Flowing Blood

Blood velocity near the boundary of a vessel is much lower than the average value due to the shear stress. Thus, an effective approach to maintain swarm stability in dynamic flow is to reduce the impact of fluidic drag by navigating RCMSs near boundaries. The distribution of flow in a branching tube is simulated, as shown in Fig. 9.12a, b. The flow velocity along the z-axis is approximately a parabolic profile, and the flow velocity is significantly lower near the boundary compared with the average velocity. As analyzed in Fig. 9.2, the gradient of the magnetic field exists along both the horizontal and vertical directions, yielding magnetic attraction force on nanoparticles in parallel and vertical directions. The nanoparticles are closer to the boundary and are more closely gathered, which increases the pattern stability in the bloodstream.

The primary influence on the RCMS in flowing fluids is the fluidic drag, which affects the inner hydrodynamic interactions as well. Such hydrodynamic interactions can be maintained in environments with relatively low flow rates (Table 9.1). The comparison between the area ratio of swarms in flowing (S_f) and stagnant blood (S_{stag}) shows that most nanoparticles are successfully gathered into the RCMS under the flow rate of 1–3 mL/min; meanwhile, the control parameters d_{ms} and f can affect formation of the microswarm in flowing fluids. The comparison between cases III and IV indicates that insufficient agent-agent magnetic interactions result in pattern disruption or the failure of swarm formation. By increasing the input frequency from 6 to 8 Hz, the microswarm with a more stable pattern can be obtained (cases IV and V), which is caused by the enhanced inward trapping force (Lecuona et al. 2002). The control parameters should satisfy two requirements to realize effective formation and navigation of RMCSs in flowing fluids. The first is swarm formation in stagnant blood and environments with low flow rates (1–4 mL/min). A reference range of d_{ms} has been determined according to the phase diagram of swarm formation in stagnant fluids (Fig. 9.4b) and the experimental results in environments with low flow rates (Table 9.1). The second is effective upstream and downstream navigation in flowing blood. The magnetic field with insufficient field gradients may cause disruption of microswarms and reduce the access rate of nanoparticles. According to the relationship between field gradient and d_{ms}, the range of d_{ms} at different frequencies is selected and further validated, in which a RCMS is first formed in stagnant blood and then navigated in flowing fluids. Finally, the optimized field parameters for generating and navigating microswarms in different bloodstream environments are determined (Fig. 9.12c).

Table 9.1 Swarm formation tests in flowing blood inside a circular tube with a diameter of 2.6 mm

	Flow rate (mean velocity)	Field parameters (d_{ms})	Formation time	Area ratio (S_f/S_{stag})
Case I	1 mL/min (3.1 mm/s)	30 mm, 6 Hz	10 ± 2 s	94 ± 3%
Case II	2 mL/min (6.3 mm/s)	30 mm, 6 Hz	10 ± 4 s	95 ± 4%
Case III	3 mL/min (9.4 mm/s)	30 mm, 6 Hz	×	×
Case IV	3 mL/min (9.4 mm/s)	20 mm, 6 Hz	16 ± 5 s	87 ± 6%
Case V	3 mL/min (9.4 mm/s)	20 mm, 8 Hz	12 ± 3 s	92 ± 4%
Case VI	4 mL/min (12.6 mm/s)	20 mm, 8 Hz	15 ± 4 s	82 ± 8%

9.4 Swarm Formation and Navigation in Flowing Blood

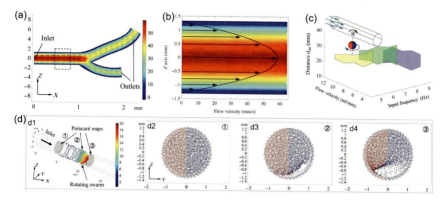

Fig. 9.12 Simulation results of navigating microswarms in flowing fluids. **a** The bloodstream in a branching tube with a diameter of 2.6 mm. The input flow rate is 10 mL/min. **b** Distribution of flow in the tube along the z-axis, corresponding to the dashed rectangular area in (**a**). **c** Field parameters (d_{ms} and f) during swarm navigation in flowing fluids. **d** Simulation of RBCs flowing through a RCMS. (d2)–(d4) are cross sections that have distances of 6 mm (left), 1 mm (left), and 1 mm (right) to the swarm center, corresponding to ①, ②, and ③ in (d1). The input frequency is 6 Hz, and the input flow rate is 10 mL/min

A RCMS in flowing blood disturbs the flow profiles and induces the Doppler effect, providing the fundamental mechanism for DUS tracking. We first investigate the influence of a RCMS on flowing RBCs using simulation. Simulated RBCs (6 μm-diameter microparticles) are released from the inlet (Fig. 9.12d1). The particles are divided into two groups according to the y-coordinate in the Poincaré map that represents the location of particle at its initial position. Under the laminar flow environment, the microparticles remain in the same position before interacting with the RCMS (Fig. 9.12d2). The initial position of 6.5% of microparticles changes after being influenced by the swarm-induced flow. 12.4% of microparticles are affected by the RCMS and move to another half-plane when flowing over it (Fig. 9.12d3, d4), indicating that the normal motion of RBCs is disturbed before contacting the swarm pattern. This disturbance induces a Doppler shift when pulses are emitted into the swarm region. Red and blue Doppler signals are generated around the RCMS navigating in flowing blood (Fig. 9.13a), and the background color representing blood flow is consistent with one of them. Increasing the input frequency can enhance the obtained signal (Fig. 9.13a1). Under the guidance of DUS imaging, RCMS is capable of performing upstream and downstream locomotion (Fig. 9.13a2, a3). Controlled by the magnet-based actuation system, the swarm exhibits locomotion at a mean velocity of 1 mm/s. However, since the detected signal originates from the interference of the swarm-induced flow to the surrounding normal blood flow, this method may not be applicable in the environment of high-rate flow. The signal can be enhanced by increasing the nanoparticle dose and the input frequency, enabling applications in environments with relatively high flow rates. The minimum frequency required to track RCMS increases with flow rate and decreases with the dose of nanoparticles (Fig. 9.13b). For example, RCMS formed by 3 μL of nanoparticles is difficult to

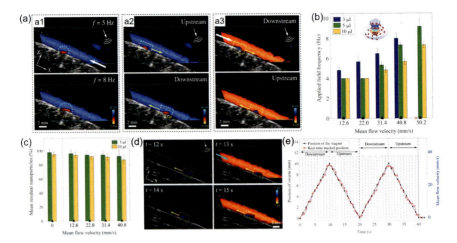

Fig. 9.13 **a** Navigation of a RCMS in flowing blood with an average flow velocity of 31.4 mm/s. White and yellow arrows indicate directions of the flow and swarm movement, respectively. **b** Minimal frequencies required to track RCMSs under different flow rates. PRF: 1.0 kHz (12.6 mm/s), 1.25 kHz (22.0–31.4 mm/s), 1.5 kHz (40.8 mm/s), and 1.75 kHz (50.2 mm/s). Each error bar denotes the standard deviation from three trials. **c** Recycled residual nanoparticles after navigation under different flow rates. Each error bar denotes the standard deviation from five trials. **d** Navigation of a RCMS in a pulsatile flow. The blue and yellow arrows indicate directions of the flow and swarm movement, respectively. **e** Comparison between the real-time tracked position of the RCMS and the position of the magnet. The blue dashed line indicates the flow profile

detect by DUS imaging in an environment with an average flow rate of 50.2 mm/s because the rotation of the swarm cannot obviously affect the blood flow. Furthermore, increasing the PRF results in reduced sensitivity to low-rate flows, failing the real-time localization of swarms. In this case, increasing the nanoparticle dose from 3 μL to 5 μL enables the swarm to be tracked again at an input frequency of 9 Hz.

The DUS signal varies dynamically with no obvious periodicity while navigating the microswarm in flowing blood. The following three steps are designed in the tracking algorithm to improve the tracking effectiveness. The first step is to define a dynamic region of interest (ROI). Since the RCMS follows the movement of the magnet during navigation, it should be restricted in a confined region near the magnet. The calibration of position between the center of the permanent magnet and the end effector of the robotic arm is conducted before the experiments. A dynamic square ROI centered with the x-coordinate of the magnet is defined, which follows the magnet during the navigation process. The second step includes color extraction and superimposition. The red color areas in ROIs are extracted, and the extraction results of n frames are superimposed ($n = 5$–15), avoiding the influence of noise signals and increasing the tracking effectiveness. The third step is to calculate the center of the superimposed Doppler color. A Kalman filter is applied to reduce fluctuations in tracked positions. Because the DUS signal generated by RCMS is evenly distributed around and above the swarm pattern, the center of the superimposed Doppler color

9.4 Swarm Formation and Navigation in Flowing Blood

can be regarded as the center of the swarm (tracked position). Considering the position of the magnet as the theoretical position, the errors of the two positions are 0.37 and 0.41 mm during the upstream and downstream locomotion, respectively. The average error is 0.34 ± 0.19 mm (21% of the RCMS's diameter), illustrating that this approach can effectively track the position of the microswarm in dynamic environments.

The penetration of ultrasound endows it with the ability to be used for deep-tissue imaging. The imaging depth of the ultrasound system is adjusted according to the position of the RCMS, and the RCMS is successfully imaged underneath porcine tissues with different thicknesses (Fig. 9.14a). The reduction of nanoparticle dose can cause image processing methods to fail to localize the swarm. The minimum nanoparticle dose required to localize RCMS increases with depth (Fig. 9.14b). RCMSs that can be localized at a depth of 4 cm need to be formed from about 4 μL of nanoparticles. The nanoparticles are collected, washed, hot air dried, and weighed after the navigation process. The results in Table 9.1 and Fig. 9.13c show that RCMSs are able to avoid disruption from flowing blood at relatively high flow rates, indicating that the interactions between nanoparticles play a crucial role in maintaining pattern stability of microswarms in dynamic environments. Around 90% of nanoparticles are gathered into the RCMS with relatively low average flow rates (1–3 mL/min). As a comparison, a RCMS exhibits locomotion in flowing blood with high access rates (~90%), although the flow rate reached over 10 mL/min.

Swarm navigation is implemented in pulsatile flowing fluids to investigate the feasibility of our approach. Pulse wave Doppler (PWD) imaging is applied to measure the flow velocity (Fig. 9.15) (Zafar et al. 2014). A mixed tracking algorithm is applied to real-time track the RCMS. In the low-flow region, the DUS signal is extracted and merged using ten continuous frames, and then the position is determined by analyzing the center of the processed signal. The algorithm in the high-flow region is like that in the constant-flow cases. The RCMS performs two downstream-upstream navigation cycles, and the tracking results show that the overall error in three trials is 0.44 ± 0.21 mm, i.e., 28% of the swarm body length (Fig. 9.13e). Although high-flow rates may interfere with DUS imaging, the tracking algorithm is still effective in the low-velocity area of the pulsating flow profile. The tracking method is tested by increasing

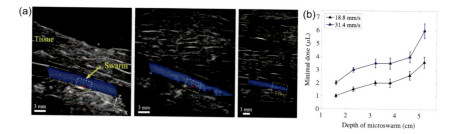

Fig. 9.14 **a** Imaging of a RCMS through porcine tissue with different thicknesses (from left to right: 1 cm, 1.8 cm, and 3 cm) under a flowing environment. The mean flow velocity is 31.4 mm/s. Parameters are $f = 6$ Hz, PRF $= 1.25$ kHz. **b** Minimal dose requirements for tracking a RCMS at different depths and flow velocities

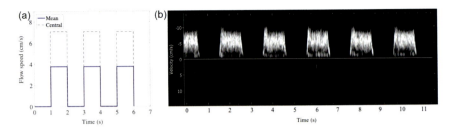

Fig. 9.15 Velocity of the pulsatile blood flow. **a** Simulation results of the mean flow velocity and the central velocity. The input bloodstream has a rate of 12 mL/min with an interval of 1 s. **b** Measurement of bloodstream in continuous 11 s using pulse wave Doppler imaging

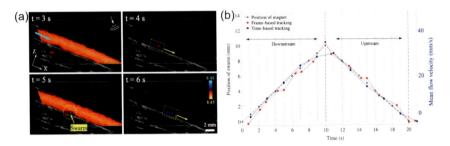

Fig. 9.16 Real-time navigation and tracking of the RCMS in the pulsatile flow. **a** Navigation of a RCMS in continuous 4 s. The input frequency is 8 Hz, the PRF is 1.75 kHz, and the input flow velocity is 14 mL/min. **b** The comparison between time-based and frame-based tracking results

the flow rate to 14 mL/min and using 3 μL of nanoparticles. The results show that RCMS cannot be tracked in the high-velocity area but can be localized in the low-velocity area according to the DUS signal. Two tracking algorithms are adopted to track the RCMS, i.e., frame-based ($n = 10$) and time-based tracking methods (the time interval is 1 s). The overall position errors of using frame- and time-based methods are 0.56 ± 0.26 and 0.32 ± 0.22 mm, respectively (Fig. 9.16). The RCMS with a low particle dose can be localized in high-flow velocity environments, although the tracking frequency became lower, i.e., $\sim$60% of the 12 mL/min case.

9.5 Real-Time Swarm Formation and Navigation in Porcine Coronary Artery Ex Vivo

To validate the feasibility of the delivery strategy, the formation and navigation of RCMSs are conducted in porcine coronary artery ex vivo (Fig. 9.17a). The whole artery is observed by using configuration I first, and the target region for nanoparticle release is selected. A catheter is used to release nanoparticles inside the artery. Meanwhile, the ultrasound probe is aligned with the catheter to observe and guide the release of nanoparticles via B-mode ultrasound imaging (Fig. 9.17b1, b2). During swarm formation, the configuration is switched to III because the 2D ultrasound images cannot give a complete picture of the curved blood vessels (Fig. 9.17b3, b4), and the cross section of the vessel is also observed using configuration III. Dynamic DUS signal is detected on the cross section of the vessel after a RCMS is formed (Fig. 9.17b4). The RCMS is then navigated by controlling the magnet, and the ultrasound probe moves along the x-axis direction. DUS signals on the cross sections (y–z plane) are recorded simultaneously (Fig. 9.17c). The y- and z-coordinates of the RCMS are extracted from the ultrasound images, while the x-coordinates are read from the manipulator. A frame-merging algorithm is used to track the RCMS in the blood vessel, as shown in Fig. 9.17d. Real-time localization of the RCMS is realized by using the ultrasound probe to scan the whole artery.

The bloodstream is restored after the RCMS is formed. Then, the swarm is navigated in flowing blood under the guidance of the induced DUS signal (Fig. 9.17e). Swarm navigation in vessels with a pulsatile flow is also performed. Similar to previous results, the RCMS is directly tracked by the induced DUS signal in the low-velocity area. The induced red color can also be distinguished from the blue color induced by the bloodstream when the RCMS is navigated in the high-velocity area. These results validate the feasibility and effectiveness of the proposed strategy in stagnant and flowing fluids ex vivo. Switchable configurations facilitate swarm formation and navigation for a variety of scenarios. For example, configuration II is suitable for guiding the release of nanoparticles due to the large observation area, while configuration III is suitable for tracking swarms in tortuous blood vessels.

9.6 Discussion and Conclusion

Micro/nanorobot swarms consisting of numerous simple building blocks are able to perform various tasks, such as targeted delivery and micromanipulation (Ahmed et al. 2017; Kokot and Snezhko 2018). Individual robots at the micro/nanoscale have very limited capabilities, while swarms formed from them can act as a more functional entity. For example, microswarms are capable of delivering more drugs and cargo and provide superior imaging contrast in deep tissues. We have covered the formation and imaging of microswarms in various bio-fluids in previous chapters, but swarm control in complex dynamic environments is more challenging, and more fundamental investigations are required. Different from the navigation of single micro/nanorobots in dynamic environments (Alapan et al. 2020; Jeong et al. 2016), the microswarm applied to dynamic environments needs to maintain the pattern stability in flowing fluids and has sufficient propulsion to achieve upstream motion. Therefore, the interactions between building blocks should be enhanced while preventing irreversible aggregation (Xu et al. 2020). Here, we introduce a permanent magnet-based microswarm-actuation method that enables swarm formation, reversible gathering, and navigation in flowing blood. Paramagnetic Fe_3O_4 nanoparticles form the RCMS under rotating fields generated by the permanent magnet system, and both upstream

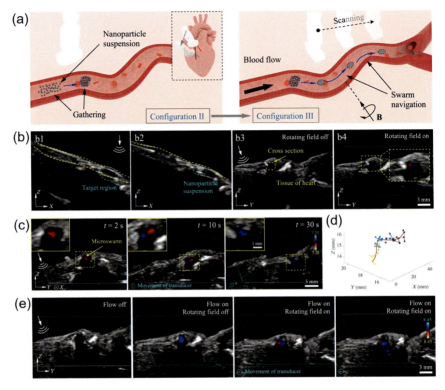

Fig. 9.17 Real-time navigation of RCMSs in porcine coronary artery ex vivo. **a** Schematic diagram of the nanoparticle release using configuration II, followed by formation and navigation of the RCMS using configuration III. **b** Nanoparticle release and swarm formation in the artery. The dashed yellow curves in (b1) and (b2) show the contour of the artery. (b3) and (b4) are ultrasound images before and after applying the rotating magnetic field. (b1) and (b2) are observed using configuration II, (b3) and (b4) are observed using configuration III. **c** Navigation of the RCMS in the vessel with stagnant blood. The insets show enlarged images of the region marked by the dashed yellow rectangles. The ultrasound probe is moved along the *x*-axis. **d** Real-time tracked position of the RCMS. Dots and the red line represent the tracked position and the fitted curve, respectively. **e** Real-time navigation of the RCMS in the vessel with flowing blood. PRFs are 1.5 kHz in (**c**), (**d**) and 1.75 kHz in (**e**)

and downstream locomotion have been realized. The RCMS is capable of spreading and regathering by adjusting the angle between the magnetic dipoles and the s-plane, which promises potential in hyperthermia (Perigo et al. 2015). The range of magnetic field parameters for RCMS formation in the blood is obtained from detailed experimental results (Figs. 9.4b, c, and 9.8), i.e., the magnetic field strength is 8 to 30 mT, and the rotation frequency is 4 to 10 Hz, which provides guidance for the swarm generation process using permanent magnet-based systems.

One of the key challenges in targeted delivery after the formation of microswarms is real-time localization and tracking, which requires imaging feedback with high temporal resolution and signal-to-noise ratio (Yang et al. 2018). This chapter introduces the real-time tracking of RCMS from multiple view angles using DUS imaging

9.6 Discussion and Conclusion

(Fig. 9.1). DUS imaging is a non-invasive imaging modality originally designed for estimating blood flow and is well compatible with magnetic fields. RCMS can be tracked in real-time in both stagnant and flowing blood, which is almost impossible for a single micro/nanorobot. Even RCMSs formed by low doses of nanoparticles are sufficient to generate detectable DUS signals in deep tissues with flowing fluids. However, high-velocity flows (e.g., in arteries) may weaken the swarm-induced flow or even lead to failure of DUS signal-based localization, which can be counteracted by increasing the failure dose or frequency. For pulsatile flow, swarm-induced DUS signals can be detected in the low-velocity region of the blood flow. Although the frequency of position feedback is reduced in this case, swarm navigation can also be achieved based on discontinuous DUS signals. Besides, multiple viewing configurations facilitate different application scenarios. Configuration I is used to fundamentally investigate the mechanism by which RCMS generates DUS signals in three dimensions, in which the imaging plane is adjusted. Configuration II is suitable for imaging in relatively straight tubes and can provide a relatively large field of view, while configuration III is suitable for imaging in tortuous tubes. Different configurations can be switched reversibly by adjusting the position of the ultrasound probe, which is convenient for tracking RCMS in different environments. In addition to ultrasound imaging based on backscattered waves, light-excited ultrasound waves have been reported to apply to real-time localization and tracking of micro/nanorobots (Aziz et al. 2019). This technique may benefit swarm localization in deep-tissue environments by combining the advantages of imaging depth and the high discrimination of microrobotic swarms from the surrounding living tissues. This technique may facilitate swarm localization in deep tissues by combining the advantages of imaging depth with high discrimination of microswarms from surrounding environments.

Passive delivery of nanoparticles using the blood circulatory system has disadvantages such as low access rates, low specificity, and potential side effects on organs. The active delivery of nanoparticles in the swarming form can reduce the total dose while maintaining a high local particle concentration, which significantly reduces the loss and adverse effects on other organs and tissues, and thus is considered as a safe and promising target therapy method. Studies have shown that nanoparticles have an access rate of more than 90% after moving in swarms for a distance of 20 mm in the flowing blood. However, moving over long distances takes a long time, and particles are inevitably lost continuously. Therefore, we propose the use of conventional medical catheters to assist the targeted delivery process. The nanoparticles are released after the catheter is guided near the target area, and a magnetic field is then applied to form the microswarm. It is noted that the catheter reduces the flow rate, providing a favorable environment for swarm formation. Then, actuated by a magnetic field and guided by DUS imaging, the microswarm can reach the target location through tortuous and narrow blood vessels, which is beyond the capability of catheters (Kim et al. 2019). This delivery strategy has a large parameter adjustment range and is expected to be applied to the control of a variety of functional magnetic micro/nanorobots (Karimi et al. 2016; Wang et al. 2018; Min et al. 2015; Shin et al. 2015; Lu et al. 2018). It is worth mentioning that the interaction between microswarms and blood vessels also needs to be evaluated before delivery, such as the effect of endothelial cells and the smoothness of the inner wall of blood vessels on the microswarm.

In summary, this chapter introduces the real-time navigation and localized delivery of microswarms in the vascular system using DUS imaging. Multiple viewing configurations are designed for tracking the real-time position of the RCMS through induced 3D DUS signals. A permanent magnet-based control system and a Doppler-based imaging system enable targeted navigation of the microswarm even in dynamic environments. These results provide a strategy for medical imaging-guided targeted delivery of concentrated agents in flowing fluids, which is validated in the ex vivo environment. The integration of microswarm control and imaging paves the way for the realization of clinical medical applications.

References

Ahmed D, Baasch T, Blondel N, Läubli N, Dual J, Nelson BJ (2017) Neutrophil-inspired propulsion in a combined acoustic and magnetic field. Nat Commun 8:770

Alapan Y, Bozuyuk U, Erkoc P, Karacakol AC, Sitti M (2020) Multifunctional surface microrollers for targeted cargo delivery in physiological blood flow. Sci Robot 5:eaba5726

Aziz A, Medina-Sánchez M, Claussen J, Schmidt OG (2019) Real-time optoacoustic tracking of single moving micro-objects in deep phantom and ex vivo tissues. Nano Lett 19:6612–6620

Aziz A, Pane S, Iacovacci V, Koukourakis N, Czarske J, Menciassi A, Medina-Sánchez M, Schmidt G (2020) Medical imaging of microrobots: toward *in vivo* applications. ACS Nano 14:10865–10893

Cassin M, Quinton A (2019) What is the more common method of obtaining velocity measurements in carotid artery studies: a 60° insonation angle versus a convenient insonation angle? Sonography 6:5–9

Ceylan H, Yasa IC, Kilic U, Hu W, Sitti M (2019) Translational prospects of untethered medical microrobots. Prog Biomed Eng 1:012002

Cheng R, Huang W, Huang L, Yang B, Mao L, Jin K, ZhuGe Q, Zhao Y (2014) Acceleration of tissue plasminogen activator-mediated thrombolysis by magnetically powered nanomotors. ACS Nano 8:7746–7754

Chong K, Kelly SD, Smith S, Eldredge JD (2013) Inertial particle trapping in viscous streaming. Phys Fluids 25:033602

Felfoul O, Mohammadi M, Taherkhani S, de Lanauze D, Xu YZ, Loghin D, Essa S, Jancik S, Houle D, Lafleur M, Gaboury L, Tabrizian M, Kaou N, Atkin M, Vuong T, Batist G, Beauchemin N, Radzioch D, Martel S (2016) Magneto-aerotactic bacteria deliver drug-containing nanoliposomes to tumour hypoxic regions. Nat Nanotechnol 11:941–947

Giovanazzi S, Görlitz A, Pfau T (2002) Tuning the dipolar interaction in quantum gases. Phys Rev Lett 89:130401

Han K, Kokot G, Das S, Winkler RG, Gompper G, Snezhko A (2020) Reconfigurable structure and tunable transport in synchronized active spinner materials. Sci Adv 6:eaaz8535

Hu W, Lum GZ, Mastrangeli M, Sitti M (2018) Small-scale soft-bodied robot with multimodal locomotion. Nature 554:81–85

Jeong S, Choi H, Go G, Lee C, Lim KS, Sim DS, Jeong MH, Ko SY, Park J-O, Park S (2016) Penetration of an artificial arterial thromboembolism in a live animal using an intravascular therapeutic microrobot system. Med Eng Phys 38:403–410

Josserand C, Rossi M (2007) The merging of two co-rotating vortices: a numerical study. Eur J Mech B Fluids 26:779–794

Karimi M, Ghasemi A, Sahandi Zangabad P, Rahighi R, Moosavi Basri SM, Mirshekari H, Amiri M, Shafaei Pishabad Z, Aslani A, Bozorgomid M, Ghosh D, Beyzavi A, Vaseghi A, Aref AR, Haghani L, Bahrami S, Hamblin MR (2016) Smart micro/nanoparticles in stimulus-responsive drug/gene delivery systems. Chem Soc Rev 45:1457–1501

Kim Y, Parada GA, Liu S, Zhao X (2019) Ferromagnetic soft continuum robots. Sci Robot 4:eaax7329

Kokot G, Snezhko A (2018) Manipulation of emergent vortices in swarms of magnetic rollers. Nat Commun 9:2344

Lecuona A, Ruiz-Rivas U, Nogueira J (2002) Simulation of particle trajectories in a vortex-induced flow: Application to seed-dependent flow measurement techniques. Meas Sci Technol 13:1020–1028

Li J, Esteban-Fernández de Ávila B, Gao W, Zhang L, Wang J (2017) Micro/nanorobots for biomedicine: Delivery, surgery, sensing, and detoxification. Sci Robot 2:eaam6431

Li D, Liu C, Yang Y, Wang L, Shen Y (2020) Micro-rocket robot with all-optic actuating and tracking in blood. Light Sci. Appl. 9:84

Lu GJ, Farhadi A, Szablowski JO, Lee-Gosselin A, Barnes SR, Lakshmanan A, Bourdeau RW, Shapiro MG (2018) Acoustically modulated magnetic resonance imaging of gas-filled protein nanostructures. Nat Mater 17:456–463

Medina-Sánchez M, Schmidt OG (2017) Medical microbots need better imaging and control. Nature 545:406–408

Min KH, Min HS, Lee HJ, Park DJ, Yhee JY, Kim K, Kwon IC, Jeong SY, Silvestre OF, Chen X, Hwang Y-S, Kim E-C, Lee SC (2015) pH-controlled gas-generating mineralized nanoparticles: a theranostic agent for ultrasound imaging and therapy of cancers. ACS Nano 9:134–145

Nelson BJ, Kaliakatsos IK, Abbott JJ (2010) Microrobots for minimally invasive medicine. Annu Rev Biomed Eng 12:55–85

Oglat AA, Matjafri MZ, Suardi N, Oqlat MA, Abdelrahman MA, Oqlat AA (2018) A review of medical doppler ultrasonography of blood flow in general and especially in common carotid artery. J Med Ultrasound 26:3–13

Perigo EA, Hemery G, Sandre O, Ortega D, Garaio E, Plazaola F, Teran FJ (2015) Fundamentals and advances in magnetic hyperthermia. Appl Phys Rev 2:041302

Qiu Y, Myers DR, Lam WA (2019) The biophysics and mechanics of blood from a materials perspective. Nat Rev Mater 4:294–311

Schwarz L, Medina-Sánchez M, Schmidt OG (2017) Hybrid BioMicromotors. Appl Phys Rev 4:031301

Servant A, Qiu F, Mazza M, Kostarelos K, Nelson BJ (2015) Controlled in vivo swimming of a swarm of bacteria-like microrobotic flagella. Adv Mater 27:2981–2988

Shin T-H, Choi Y, Kim S, Cheon J (2015) Recent advances in magnetic nanoparticle-based multi-modal imaging. Chem Soc Rev 44:4501–4516

Singh H, Laibinis PE, Hatton TA (2005) Rigid, superparamagnetic chains of permanently linked beads coated with magnetic nanoparticles. Synthesis and rotational dynamics under applied magnetic fields. Langmuir 21:11500–11509

Sitti M (2017) Mobile microrobotics, MIT Press

Snezhko A, Aranson IS (2011) Magnetic manipulation of self-assembled colloidal asters. Nat Mater 10:698–703

Venugopalan PL, Esteban-Fernández de Ávila B, Pal M, Ghosh A, Wang J (2020) Fantastic voyage of nanomotors into the cell. ACS Nano 14:9423–9439

Wang B, Chan KF, Yu J, Wang Q, Yang L, Chiu PWY, Zhang L (2018) Reconfigurable swarms of ferromagnetic colloids for enhanced local hyperthermia. Adv Funct Mater 28:1705701

Wang Q, Yang L, Yu J, Chiu PWY, Zheng Y-P, Zhang L (2020b) Real-time magnetic navigation of a rotating colloidal microswarm under ultrasound guidance. IEEE Trans Biomed Eng 67:3403–3412

Wang Q, Yu J, Yuan K, Yang L, Jin D, Zhang L (2020c) Disassembly and spreading of magnetic nanoparticle clusters on uneven surfaces. Appl Mater Today 18:100489

Wang Q, Wang B, Yu J, Schweizer K, Nelson BJ, Zhang L (2020a) Reconfigurable magnetic microswarm for thrombolysis under ultrasound imaging. In: IEEE international conference on robotics and automation, IEEE, pp 10285–10291

Wu Z, Troll J, Jeong H-H, Wei Q, Stang M, Ziemssen F, Wang Z, Dong M, Schnichels S, Qiu T, Fischer P (2018) A swarm of slippery micropropellers penetrates the vitreous body of the eye. Sci Adv 4:eaat4388

Wu Z, Li L, Yang Y, Hu P, Li Y, Yang SY, Wang LV, Gao W (2019) A microrobotic system guided by photoacoustic computed tomography for targeted navigation in intestines *in vivo*. Sci Robot 4:eaax0613

Xu H, Medina-Sánchez M, Maitz MF, Werner C, Schmidt OG (2020) Sperm micromotors for cargo delivery through flowing blood. ACS Nano 14:2982–2993

Yan X, Zhou Q, Vincent M, Deng Y, Yu J, Xu J, Xu T, Tang T, Bian L, Wang YXJ, Kostarelos K, Zhang L (2017) Multifunctional biohybrid magnetite microrobots for imaging-guided therapy. Sci Robot 2:eaaq1155

Yang G-Z, Bellingham J, Dupont PE, Fischer P, Floridi L, Full R, Jacobstein N, Kumar V, McNutt M, Merrifield R, Nelson BJ, Scassellati B, Taddeo M, Taylor R, Veloso M, Wang ZL, Wood R (2018) The grand challenges. Sci Robot 3:eaar7650

Yu J, Jin D, Chan K-F, Wang Q, Yuan K, Zhang L (2019) Active generation and magnetic actuation of microrobotic swarms in bio-fluids. Nat Commun 10:5631

Zafar H, Sharif F, Leahy MJ (2014) Measurement of the blood flow rate and velocity in coronary artery stenosis using intracoronary frequency domain optical coherence tomography: validation against fractional flow reserve. IJC Heart Vasc 5:68–71

Zhao X, Kim Y (2019) Soft microbots programmed by nanomagnets. Nature 575:58–59

Chapter 10
Applications of Micro/Nanorobot Swarms in Biomedicine

Abstract Micro/nanorobot swarms promise significant potential in various fields, especially in biomedicine. To date, how to realize clinical medical applications of microswarms still remains one of the biggest challenges in this research field. This chapter first briefly introduces several representative examples of biomedical applications of microswarms, such as rapid biosensing, detecting bio-fluid properties, photothermal–chemotherapy, in vivo imaging-guided navigation, and in vivo embolization. We then highlight three different biomedical applications of magnetic microswarms in detail, including enhancing local hyperthermia by utilizing the gathering effect of vortex-like swarms, eradicating biofilm with porous Fe3O4 nanoparticle swarms and magnetic urchin-like capsule robot swarms, and accelerating thrombolysis through ribbon-like swarm-enhanced convection and shear stress.

Keywords Microswarm · Biomedical application · Hyperthermia · Biofilm eradication · Thrombolysis

10.1 Introduction

Micro/nanorobot swarms have great promise in biomedical applications and are expected to bring enormous changes to traditional biomedicine (Wu et al. 2020; Chen et al. 2021; Wang et al. 2021; Fu et al. 2022). The advantages of micro/nanorobot swarms in the biomedical field are mainly reflected in the following aspects. First, micro/nanorobot swarms consisting of numerous individuals are capable of delivering large doses of cargo (such as drugs or cells) and can realize larger-scale energy conversion (e.g., thermal effects and mechanical forces). Second, the swarm pattern can be regarded as an entirety due to the accumulative effect, providing better imaging contrast and greater resistance to external disturbances than single agents at the micro/nanoscale. Third, the flexible morphology endows the microswarm with outstanding environmental adaptability, which is beneficial for adapting to complex internal environments of the human body, such as tortuous lumens. Furthermore, the functionalization of microswarms can be realized by simply modifying the building blocks.

Various energy sources and materials can be applied to form microswarms, and most of them have good biocompatibility (Wang and Zhang 2018; Jin and Zhang 2022; Wang and Pumera 2020). Fundamental research on micro/nanorobot swarms has not only deepened the understanding of collective behaviors at the micro/nanoscale, but also greatly promoted the development of microswarms in the field of biomedicine.

In recent years, a wide range of biomedical applications using microswarms have been reported, and various complex tasks have been realized, such as micromanipulation, targeted delivery and therapy, and biosensing and cleaning (Yang et al. 2022). These biomedical applications of microswarms range from in vitro (cell-line) and ex vivo (organ outside the body) to in vivo (inside the body) (Ji et al. 2022). For example, based on ultrasonic aggregation-induced enrichment with Raman enhancement, functionalized gold nanorod swarms can achieve ultrasensitive and rapid biosensing (Fig. 10.1a) (Xu et al. 2020). The principle is that the DNA probe-modified nanorods have the ability to capture specific biomolecular, and the formation of swarms enhances this capture ability, thereby realizing ultrasensitive detection of ultratrace biomarkers in a clinical sample solution. The spreading rate and elongation rate of the vortex-like swarm are influenced by the fluidic viscosity and ionic strength (Fig. 10.1b) (Chen et al. 2022). Thus, the microswarm can serve as an indicator for sensing the physical properties of biological fluids, such as blood. Under the irradiation of near-infrared laser, the gold nanoshell swarm can generate heat through the photothermal effect to kill nearby cancer cells (Fig. 10.1c) (Wang et al. 2016a). Meanwhile, the nanoshell will also rupture and release the internal anticancer drug (doxorubicin), which further inhibits tumor growth. This combined photothermal–chemotherapy utilizing the gold nanoshell swarm is more effective than chemotherapy or photothermal therapy alone. The swarm of artificial bacterial flagella can be used for in vivo targeted delivery. Actuated by a rotating magnetic field, the near-infrared probe-modified flagella swarm successfully performs navigated locomotion in the peritoneal cavity of a mouse under the guidance of near-infrared fluorescence and the IVIS imaging system (Fig. 10.1d) (Servant et al. 2015). Magnetic nanoparticle swarms can achieve selective embolization in blood vessels for the treatment of tumors, fistulas, and arteriovenous malformations, which has been verified in microfluidic channels, ex vivo tissues, and in vivo pig kidneys (Fig. 10.1e) (Law et al. 2022). Overall, the performance and safety of microswarms have been continuously validated as clinical trials are conducted, indicating their prospective transition from experimentation to practice.

This chapter introduces several representative studies on biomedical applications of magnetic microswarms in detail, including enhancing local hyperthermia, biofilm eradication, accelerating thrombolysis, and in vivo drug delivery and imaging-guided therapy. First, the vortex-like swarm is able to increase the local nanoparticle concentration due to its gathering effect. Therefore, it is used to enhance the efficiency of local magnetic hyperthermia, including increasing the temperature and improving the heating rate (Wang et al. 2018). Second, two biofilm eradication methods based on two different types of microswarms are introduced. One consists of porous Fe_3O_4 nanoparticles (Dong et al. 2021), the other uses magnetic urchin-like capsule

10.1 Introduction

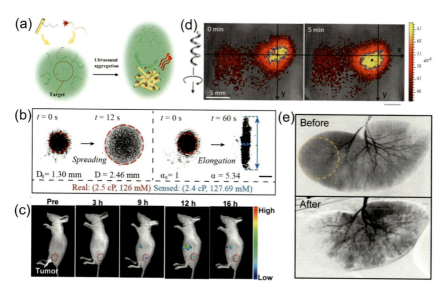

Fig. 10.1 Research on biomedical applications of micro/nanorobot swarms. **a** Ultrasonic aggregated gold nanorod swarm is used for ultrasensitive and rapid biosensing of miRNA-1246 detection in serum. Reprinted with permission from reference (Xu et al. 2020). Copyright 2020 American Chemical Society. **b** The vortex-like swarm serves as an indicator for sensing the physical properties of biological fluids. Reprinted with permission from reference (Chen et al. 2022). Copyright 2022 American Chemical Society. **c** The gold nanoshell swarm is applied to combined photothermal-chemotherapy of tumors in nude mice. Reprinted with permission from reference (Wang et al. 2016a). Copyright 2016 Elsevier. **d** The in vivo imaging of a swarm of artificial bacterial flagella in the intra peritoneal cavity of a mouse. Reprinted with permission from reference (Servant et al. 2015). Copyright 2015 John Wiley and Sons. **e** Magnetic nanoparticle swarms perform in vivo embolization in porcine kidneys. Reprinted with permission from reference (Law et al. 2022). Copyright 2022 American Association for the Advancement of Science

robots loaded with magnetic liquid metal droplets as building blocks (Sun et al. 2022). Despite the mechanism is different, both microswarm platforms exhibit the ability to efficiently eliminate stubborn biofilms in hard-to-reach regions. Third, a microswarm-assisted method for accelerating thrombolysis is introduced. The microswarm can enhance the convection and shear stress at the blood clot-liquid interface, which improves the efficiency of drugs to dissolve thrombus and thus reduce the drug dose required for thrombolysis (Wang et al. 2022).

10.2 Enhance Local Hyperthermia

10.2.1 Magnetic Local Hyperthermia

Hyperthermia is considered an effective therapeutic method for the treatment of tumors in medical oncology. Compared with radiotherapy, chemotherapy, and photothermal therapy, hyperthermia has the advantages of being able to remotely act on the deep tissues of the human body and having fewer side effects, thus has received extensive attention in recent years (Dahl and Overgaard 1995). Hyperthermia generally utilizes a temperature over 40 °C to inhibit the growth of tumor cells or even kill tumor cells, which is based on the higher thermal sensitivity of tumor cells than normal cells (Roussakow 2013). Hyperthermia is classified into three types according to the heating area, i.e., whole-body hyperthermia, partial hyperthermia, and local hyperthermia. Among the three types, whole-body hyperthermia is performed in a heat chamber, while partial hyperthermia depends on perfusion or microwaves. Compared with the first two methods, local hyperthermia has the smallest heating area, which means higher accuracy. In local hyperthermia, only tumor cells are killed by the high temperature, while most normal cells still work outside the affected area of hyperthermia. Methods to achieve local hyperthermia include alternating magnetic fields, ultrasonic waves, microwaves, and near-infrared radiation. The method of using an alternating magnetic field to make magnetic nanoparticles generate heat in a designated area for the treatment of tumors is called magnetic hyperthermia (Périgo et al. 2015). The magnetic field can affect deep tissues without obvious damage to the human body, thus having good penetrability and biocompatibility. In addition, magnetic nanoparticles have good targeting ability and biocompatibility, and their dosage, size, geometry, and functionalization can be designed according to requirements (Miller et al. 2015; Karimi et al. 2016; Huang et al. 2016). Therefore, magnetic hyperthermia is considered to be an ideal tumor treatment method.

Ferromagnetic nanoparticles with a wide range of sizes are commonly used in hyperthermia and can be synthesized by various types of methods, such as coprecipitation, hydrothermal, solvothermal, and thermal decomposition (Chalasani and Vasudevan 2011). Improving the heating rate and specific loss power (SLP) is crucial for reducing the dosage of magnetic nanoparticles used in hyperthermia. For example, the optimized magnetic core–shell nanoparticles have much higher SLP values than ordinary nanoparticles due to the exchange coupling between the hard core and soft shell (Lee et al. 2011). In the treatment experiments of mice tumors, the temperature of tumor sites reaches about 43–48 °C by using these core–shell nanoparticles. Meanwhile, the dosage is reduced to 10% of that used in traditional methods, which greatly avoids the attack from the immune system. In addition, the development of magnetothermal materials also needs to consider their interactions with biological systems (Huang et al. 2015; Li et al. 2015; Wang et al. 2016b; Zhang et al. 2013; Monopoli et al. 2012; Feliu et al. 2016). Natural and artificial polymers, such as polydopamine, polyvinylpyrrolidone, and polyethylene glycol (PEG), have been reported

10.2 Enhance Local Hyperthermia

to modify the surface of nanoparticles to improve their stability and biocompatibility (Feliu et al. 2016; Saei et al. 2017).

The improvement of thermal treatment efficiency is mainly achieved by improving the design of magnetic materials, including increasing the magnetization, changing the particle geometry, and designing special structures (such as the core–shell structure). Here, we introduce using micro/nanorobot swarms in magnetic hyperthermia to obtain enhanced heating efficiency and tunable energy dose (Wang et al. 2018). Benefitting from the gathering effect of the microswarm, the local nanoparticle concentration can be increased to 500% of the initial state, which greatly improves the heating efficiency and reduces the dose of nanoparticles. Furthermore, since the microswarm has reversible pattern transformation ability, including swelling and shrinkage, the local particle concentration can be adjusted in real time as needed.

10.2.2 Adjustment of the Nanoparticle Concentration

The magnetic nanoparticles used here are Fe_3O_4 nanoparticles synthesized by a solvothermal method based on green chemical reactions, and no toxic products are obtained (Deng et al. 2005). The Fe_3O_4 nanoparticles have a diameter in the range of 200–300 nm, and scanning electron microscopy (SEM) and transmission electron microscopy (TEM) images are shown in Fig. 10.2. The synthesized nanoparticles have a PEG capping layer, which reduces the possibility of cytotoxicity and being recognized as foreign molecules by the immune system, thus helping to increase the retention time of the particles in the body.

The comparison of the commonly used commercial thermal seeds and the as-prepared nanoparticles is shown in Fig. 10.3a. The thermal seeds have a length of ~6 mm and a diameter of ~1 mm. Figure 10.3b presents the magnetism of the nanoparticles (red curve) and thermal seeds (blue curve) measured with a vibrating sample magnetometer. It can be found from the hysteresis loops that both the nanoparticles

Fig. 10.2 Fe_3O_4 nanoparticles for magnetic local hyperthermia. **A** SEM image of the nanoparticle. **b** TEM bright field image of the nanoparticles deposited from ethanolic dispersion on amorphous carbon-coated copper grids and dried at 60 °C. **c** Enlarged TEM image of the nanoparticles

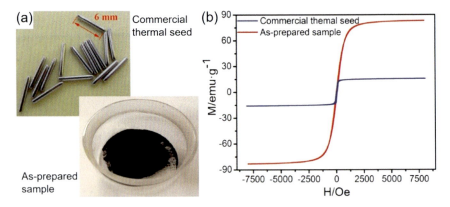

Fig. 10.3 Comparison between as-prepared nanoparticles and commercial thermal seeds. **a** Images of the as-prepared nanoparticles and commercial thermal seeds. **b** Magnetism of the as-prepared nanoparticles and the commercial thermal seeds at 300 K measured with a vibrating sample magnetometer

and thermal seeds are superparamagnetic, with residual magnetization and coercivity values around 0. The saturation magnetization of synthesized magnetic nanoparticles (83.1 emu·g^{-1}) is about five times larger than that of the commercial thermal seeds (16.0 emu·g^{-1}), indicating that the nanoparticles may have a better magnetocaloric effect than the thermal seeds.

It is well known that the most straightforward and effective way to increase the heating rate and SLP of magnetic particles is to increase the dose or concentration. In practical treatments, large doses of nanoparticles are likely to be recognized by the immune system as foreign invading substances, thereby triggering a strong allergic reaction (Maldonado et al. 2015). This makes traditional hyperthermia a trade-off between efficiency and safety, as safe doses often mean poorer therapeutic effects. Using the gathering effect of micro/nanorobot swarms can increase the concentration of nanoparticles in local areas, promising high thermal efficiency at low particle doses.

Previous research has reported that programmable spatial dynamic magnetic fields are capable of triggering the disassembly of nanoparticle clusters and swelling of microswarm patterns (Yu et al. 2017). By using this previously reported dynamic field, a nanoparticle cluster is disassembled into plenty of smaller fragments, and the area covered by nanoparticles gradually expands (Fig. 10.4a1–a5). The disassembled nanoparticles can regather into a vortex-like swarm by simply applying an in-plane rotating magnetic field (Fig. 10.4b1–b5) (Yu et al. 2018). Figure 10.5a shows the areal density of nanoparticles (the mass of nanoparticles per square centimeter, i.e., nanoparticle concentration) decreases over time during the swelling process. The higher strength of the dynamic field results in lower nanoparticle areal density. For example, a dynamic field with a strength of 8 mT can reduce the area density from 4.45 mg/cm^2 to 0.82 mg/cm^2 in 70 s, while the area density only decreases from 3.8 mg/cm^2 to 3.1 mg/cm^2 at a lower field strength (2 mT) within the same time.

10.2 Enhance Local Hyperthermia

Therefore, to disperse the nanoparticles, the high field strength is beneficial to obtain a relatively low area density (e.g., lower than 1 mg/cm^2), and the low field strength is more suitable for fine-tuning area density. On the other hand, the area of nanoparticles decreases over time in the swarm formation process. Since most of the nanoparticles are confined inside the swarm pattern, the area density gradually increases. Experimental results show that the initial concentration of nanoparticles before applying the rotating magnetic field has an influence on the formation time of the vortex-like swarm (Fig. 10.5b). It takes a longer time to generate the swarm at a lower initial concentration due to weaker attraction between vortices, which is consistent with the results we introduced in Chap. 2. Increasing the field strength makes the length of nanoparticle chains become longer, which expands the affected area of the vortex and shortens the swarm formation time. This effect is negligible for larger initial nanoparticle concentrations (>1 mg/cm^2) since the vortex-like swarm can form rapidly in these cases. In addition, the final nanoparticle concentration in the vortex-like swarm increases with the rotating frequency of the magnetic field, as shown in Fig. 10.5c. After the frequency increases to more than 10 Hz, the nanoparticle concentration does not continue to increase but remains at about 4 mg/cm^2.

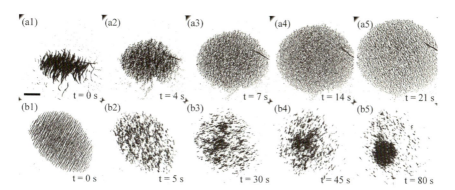

Fig. 10.4 Reversible pattern reconfiguration of the magnetic nanoparticles. **a** The swelling process under a dynamic field. **b** The shrinkage process under a rotating field

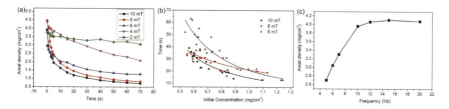

Fig. 10.5 Area density of the nanoparticles. **a** The change of area density during the swelling process of nanoparticles. **b** The relationship between formation time of the vortex-like swarm and initial concentration of the nanoparticles. **c** The relationship between final nanoparticle concentration and the rotating frequency

10.2.3 Photothermal Effect of the Nanoparticle Swarm

Under external radio frequency magnetic fields, the nanoparticles are able to convert electromagnetic energy into thermal energy, and the heating efficiency is directly related to the density of nanoparticles. In the experiment, an alternating magnetic field with a frequency of 180 kHz is used to heat the Fe_3O_4 nanoparticles. The change of temperature in the central region of the nanoparticles with time is measured, and the results are shown in Fig. 10.6a. The central temperature of nanoparticles with different area densities all increases rapidly after applying the alternating magnetic field and then gradually becomes stable. After applying the magnetic field for half an hour, the temperature reaches a relatively stable value. Figure 10.6b, c shows the final temperature that the nanoparticles can reach and the relationship between the heating rate and areal density, respectively. It can be found that the larger the areal density, the faster the temperature rises and the higher the final temperature. In general, temperatures above 40 °C are effective in killing tumor cells. According to Fig. 10.6b, area densities less than ~0.75 mg/cm^2 cannot heat localized areas above 40 °C. The high areal density is able to shorten the temperature rise time, which significantly improves the efficiency of hyperthermia (Fig. 10.6c).

During the heating process, the nanoparticle-enriched region produces a strong thermal effect. Figure 10.6d shows the temperature at different positions from the nanoparticle swarm pattern (the area density is 20 mg/cm^2). The temperature decays

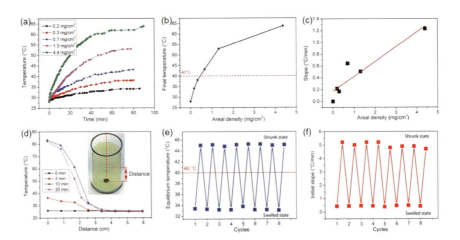

Fig. 10.6 Adjustment of the nanoparticle concentration. **a** The change of temperature in the central region of the nanoparticles with different area densities. **b** The final temperature of nanoparticles with different area densities. **c** The relationship between the heating rate and areal density. **d** The temperature at different positions from the nanoparticle swarm pattern in agar gel. The area density of the nanoparticles is 20 mg/cm^2. **e** The final equilibrium temperature of the nanoparticle swarm during eight swelling–shrinkage cycles. **f** The initial heating rate of the swarm in shrunk state and swelling state. The frequency of the alternating magnetic field is 180 kHz, and the coil current is 40 A

10.2 Enhance Local Hyperthermia

exponentially with increasing distance from the swarm center. When the distance is larger than 2 cm, the temperature is substantially less than 40 °C, so the effective area of local hyperthermia in this case has a radius of about 2 cm. The swelling and shrinkage of the swarm pattern affect the density of nanoparticles and thus have an influence on the local temperature. Figure 10.6e shows the equilibrium temperature of the nanoparticle swarm during eight swelling–shrinkage cycles. It can be found that shrinkage of the swarm results in a higher temperature (>40 °C), and the swelling leads to a lower temperature (<40 °C). Therefore, the shrunk state of the nanoparticle swarm is more favorable for local magnetic hyperthermia. The shrunk and swelled states of the swarm also have an effect on the heating rate. Figure 10.6f shows that the average heating rate in the shrunk state is 5 °C/min, while it is only 0.4 °C/min in the swelled state. In addition, it can also be found from Fig. 10.6e, f that after the nanoparticles undergo multiple swelling–shrinkage cycles, their heat-generating capacity has no obvious decrease.

10.2.4 Enhanced Local Hyperthermia

In vitro experiments are conducted to demonstrate the capability of microswarms to perform enhanced local hyperthermia. First, the vortex-like swarm is formed under the rotating magnetic field. Then, the swarm is navigated to a chamber through a long straight channel (Fig. 10.7a) or turns above and enters another chamber (Fig. 10.7b). These chambers are cultured with HepG2 cells (a commonly used tumor cell). After reaching the target area and completing the magnetic hyperthermia treatment, the swarm pattern can also swell to obtain a low area density of nanoparticles.

In the in vitro experiments, the nanoparticles have an initial concentration of 0.5 mg/cm^2. According to the results in Fig. 10.6b, this concentration is not able to raise the local temperature to the 40 °C required to kill cancer cells. The formation of the vortex-like swarm can lead to a significant increase in the concentration of

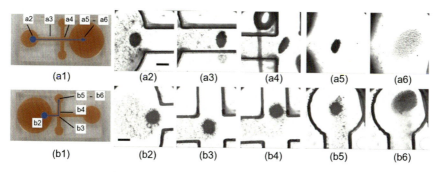

Fig. 10.7 Delivery process of the microswarm. The swarm **a** enters a chamber after passing through a long straight channel or **b** turns above and enters another chamber. The blue arrows indicate the trajectories of the swarm. The scale bars are 1 mm

nanoparticles. Figure 10.8 shows the schematic of the local hyperthermia of tumor cells using the nanoparticle swarm with tunable pattern and directional movement capabilities. In the magnetic hyperthermia process, an alternating magnetic field with a frequency of 180 kHz and a coil current of 40 A is applied for half an hour. The tumor cells covered by the nanoparticle swarm are heated to a high temperature, as indicated by the circular region in Fig. 10.8b. After the heating process, cells in the chamber are stained with propidium iodide (PI) and calcein-AM (calcein-AM for live cells and PI for dead cells). Figure 10.9 shows the fluorescence microscope image of the red rectangular area in Fig. 10.8b, where Fig. 10.9a–c is the calcein-AM channel, the PI channel, and the combined images of the calcein-AM and PI channels, respectively. The results show that HepG2 cells heated by the swarm have died, while cells outside the area covered by the swarm are still alive because they are not affected by the high temperature. There is a clear boundary between dead cells and living cells, and enlarged images of the boundary area are shown in Fig. 10.9d, e. It can be found that the interpenetration of live cells and dead cells is only a few tens of micrometers, indicating that the hyperthermia area can be precisely controlled by adjusting the position of the swarm pattern. In the control group, hyperthermia is performed with the same dose of magnetic nanoparticles under the same alternating magnetic field. Because there is no gathering effect from swarms, the nanoparticles are uniformly distributed on the cellular substrate. After half an hour of hyperthermia, most of the HepG2 cells are viable with only a small amount of death (Fig. 10.9f), indicating insufficient heat generation from dispersed nanoparticles. These results demonstrate that nanoparticle swarms with reconfigurable patterns can effectively increase the nanoparticle concentration to achieve high-efficiency local hyperthermia of tumor cells.

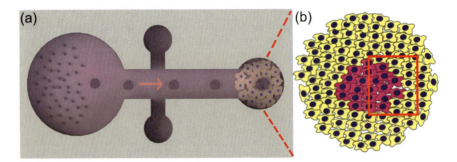

Fig. 10.8 Schematic illustration of enhanced magnetic hyperthermia using the nanoparticle swarm. **a** Schematic of in vitro thermal generation using the swarm in shrunk state after delivery. **b** Schematic of the tumor cells after the magnetic hyperthermia. The red cells are covered by the swarm during the hyperthermia process

10.3 Biofilm Eradication

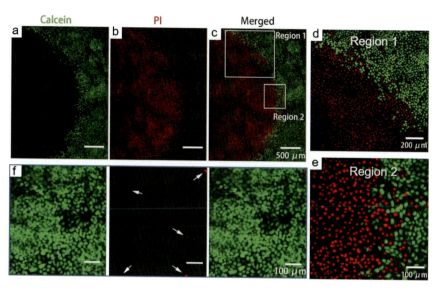

Fig. 10.9 Fluorescence microscopic images of the tumor cells after hyperthermia. **a** The calcein-AM channel, **b** PI channel, and **c** the merged fluorescence images of the boundary area of dead and live cells (the red rectangular area in Fig. 10.8b). **d, e** The merged fluorescence images of region 1 and region 2 in "(**c**)". **f** The control group in which the hyperthermia is conducted by using the same dose of nanoparticles in a swelling state. The white arrows show the position of dead cells. The frequency of the alternating magnetic field is 180 kHz, and the coil current is 40 A

10.3 Biofilm Eradication

10.3.1 Introduction to Biofilms

Biofilms are typically composed of bacterial cells enmeshed in extracellular polymeric substances (EPS) (Whitchurch et al. 2002; Flemming et al. 2016). The EPS significantly enhances the resistance of bacterial cells to conventional antibiotics and the immune system while improving the mechanical strength of biofilms (Liu et al. 2019a; Flemming and Wingender 2010). Biofilms are widely distributed in different parts of the human body, including the surfaces of natural organs (e.g., teeth, bones, and urethra) and implanted equipment (e.g., contact lenses, indwelling stents, and medical catheters), which may cause many infectious diseases (Liu et al. 2019a; Behlau and Gilmore 2008; Koo et al. 2017). To date, the commonly used biofilm prevention and treatment methods are mainly divided into two categories, namely passive prevention and active removal (Cheeseman et al. 2020a). Passive prevention mainly reduces the possibility of biofilm formation by surface treatment of biofilm-prone sites, such as adding antibacterial agents/inhibitors, or modifying nanostructured topography with antibacterial effects (Kälicke et al. 2006; Chung et al. 2007; Pogodin et al. 2013). This method has the disadvantage of not being able to remove biofilms that have already formed (Banerjee et al. 2011). The general method used in the active removal of biofilms is the systemic administration of high concentrations of antibiotics, which can easily lead to severe drug resistance and the destruction of normal microbiota in the human body (Flemming et al. 2016). In addition, new antimicrobial agents like artificial enzymes with reactive oxygen species (ROS) generation ability have also been used to remove biofilms (Chen et al. 2018; Gao et al. 2016; Liu et al. 2019b). No matter which method is used to actively remove biofilms, there are four key points that must be taken into account. First, the antimicrobial agent should have the ability to break through the EPS barrier and kill the bacterial cells inside. Second, antibacterial agents need to be delivered precisely to the biofilm location to reduce side effects and dose of drugs. Third, since the bacteria in the biofilm are not just one type, the antibacterial agent should have broad-spectrum antibacterial ability. Finally, considering that a part of the free bacteria is released after the biofilm is destroyed, the antibacterial agent should be able to kill planktonic bacteria to prevent the biofilm from regenerating.

Micro/nanorobot swarms have received extensive attention in the field of biomedical applications due to remote actuation and localization capability and enhanced functions from swarming effects. Microswarms can carry large doses of antimicrobial agents and generate stronger fluid flow and mechanical force, which is conducive to penetrating and destroying biofilms. Besides, magnetically driven microswarms have the characteristics of multimodal transformation, high controllability, and good biocompatibility, which are very suitable for the eradication of biofilms. In addition to the actuating method, building blocks of the microswarm also play a crucial role in biofilm removal. In order to minimize the drug resistance of biofilms and improve the removal efficiency, a variety of new antibacterial agents have been developed, including metal ions (Soto et al. 2020), antimicrobial peptides (Yu et al. 2020), lysozyme (Kiristi et al. 2015), and liquid metals (Elbourne et al. 2020). Edward Steager and Hyun Koo et al. use iron oxide nanoparticles as building blocks to construct small-scale catalytic antimicrobial robots (CARs) for biofilm eradication (Hwang et al. 2019). Two distinct CAR platforms are developed, which are able to kill and clean biofilms in hard-to-reach area or on highly confined anatomical surfaces, such as the interior of human teeth, through combined catalytic and mechanical action. Martin Pumera and colleagues have developed a series of micro/nanorobots for biofilm removal, including aqua sperm micromotors (Mayorga-Martinez et al. 2021), light-powered ZnO micromotors (Ussia et al. 2021), enzyme-photocatalyst tandem microrobot based on urease-immobilized TiO_2/CdS nanotube bundles (Villa et al. 2022), and so on. Besides, researchers have made many other efforts in using micro/nanorobots and their swarms to eliminate biofilms (Deng et al. 2022; Zhang et al. 2022; Xie et al. 2020; Liu et al. 2018; Oh et al. 2022). In this section, we focus on two types of microswarms for biofilm eradication, which are composed of different building blocks, i.e., porous Fe_3O_4 (p-Fe_3O_4) nanoparticles and magnetic urchin-like capsule robots loaded with magnetic liquid metal droplets (MUCR@MLMDs).

As introduced in the previous chapters, paramagnetic Fe_3O_4 particles are ideal building blocks for constructing various types of microswarms. Meanwhile, the Fe

10.3 Biofilm Eradication

element in it can also generate free radicals through the Fenton reaction, demonstrating the ability to kill bacterial cells (Naha et al. 2019; Wang et al. 2014). Fabricated by facile acid etching of Fe_3O_4 nanoparticles, the p-Fe_3O_4 nanoparticles have a larger specific surface area and more active sites, as shown in Fig. 10.10a. The microswarm formed by p-Fe_3O_4 nanoparticles (p-Fe_3O_4 swarm) under a rotating magnetic field has the ability to efficiently remove biofilms, which is mainly reflected in the following aspects. On the one hand, the porous structure and rough surface of the p-Fe_3O_4 nanoparticles enable them to catalyze the generation of more free radicals and enhance the mechanical friction between the particles and the biofilm. On the other hand, the formation of microswarm further enhances these effects. To be specific, the swarming motion speeds up the Fenton reaction, resulting in local enrichment of free radicals. The p-Fe_3O_4 swarm can also generate stronger fluid flow and mechanical forces for breaking the EPS barrier, facilitating the penetration of free radicals into the interior of the biofilm to kill bacteria and degrade the EPS (Fig. 10.10b). Benefiting from the capabilities of simultaneously degrading the EPS barrier and killing the bacteria cells, p-Fe_3O_4 swarm exhibits high biofilm elimination efficacy.

Another microswarm platform for biofilm eradication utilizes natural urchin-like sunflower pollens as building blocks. The magnetic urchin-like capsule robots (MUCR) load and deliver magnetic liquid metal droplets (MLMD). Liquid metals have been applied to broad-spectrum treatment of bacterial biofilms due to properties of deformation, acceptable cell toxicity, and biodegradability (Cheeseman et al. 2020b; Lin et al. 2021, 2020; Li et al. 2021). Actuated by the rotating magnetic field, the MUCR@MLMD swarm moves toward the biofilm and physically breaks the EPS barrier using the inherited microspikes of sunflower pollens. Then, the MLMD is released from the MUCR and kills bacterial cells using its sharp edges. The synergistic effect of MUCR and MLMD greatly enhances the efficiency of biofilm removal. Moreover, the MUCR can also be loaded with a chemotactic agent (L-aspartic acid) to attract, capture, and kill planktonic bacteria released from the biofilm during the removal process. In addition to being effective in removing biofilms in two-dimensional planes, the MUCR@MLMD swarm is able to eradicate biofilms in some hard-to-reach sites in the human body. For example, the surface of biliary stents is prone to adhesion of bacteria and the formation of biofilms containing various types of bacteria (e.g., *Escherichia coli*, *Enterococcus faecalis*, *Klebsiella*, *Bacillus*, and *Candida tropicalis*), which may further lead to clogging of the stent and even threaten the patient's life (Vaishnavi et al. 2018; Sung et al. 1993; Leung et al. 1988, 2000; Berkel et al. 2005). For patients requiring long-term placement of biliary stents, frequent replacement to avoid stent obstruction is painful and expensive. The MUCR@MLMD swarm can be delivered to the vicinity of the biliary stent by endoscope, and then be navigated into the stent under the guidance of fluoroscopic imaging for biofilm eradication (Fig. 10.11).

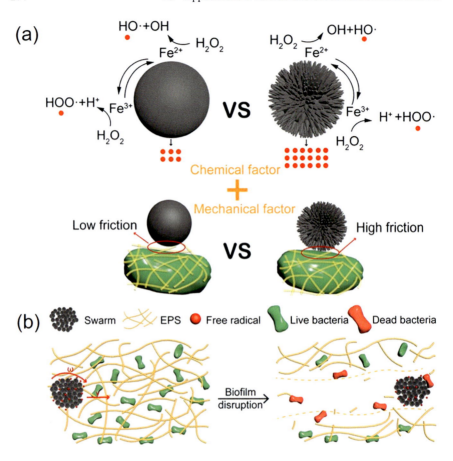

Fig. 10.10 Schematic illustration of the p-Fe$_3$O$_4$ swarm for biofilm elimination. **a** Comparison of the chemical and mechanical factors of the bacteria killing of Fe$_3$O$_4$ and p-Fe$_3$O$_4$ nanoparticles. **b** Schematic of the biofilm removal process using the p-Fe$_3$O$_4$ swarm

10.3.2 The p-Fe$_3$O$_4$ Swarm with Catalytic and Antimicrobial Capabilities

In order to improve the catalytic activity of Fe$_3$O$_4$ nanoparticles, oxalic acid solution is used for etching at room temperature for one hour. Figure 10.12 shows the microscopic images of the particles before and after etching. It can be seen from the SEM images (Fig. 10.12a, b) that the etched particles have porous structures and rough surfaces. The magnified TEM image (Fig. 10.12d) shows that the interior structure of the particles is also full of slender pores. In addition, the etching also leads to a decrease in the diameter of Fe$_3$O$_4$ nanoparticles, and the mean diameters before and after etching are 362.79 nm and 321.87 nm, respectively. Therefore, p-Fe$_3$O$_4$ nanoparticles have larger specific surface area and more rich catalytic sites than

10.3 Biofilm Eradication

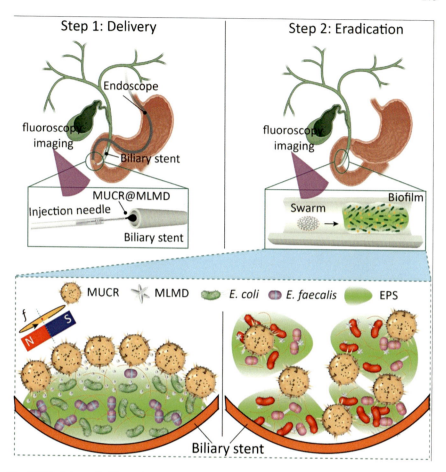

Fig. 10.11 Schematic illustration of the eradication of biofilms in the biliary stent by using the MUCR@MLMD swarm

Fe$_3$O$_4$ nanoparticles. According to the Brunauer–Emmett–Teller (BET) test results, the specific surface area, pore size, and porosity of p-Fe$_3$O$_4$ nanoparticles are 48.34 m^2/g, 9.81 nm, and 0.119 cm^3/g, respectively, which are significantly better than those of Fe$_3$O$_4$ nanoparticles (21.61 m^2/g, 5.37 nm, and 0.029 cm^3/g). The magnetic saturation of the p-Fe$_3$O$_4$ nanoparticles (71 emu/g) is similar to that of the untreated Fe$_3$O$_4$ nanoparticles (82 emu/g), indicating that they still have stable magnetism.

Actuated by the rotating magnetic field generated by the three-axis Helmholtz coil system, the p-Fe$_3$O$_4$ nanoparticles form a vortex-like swarm with a stable circular pattern, as shown in Fig. 10.13a. Considering the practical environment of biofilm removal, the average moving speeds of p-Fe$_3$O$_4$ swarm in phosphate-buffered brine (PBS) and liquid culture medium (LCM) solutions at different times are measured (Fig. 10.13b). The moving velocity of the swarm in LCM solution (~69.1 μm/s) decreases by about 25% compared to that in PBS solution (~92.4 μm/s) due to the

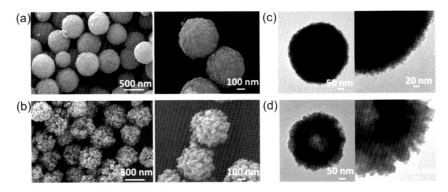

Fig. 10.12 Microscopic images of the nanoparticles before and after etching. **a, b** SEM images of Fe$_3$O$_4$ and p-Fe$_3$O$_4$ nanoparticles. **c, d** TEM images of Fe$_3$O$_4$ and p-Fe$_3$O$_4$ nanoparticles

increase in liquid viscosity. In the LCM solution, the speed of the swarm can remain stable for a long time. Figure 10.13c shows that the area of the swarm decreases over time and then tends to stabilize. In addition, an increase in the magnetic field strength resulted in a decrease in the swarm area, and the particles in the swarm are more tightly bound. The above results indicate that p-Fe$_3$O$_4$ nanoparticles can form swarm under a uniform rotating magnetic field and have good controllability and stability in LCM solution.

In addition to the three-axis Helmholtz coil system, a permanent magnet system (Fig. 10.13d) consisting of a spherical magnet, a rotating motor, and a movable chassis is also able to provide the rotating magnetic field for the actuation of the p-Fe$_3$O$_4$ swarm. Figure 10.13e shows the magnetic flux distribution density in the working space of the permanent magnet system, and the magnetic field strength can be adjusted by changing the distance between the working space and the spherical magnet. The average velocity of the swarm driven by the permanent magnet system in the *E. coli* biofilm (~264 μm/s) decreased by about 44% compared to that in the PBS solution (~473 μm/s). In total, both magnetic actuation systems can generate the rotating magnetic fields needed to form and control the p-Fe$_3$O$_4$ swarm, and both enable the swarm to be delivered with high access rates. The access rates of the swarm driven by the electromagnetic coil system and the permanent magnet system are 80.6% and 89.5%, respectively. The higher access rate of the latter one is because the gradient created by the permanent magnet has an additional attractive effect on the particles in the swarm. In order to obtain faster moving speeds, higher access rates, and larger working space, a permanent magnet system is used to form and actuate the p-Fe$_3$O$_4$ swarm in the biofilm removal experiments.

The catalytic performance of p-Fe$_3$O$_4$ swarms enhanced by the porous structure and collective behavior is evaluated. As shown in Fig. 10.14, the p-Fe$_3$O$_4$ swarm first forms on the left side of the channel and passes through the narrow channel to the right region where 3,3',5,5'-tetramethylbenzidine (TMB) is added as a colorimetric agent. The p-Fe$_3$O$_4$ swarm is capable of catalyzing TMB in the presence of H$_2$O$_2$,

10.3 Biofilm Eradication

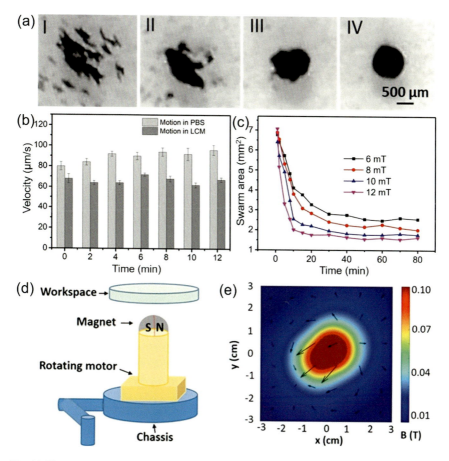

Fig. 10.13 Microswarm formed by p-Fe₃O₄ nanoparticles. **a** Formation process of the p-Fe₃O₄ swarm under a rotating field generated by the electromagnetic coil system. **b** Moving velocity of the p-Fe₃O₄ swarm in PBS and SLM solution at different times. **c** The change of swarm area with time under different magnetic field strengths. **d** Schematic of the permanent magnet system for generating the rotating field. **e** Magnetic flux density at a distance of 1 cm from the tip of the spherical permanent magnet

and the generated product is blue. After reaching the targeted area, the swarm induces a catalytic reaction to generate hydroxyl radicals that turn the surrounding liquid blue (Fig. 10.14c). The catalytic reaction rate in the adjacent region of the swarm is faster than that in the distal region, which is due to the fact that the reaction occurs on the surface of p-Fe₃O₄ nanoparticles and the swarm has the effect of trapping chemicals and enhancing convection. The blue color of the liquid in the right region gradually becomes uniform and darker with the reaction time. This experiment validates the active delivery and target catalysis capabilities of the p-Fe₃O₄ swarm. The catalytic activity of different nanoparticles is compared by the degradation performance of methylene blue, and four groups are included in the experiments, i.e., static Fe₃O₄, static p-Fe$_3$O$_4$, Fe$_3$O$_4$ swarm, and p-Fe$_3$O$_4$ swarm. The results in Fig. 10.14d, e show that the p-Fe$_3$O$_4$ swarm has the highest peroxidase-like activity compared with other groups, which is also ascribed to the porous structure and swarming effect. The pH of the surrounding solution has an influence on the catalytic activity of the p-Fe$_3$O$_4$ swarm. The p-Fe$_3$O$_4$ swarm exhibits catalytic activity at pH 3–8 and has the best catalytic activity at pH 3–4, which is very suitable for the acidic microenvironment in biofilms (Liu et al. 2019a). In addition, the p-Fe$_3$O$_4$ nanoparticle can be recycled, and it still exhibits good reusability and stability after undergoing over eight catalytic degradation cycles in one day or over seven catalytic degradation cycles in 30 days.

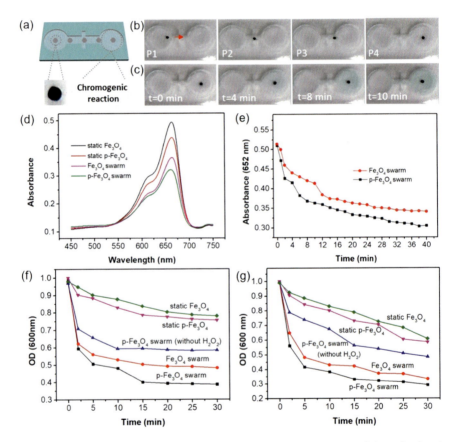

Fig. 10.14 Catalytic and antimicrobial performances of the p-Fe$_3$O$_4$ swarm. **a** Schematic of navigating the p-Fe$_3$O$_4$ swarm in the channel for targeted catalysis of TMB. **b** Navigation process of the p-Fe$_3$O$_4$ swarm in the channel. **c** Targeted catalysis process of TMB with the p-Fe$_3$O$_4$ swarm. **d** Absorbance spectra of methylene blue after treating for 30 min. **e** The change in absorbance of methylene blue with time during the degradation process. **f** The change in OD$_{600}$ with time for *E. coli*. **g** The change in OD$_{600}$ with time for *B. cereus*. 1 mL of 5 ppm methylene blue solution contains 5 μg of particles and 100 mM H$_2$O$_2$ in (**d**) and (**e**)

The drug-resistant Gram-negative *E. coli* and Gram-positive *B. cereus* are incubated to characterize the antibacterial ability of the p-Fe$_3$O$_4$ swarm. The optical density at 600 nm (OD$_{600}$) of the LCM solution, which is positively correlated with bacterial activity, is presented in Fig. 10.14f, g. The experiments include five groups, i.e., static Fe$_3$O$_4$, static p-Fe$_3$O$_4$, p-Fe$_3$O$_4$ swarm (without H$_2$O$_2$), Fe$_3$O$_4$ swarm, and p-Fe$_3$O$_4$ swarm. The OD$_{600}$ decreases with the treatment time in all the groups, which indicates a gradual death of bacteria. The p-Fe$_3$O$_4$ swarm shows the best antibacterial effect, which is caused by various factors. First, higher catalytic efficiency leads to more generation of bactericidal hydroxyl radicals. Second, the sharp edges of p-Fe$_3$O$_4$ can mechanically disrupt bacterial membranes and kill bacterial cells. Therefore, even in the absence of H$_2$O$_2$, the p-Fe$_3$O$_4$ swarm still exhibits a certain bactericidal ability. The statistical results of the killing efficiency of bacteria show that the relative bacterial viabilities of *E. coli* and *B. cereus* are reduced to 0.3% and 0% after the treatment of the p-Fe$_3$O$_4$ swarm, respectively.

10.3.3 Biofilm Elimination by the p-Fe$_3$O$_4$ Swarm

Bacteria inside the biofilm are more difficult to kill than planktonic bacteria due to the protection of the EPS barrier. The mechanical stability of the biofilm is also significantly enhanced by the EPS. Benefiting from high catalytic activity, good controllability, and ability to generate strong convection and mechanical forces, the designed p-Fe$_3$O$_4$ swarm has great potential in eliminating biofilms. The biofilm removal performance of the p-Fe$_3$O$_4$ swarm is evaluated using Gram-negative *E. coli* biofilms cultured on 24-well plates. The viability of bacterial cells in biofilms is characterized by fluorescent staining, with green and red representing live and dead cells, respectively. Figure 10.15a shows 3D fluorescence microscopy images of *E. coli* biofilms for two different treatment groups, i.e., static p-Fe$_3$O$_4$ nanoparticles mixed with H$_2$O$_2$ and p-Fe$_3$O$_4$ swarm in H$_2$O$_2$ solution. The results show that the thickness of the biofilm and the activity of the bacteria decrease with time. Biofilms treated with the p-Fe$_3$O$_4$ swarm are thinner and have fewer viable bacterial cells than biofilms treated with static p-Fe$_3$O$_4$ nanoparticles. Figure 10.15b shows that the thickness of the biofilm decreases from nearly 60 μm to less than 5 μm after being treated with the p-Fe$_3$O$_4$ swarm for 30 min. The proportion of viable cells in the biofilm is measured using colony-forming units (CFU) counting, and the results show that the p-Fe$_3$O$_4$ swarm kills more than 99.99% of the bacterial cells in the biofilm (Fig. 10.15c).

Eight groups of comparative experiments are conducted to demonstrate the performance of p-Fe$_3$O$_4$ swarms to efficiently remove biofilms. The 3D fluorescence images of biofilms and the number of living cells after 30 min of treatment are shown in Fig. 10.16a, b, respectively. The comparison between groups 7 and 8 shows that the biofilm removal rate and the killing efficiency of the p-Fe$_3$O$_4$ swarm are higher than those of the p-Fe$_3$O$_4$ swarm, which proves that the p-Fe$_3$O$_4$ nanoparticles have better

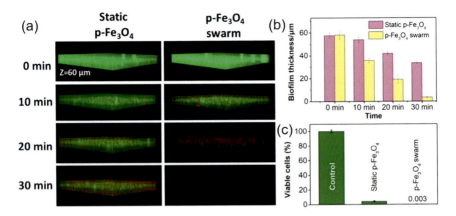

Fig. 10.15 Removal of *E. coli* biofilms using the p-Fe$_3$O$_4$ swarm. **a** 3D fluorescence microscopy images of *E. coli* biofilms. Green and red represent live and dead cells, respectively. **b** The change in the thickness of the biofilm during the treatment process. **C** The proportion of viable cells in the biofilm measured by using CFU counting. Each error bar denotes the standard deviation from three trials

catalytic activity and stronger mechanical effect than ordinary p-Fe$_3$O$_4$ nanoparticles. In group 6, the p-Fe$_3$O$_4$ nanoparticle chains are driven by a rotating magnetic field perpendicular to the substrate and perform a simple tumbling motion, and the destruction of the biofilm is smaller than when the particles perform a swarming motion. This suggests that the formation of a swarm concentrates the particles in a smaller area, which elevates the local hydroxyl radical's concentration and generates strong convection for more effective biofilm removal. In group 2, the pure H$_2$O$_2$ solution performs poorly in eliminating biofilms, further demonstrating the importance of the p-Fe$_3$O$_4$ swarm in biofilm elimination. The p-Fe$_3$O$_4$ swarm without H$_2$O$_2$ solution (group 4) also shows a certain ability to eliminate biofilms, which is consistent with the results of antibacterial experiments. In this case, the elimination of the biofilm is mainly achieved by pure physical interaction, that is, the strong fluid flow generated by the p-Fe$_3$O$_4$ swarm and the nanostructure on the surface of p-Fe$_3$O$_4$ particles play the role of destroying EPS and killing bacteria. Overall, the p-Fe$_3$O$_4$ swarm can destroy EPS and kill bacteria through both chemical catalysis and mechanical reactions, showing an excellent ability to eliminate biofilms.

The p-Fe$_3$O$_4$ nanoparticles after removing the biofilm remain intact and can be recovered and reused by ultrasonic cleaning with ethanol (75%) solution, providing more possibilities for practical applications. According to the results of 3-(4,5-dimethylthiazol-2-yl)-5-(3-carboxymethoxyphenyl)-2-(4-sulfophenyl)-2H-tetrazolium (MTS) cell viability assay, when the concentration is as high as 100 μg/mL, the cell viability of static p-Fe$_3$O$_4$ and p-Fe$_3$O$_4$ swarm to normal cells (3T3 cells) are both higher than 70%, indicating that p-Fe$_3$O$_4$ nanoparticles have no obvious toxicity to normal cells. Targeted removal of biofilms using p-Fe$_3$O$_4$ swarm along a predetermined route is demonstrated, and the schematic

10.3 Biofilm Eradication

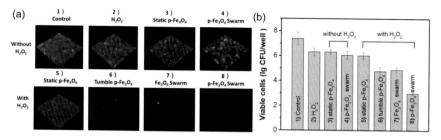

Fig. 10.16 Eight groups of comparative experiments of biofilm removal. **a** 3D fluorescence microscopy images of biofilms in eight different treatment groups. **b** The number of viable cells in the biofilm measured by using CFU counting. Each error bar denotes the standard deviation from three trials

diagram is shown in Fig. 10.17a. The p-Fe$_3$O$_4$ swarm is navigated along a square trajectory on the biofilm, and the optical images show that the biofilm on the trajectory is eliminated (Fig. 10.17b). This indicates that the swarm is able to remove biofilms at specified locations with high precision without damaging nearby normal tissue, and its size can be tuned by changing the dose of added particles. The fluorescence images (Fig. 10.17c) show that bacterial cells in the treated area are all killed, and no bacterial regrowth is observed after an additional 24 h of incubation, which is important for biofilm treatment. In addition to open flat surfaces, the p-Fe$_3$O$_4$ swarm can also perform biofilm removal in hard-to-reach cavities and confined spaces (e.g., tortuous lumens and tiny crevices) where biofilms are easier to grow. The p-Fe$_3$O$_4$ swarm is used to remove biofilm in a tiny U-shaped tube with an inner diameter of 0.9 mm. The biofilm grows on the curved part of the tube, as shown in Fig. 10.17d. Actuated by the rotating magnetic field, the p-Fe$_3$O$_4$ swarm moves back and forth for 9.5 min in the U-shaped tube to remove the biofilm on the inner wall, and the biofilm residue left in the tube after removal can be easily removed by flushing (Fig. 10.17e). The p-Fe$_3$O$_4$ swarm moving inside the tube can enhance the fluid convection in the whole 3D space, as shown in Fig. 10.17f. When the p-Fe$_3$O$_4$ swarm is close to the biofilm that blocks the tube (to the left of the cyan boundary line in Fig. 10.17f), the local high flow rate and strong fluctuations enhance the interaction with the biofilm and improve the efficiency of biofilm removal. Overall, this p-Fe$_3$O$_4$ swarm-based biofilm elimination method takes full advantage of the nanoenzyme catalytic performance of p-Fe$_3$O$_4$ nanoparticles as well as the swarm's targeting properties and ability to enhance chemical and physical reactions, which is expected to be a facile and efficient strategy for the treatment of stubborn biofilms in medical devices or industrial settings.

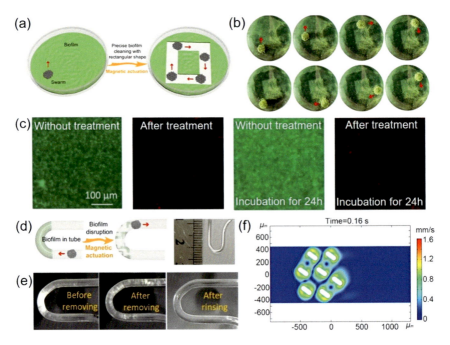

Fig. 10.17 Biofilm removal on flat surfaces and in curved tubes. **a** Schematic diagram of controlled biofilm removal using the p-Fe$_3$O$_4$ swarm. **b** Experimental images of the biofilm during the removal process. The red arrows represent the moving direction of the swarm. **c** Fluorescence images of biofilms with and without treatment using the p-Fe$_3$O$_4$ swarm. The regrowth ability of biofilms in the treated area is tested by additionally incubating for 24 h. **d** Schematic diagram of biofilm removal in a U-shaped tube. **e** Experimental images of the biofilm before and after removing. **f** Simulation result of fluid velocity in the tube when the p-Fe$_3$O$_4$ swarm is approaching the biofilm

10.3.4 The MUCR@MLMD and Its Swarming Behavior

Another microswarm platform for biofilm removal uses MUCR@MLMDs as building blocks. The MUCR@MLMDs are fabricated by loading liquid metal droplets with a core–shell structure into the inner cavity of a capsule microrobot, the specific fabrication process is shown in Fig. 10.18a. First, eutectic gallium-indium alloy and iron powder are mixed and ground in a mass ratio of 10:1. The liquid metal alloy is then added to anhydrous ethanol and sonicated to obtain MLMD (Elbourne et al. 2020). The TEM image shows that the obtained MLMDs are essentially spherical (Fig. 10.18b), with diameters concentrate below 1 µm and the aspect ratio of about 1 without magnetic fields. On the other hand, MUCR is obtained by sequentially subjecting raw sunflower pollen grains to phosphoric acid digestion and chemical bath deposition. The internal cytoplasm of native pollen is removed by phosphoric acid, while its external extexine is retained. The chemical bath deposition plays a role in modifying Fe$_3$O$_4$ nanoparticles to the surface of the enucleated pollen, which makes the spines on the pollen surface become rougher, as shown in Fig. 10.18c.

10.3 Biofilm Eradication

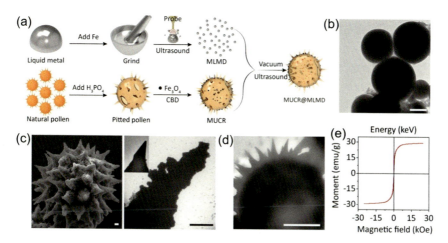

Fig. 10.18 Fabrication and characterization of the MUCR@MLMD. **a** Schematic diagram of the fabrication process of the MUCR@MLMD. **b** TEM image of the as-prepared MLMD. The scale bar is 100 nm. **c** SEM images of the as-prepared MUCR (left) and the magnified TEM image of the individual spines of MUCR (right). The scale bars are 1 μm. **d** TEM image of the as-prepared MUCR@MLMD. The scale bar is 10 μm. **e** Magnetic hysteresis loop of the MUCR@MLMD

Finally, the prepared MLMD is mixed with dry MUCR and then treated by vacuum loading and sonication, which enables the successful loading of MLMDs into the cavities of MUCR. The TEM image in Fig. 10.18d indicates the successful loading of the nanoscale MLMDs into the interior of the MUCR. The hysteresis loop at 300 K (red curve in Fig. 10.18e) shows the as-prepared MUCR@MLMD has good magnetic performance, and the results of quantitative vibrating sample magnetometry measurements show that its saturation magnetic susceptibility is 29.3 emu/g. In addition, according to the thermogravimetric analysis, the contents of liquid metal, Fe, Fe_3O_4, and enucleated pollen in MUCR@MLMD are about 20%, 2%, 47%, and 31%, respectively.

According to the self-terminated Cabrera–Motte oxidation mechanism, the MLMD consists of a nanometer-thick oxidized shell and a liquid core. The morphology of the MLMD changes under the influence of a magnetic field due to its liquid nature, and the increased surface area through deformation promotes oxidation of the surface and maintains a specific shape of the MLMD. Figure 10.19a shows that the MLMD changes from spherical to three different shapes under the magnetic field, i.e., jagged spheres, rods, and spines, with proportions of 52%, 19%, and 29%, and aspect ratios of 1, 4, and 1, respectively. TEM images indicate that such deformations enable the MLMD to form sharp edges at the nanoscale (Fig. 10.19b). The energy-dispersive X-ray spectroscopy result (Fig. 10.19c) shows that the post-magnetized liquid metal droplets have the same composition as the premagnetized one. The individual MUCR@MLMD exhibits three different motion modes under the actuation of rotating magnetic fields, i.e., spinning-fixed, spinning-translation, and rolling, as shown in Fig. 10.19d. When the pitch angle of rotating field is zero (i.e., the field is in the *x–y* plane), the MUCR@MLMD performs spinning-fixed motion, that is, in situ rotation instead of translation motion. When the pitch angle is 90°, the MUCR@MLMD will roll forward (rolling motion). When the pitch angle is between 0 and 90°, the MUCR@MLMD produces displacement while rotating (spinning-translation motion). The moving velocity of MUCR@MLMD performing rolling motion increases with the input frequency of the rotating field (Fig. 10.19e). If the input frequency exceeds the step-out frequency of the MUCR@MLMD, the velocity gradually decreases due to asynchronous rotation.

MUCR@MLMDs form swarm with a circular pattern under the actuation of in-plane rotating magnetic fields due to the balance of magnetic and hydrodynamic interactions. Similar to the p-Fe$_3$O$_4$ swarm, the MUCR@MLMD swarm is driven by a permanent magnet system consisting of a spherical permanent magnet, a 6-DOF manipulator, and a rotating motor. The system has a spacious working space (1800 mm diameter) and a high-intensity magnetic field (0–600 mT). The schematic diagram and magnetic flux density distribution of the permanent magnet system are shown in Fig. 10.20a. The MUCR@MLMD swarm can move through a complex maze to a specified position (Fig. 10.20b). Since the channel width of this maze (~4 mm) is smaller than the diameter of the MUCR@MLMD swarm, the swarm pattern automatically changes from circular to fusiform as it moves through the channel. That is, the MUCR@MLMD swarm remains stable under the influence of

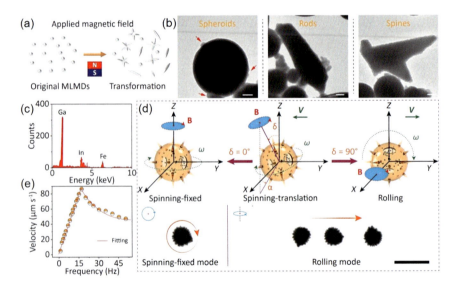

Fig. 10.19 Magnetic response of the MLMD and individual MUCR@MLMD. **a** Schematic diagram of morphology deformation of the MLMD in magnetic fields. **b** TEM images of different shapes of the MLMD in magnetic fields: jagged spheres, rods, and spines. The red arrows indicate sharp edges. The scale bars are 50 nm. **c** EDX spectrum of post-magnetized MLMD. **d** Schematic illustration of three different motion modes of individual MUCR@MLMD and the superimposed snapshots of spinning-fixed and rolling modes. The scale bar is 50 μm. **e** The relationship between moving velocity of the MUCR@MLMD (in rolling mode) and input rotating frequency

10.3 Biofilm Eradication

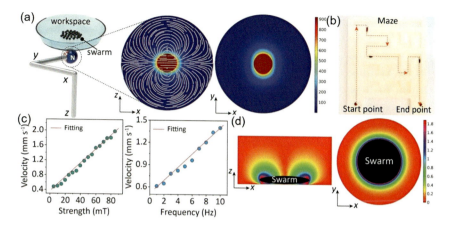

Fig. 10.20 Characterization of the MUCR@MLMD swarm. **a** Schematic diagram and magnetic flux density distribution of the permanent magnet system. **b** Snapshots of the MUCR@MLMD swarm passing through a complex maze. **c** Moving velocity of the MUCR@MLMD swarm varies with the field strength and rotating frequency. **d** Simulation results of flow velocity induced by the MUCR@MLMD swarm (unit: mm/s). Each error bar denotes the standard deviation from three trials

the inner wall of the channel, and there is almost no particle loss during the motion. The moving velocity of the MUCR@MLMD swarm is proportional to the magnetic field strength and rotating frequency in the working space, and its maximum moving velocity is about 2 mm/s (Fig. 10.20c). Simulation results show that the swarm-induced flow velocity is much larger than that induced by a single MUCR@MLMD, which is up to 100 µm/s (Fig. 10.20d). The strong fluid flow generated by the swarm has a stronger fluid impact on the bacterial community, which is beneficial to the removal of biofilms. Taken together, the MUCR@MLMD exhibits excellent magnetic responsiveness, and the MUCR@MLMD swarm is expected to serve as ideal mobile platforms for targeted drug delivery and biofilm eradication.

10.3.5 Biofilm Eradication Using the MUCR@MLMD Swarm

Benefiting from inherent microspikes, magnetically induced deformation, and abilities to perform controllable locomotion and generate strong fluid convection, the MUCR@MLMD swarm is promising for biofilm eradication. Gram-negative multidrug-resistant *E. coli* (MREC) biofilms are cultured to evaluate the biofilm removal performance of the MUCR@MLMD swarm. Five different groups are included in the experiments, i.e., the control group, the static MUCR@MLMD, the MUCR swarm, the MLMD swarm, and the MUCR@MLMD swarm. After 30 min of treatment, the viability of bacteria in the biofilm is observed using the 3D fluorescence microscope (Fig. 10.21a). The results show that MUCR@MLMDs without the magnetic field applied (static MUCR@MLMD) have minimal effect on the biofilm. The dynamic swarms can significantly improve the removal efficiency of biofilms, among which the MUCR@MLMD swarm performs the best. Figure 10.21b shows the average thickness of biofilms treated in different ways for 30 or 60 min (the original thickness is 19.3 ± 0.6 μm). The thickness reduction of the biofilm treated by the MUCR@MLMD swarm is significant. For example, when the treatment time is 30 min, the thickness of biofilm treated by the static MUCR@MLMD, the MUCR swarm, the MLMD swarm, and the MUCR@MLMD swarm is reduced by 2.9, 3.9, 5.6, and 5.0 μm, respectively. According to bacterial viabilities of residual biofilm fragments evaluated by CFU counting, the MUCR@MLMD swarm is capable of killing more than 99% of bacterial cells in biofilms (Fig. 10.21c). The above results indicate that the swarming behavior has the effect of improving the biofilm removal efficiency. In addition, the MUCR@MLMD swarm has better biofilm removal performance than the MUCR swarm or MLMD swarm, because the MUCR@MLMD swarm combines the biofilm-destroying abilities of magnetic urchin-like pitted pollen and magnetic liquid metal droplets. According to Stokes' law, the magnetic forces of a single MUCR and MLMD are estimated to be approximately 450 pN (targeting EPS) and 7.5 pN (targeting embedded bacterial cells), respectively. Thus, MUCR generates enough force to destroy EPS, while MLMD induces adequate power to disrupt bacterial cell membranes (Flemming and Wingender 2010; Elbourne et al. 2020). The mechanism of biofilm eradication by the MUCR@MLMD swarm is verified by the SEM images in Fig. 10.21d. First, the spikes of MUCR penetrate the EPS barrier and disrupt the structure of the biofilm (Fig. 10.21d, left). Then, MLMDs enter the interior of the biofilm, pierce, and kill the embedded bacterial cells using spikes generated in the magnetic field (Fig. 10.21d, right).

The MUCR@MLMD swarm has a similar ability to controllably remove biofilms at specified area as the p-Fe_3O_4 swarm. Figure 10.22a shows the biofilm removal process using a MUCR@MLMD swarm with a diameter of about 4 mm guided by an external magnetic field along a predesigned curved trajectory. MUCR@MLMD swarms can also remove biofilms in hard-to-reach and narrow sites. MUCR@MLMDs are injected into a U-shaped tube (inner diameter: 2 mm) with biofilms cultured inside and form swarm under a rotating magnetic field. The swarm pattern transforms to fusiform due to the limitation of the tube wall, which enables the swarm to better fit the inner wall and remove the biofilm on it (Fig. 10.22b). Finally, the swarm completely eradicates biofilms growing inside the U-shaped tube by reciprocating motion, indicating that the MUCR@MLMD swarm is capable of performing targeted biofilm eradication in narrow and tortuous tubes such as biliary stents.

During the growth and removal of biofilms, a large number of bacteria in planktonic state will be produced, which can easily lead to the regeneration of biofilms (Kaplan, J.á. 2010; Kostakioti et al. 2013). MUCR loaded with the chemotactic agent L-aspartic acid (LAA) can attract and capture these free bacteria. The loading efficiency of MUCR for LAA is about 10.6%, and the releasing process can be completed

10.3 Biofilm Eradication

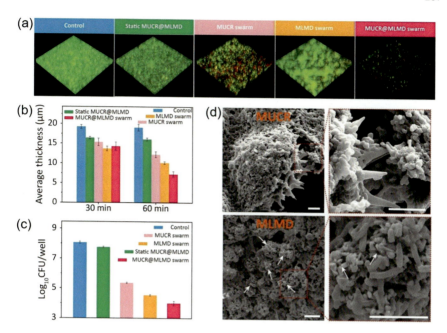

Fig. 10.21 MUCR@MLMD swarm for biofilm eradication. **a** 3D fluorescence microscopy images of the MREC biofilm with different treatments for 30 min. The green and red signals indicate living and dead bacteria, respectively. **b** The average thickness of biofilms treated in different ways for 30 or 60 min. **c** The $\log_{10}$ of CFU results of bacterial residues after different treatments. **d** SEM images showing the biofilm eradication mechanism of the MUCR@MLMD swarm. The scale bars are 5 μm. Each error bar denotes the standard deviation from three trials

Fig. 10.22 Demonstration of biofilm eradication. **a** Controlled removal of the biofilm grown on a flat surface using the MUCR@MLMD swarm. **b** Disruption of the biofilm in a U-shaped tube (inner diameter: 2 mm) by the MUCR@MLMD swarm. All the Scale bars are 5 mm

within 20 min. The schematic diagram shows chemotaxis released from the MUCR by diffusion attracts bacteria (Fig. 10.23a). Natural pollen, MUCR@MLMD, and MUCR@MLMD@LAA are added to the *E. coli* suspensions with the same concentration, respectively. After 60 min, it can be found that a large number of *E. coli* are enriched around MUCR@MLMD@LAA, while the attraction of natural pollen and MUCR@MLMD to *E. coli* is very small (Fig. 10.23b). Notably, the cavity of MUCR acts as a structural trap to capture some of the attracted *E. coli*. The simulation results of the diffusion process show that LAA is continuously released and diffused from MUCR, and the concentration is high at the germination pores (Fig. 10.23c). The enrichment of *E. coli* is caused by directional swimming toward the released LAA, which is quantified in a Y-shaped microfluidic chip (Fig. 10.23d). The Y-shaped channel is connected to three reservoirs filled with *E. coli* suspension, PBS solution, and LAA solution, respectively. After 60 min, the LAA reservoir contains four times as many bacteria as the PBS reservoir.

Gram-positive *E. faecalis* as well as Gram-negative MRECs are cultured for antimicrobial performance evaluation. PBS solution (control group), MUCR dispersion, MLMD dispersion, MUCR@MLMD dispersion, and MUCR@MLMD@LAA dispersion are added to the solution containing a certain amount of bacterial cells, and the OD_{600} is measured after a period of treatment. The results show that the bacterial survival rates of PBS solution and static MUCR@MLMD@LAA are higher than 80%. In contrast, viabilities of bacteria treated with dynamic swarms are less than 60%. Among them, viabilities of bacteria treated with the MUCR@MLMD swarm and the MUCR@MLMD@LAA swarm are less than 30%, which is attributed to the dual magnetic response of MUCR@MLMD. Notably, the MUCR@MLMD@LAA

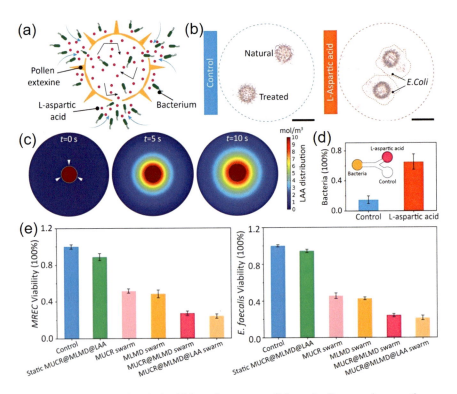

Fig. 10.23 Evaluation of antimicrobial performance. **a** Schematic diagram shows pathogens attracted by chemoattractant LAA and approach the hollow MUCR. **b** Attraction of natural pollen, MUCR@MLMD, and MUCR@MLMD@LAA to *E. coli*. The scale bars are 30 μm. **c** Simulation of the release of LAA around germination pores of the MUCR. **d** Bacteria accumulate in the LAA reservoir in the Y-shaped microfluidic chip. **e** Viabilities of bacteria after being treated in different ways. Each error bar denotes the standard deviation from three trials

10.3 Biofilm Eradication

swarm has better antibacterial properties than the MUCR@MLMD swarm, which is mainly due to the attractive effect of LAA on bacteria. In brief, these experiments confirm that the MUCR@MLMD@-based swarm has excellent antibacterial activity.

To further validate the feasibility and efficiency of the MUCR@MLMD swarm for biofilm eradication in medical implants, experiments are conducted in biliary stents collected from the patients (the indwelling time is about 2–3 months). Although magnetically actuated microswarms are suitable for adaptive locomotion in confined spaces, they are less efficient for long-distance transportation. Therefore, the MUCR@MLMDs are firstly delivered to the vicinity of the biliary tract quickly and precisely by using an endoscope, and then the swarm is actuated by the magnetic field to enter the hard-to-reach biliary stent and perform biofilm eradication. As shown in Fig. 10.24a, an injection needle (internal diameter: 1.5 mm) reaches the biliary stent entrance through the esophagus, stomach, and duodenum. Then, the MUCR@MLMDs are delivered into the biliary stent by injection under endoscopy monitoring and fluoroscopy imaging. The total delivery distance is about 100 cm and can be completed in 10 min. The fluoroscopy imaging is adopted to monitor the microswarm after entering the biliary stent where endoscope cannot track and observe (Fig. 10.24b). The contrast of fluoroscopy imaging is significantly enhanced by MLMD inside the microswarm.

Under the real-time navigation of fluoroscopy images, the MUCR@MLMD swarm moves back and forth within the bile duct stent to scrape the biofilm on the inner wall. Fluid flow induced by swarm movement exerts shear and pressure on the inner wall of the stent, which is beneficial for biofilm removal. Notably, the biofilm grew around 360° within the biliary stent. Therefore, there are two ways to

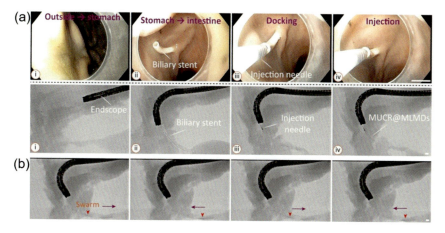

Fig. 10.24 Biofilm eradication of the biliary stent. **a** The MUCR@MLMDs are delivered into the biliary stents by an injection needle equipped on the endoscope. Images on the top are taken by endoscopy, and images at the bottom are taken by fluoroscopy imaging. **b** Controlled navigation of the microswarm within the biliary stents under real-time guidance of fluoroscopy imaging. The red and purple arrows indicate the position and moving direction of the swarm, respectively. The scale bars are 5 mm

ensure that the biofilm inside the stent is completely removed. The first method is to keep the patient still, and use the 6-DOF robotic arm of the permanent magnet system to flip the spherical permanent magnet 180° to the opposite side to remove the biofilm on the other side of the stent. Another way is to flip the biliary stent 180°, i.e., flip the patient's position. After 30 min of treatment, crystal violet staining, live/dead fluorescent staining imaging, and SEM imaging are performed on the inner wall of the biliary stent (Fig. 10.25). Optical images show that the interior of the untreated biliary stent (control group) is yellow and becomes purple after stained with crystal violet, indicating that there are many bacteria. The interior of the microswarm-treated biliary stent is obviously cleaner and is not stained to be purple by crystal violet, indicating that the biofilm had been scraped off. Fluorescence microscopy images indicate many live bacteria on the untreated biliary stents, but no live bacteria on the treated stents. SEM images of the interior and cross section of biliary stents show that there are no bacterial biofilm components on the treated stents. In addition, most of the bacterial cells in the biofilm fragments scraped from the biliary stent have died. Overall, the MUCR@MLMD swarm combines the active delivery properties of microswarms with the broad-spectrum antibacterial properties of magnetic liquid metal droplets and natural pollen to achieve targeted eradication of biofilms, demonstrating great potential for eradicating bacterial biofilms on medical implants such as biliary stents.

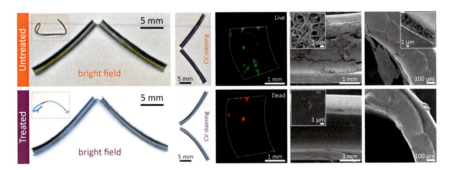

Fig. 10.25 Images of the inner surface of untreated biliary stent and that treated with the MUCR@MLMD swarm. The comparisons are made by using the bright field, crystal violet staining, live/dead staining, SEM imaging of sliced open stents, and cross section of stents, respectively

10.4 Accelerate Thrombolysis

10.4.1 Introduction to Thrombolysis

Thrombus is caused by blood clotting and can form in both veins and arteries, such as deep vein thrombosis and pulmonary embolism. Thrombus disrupts the normal circulation of the human vascular system, which may induce many life-threatening problems, such as coronary infarction and ischemic stroke (Heit 2015). Thrombosis usually begins with platelet aggregation, where activation of coagulation factors causes fibrin to accumulate around platelets to form a grid-like structure that traps more red blood cells (RBCs) and platelets. Traditional thrombus treatment methods include injection of thrombolytic drugs and the use of catheters for mechanical resection, but they all have some drawbacks and safety concerns (Zenych et al. 2020; Wan et al. 2020). For example, thrombolytic drugs are used in larger doses to effectively remove blood clots, but they are prone to cause side effects such as bleeding and blood pressure instability (Thomalla et al. 2006; Juenet et al. 2018; Chen et al. 2019). The use of catheters requires complex clinical procedures and meticulous manipulations and may lead to vascular damage, especially for vessels with relatively small diameters (Wechsler 2011; Korin et al. 2012). Recently, the rapid development of small-scale robots has provided new ideas for thrombus removal (Khalil et al. 2018; Tasci et al. 2017; Leclerc et al. 2020). For example, millirobots can mechanically disintegrate thrombus or serve as a tool to deliver thrombolytic drugs. However, the application of millimeter-scale robots is very limited, because blood clots usually form in narrow and confined areas. On the other hand, micro/nanorobots have limited impact on thrombus, and their small size makes real-time in vivo imaging more challenging. The use of micro/nanorobot swarms to remove thrombus is an effective way to solve the above problems, with the following specific advantages. First, microswarms have good environmental adaptability. Their transformable morphology is beneficial to adapt to the complex vascular network and avoid mechanical damage to the blood vessels. Second, the microswarm can serve as a targeted delivery tool to precisely remove the thrombus at the specified site. Third, the microswarm enables larger dose drug delivery, and the induced flow may accelerate mass transfer near the clot-fluid interface. Finally, the gathering effect of microswarms can increase the concentration of building blocks, thereby improving the contrast of medical imaging, which is beneficial for real-time tracking in the human body (Yu et al. 2019). To date, various imaging techniques have been applied to localize microswarms, including magnetic resonance imaging (Yan et al. 2017), positron emission tomography/computed tomography (Vilela et al. 2018), and infrared fluorescence imaging (Servant et al. 2015). Among these imaging techniques, ultrasound imaging has the advantages of low cost, non-radiation, deep imaging depth, and high temporal resolution, which makes it suitable for real-time localization of thrombus and microswarm.

This section introduces a strategy to accelerate tissue plasminogen activator (tPA)-mediated thrombolysis using the Fe_3O_4 nanoparticle ribbon-like swarm under the

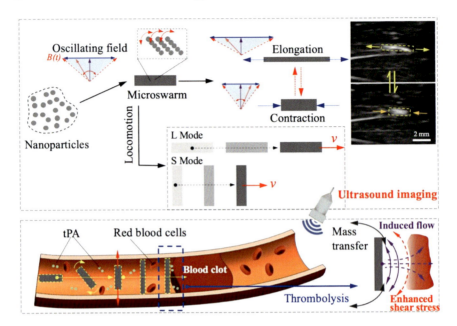

Fig. 10.26 Schematic illustration of using the ribbon-like swarm to accelerate tPA-mediated thrombolysis. The ribbon-like swarm is navigated to the targeted area in two different locomotion modes in blood and then adapts to the blood clot by pattern transformation

guidance of ultrasound imaging. tPA is a commonly used thrombolytic drug for ischemic stroke and is approved by the Food and Drug Administration (FDA). The schematic of this swarm-based accelerated thrombolysis strategy is shown in Fig. 10.26. After reaching the vicinity of the thrombus through two different motion modes under the guidance of ultrasound imaging, the ribbon-like swarm can adapt to the thrombus by pattern transformation (elongation of shrinkage). Subsequently, the swarm induces three-dimensional fluid convection around it, which enhances the shear stress near the clot-fluid interface and promotes the destruction of the clot. In addition, the swarm-induced fluid flow also enhances the thrombus-dissolving efficiency of tPA, thereby reducing the dose of tPA used. This strategy of using ultrasound-localized microswarm as a tool to accelerate the rate of thrombolysis holds great promise in clinical thrombus therapy.

10.4.2 Swarm-Induced Fluid Flow and Shear Stress

The fluid flow and shear stress induced by the swarm are the basis for accelerating the dissolution of thrombosis by tPA, which are elucidated by simulation here. First, the induced flow around the microswarm and the clot-fluid interface is investigated by using two finite element simulations. According to the chain length model, the

10.4 Accelerate Thrombolysis

length of nanoparticle chains in blood is set to 45 μm in an oscillating field with a frequency of 10 Hz (Fig. 10.27a). The viscosity and density of the fluid are set to 4 mPa·s and 1.05×10^3 kg/m^3, respectively, which is similar to blood. The distance between adjacent nanoparticle chains is twice the chain length (90 μm). In the first simulation (case I, Fig. 10.27a, b), the swarm oscillates 45 μm to the right of the clot to simulate fluid flow at the interface between the microswarm and the clot. In the second simulation (case II, Fig. 10.27c–e), the swarm is confined by three non-slip boundaries, which simulates the situation where the width of the swarm becomes the same as the blood clot through pattern transformation. The top view of simulation results (Fig. 10.27b, c) shows that the swarm induces convection in the horizontal plane (yellow arrows). The flow velocity is larger in the surroundings of swarm and can reach more than 100 μm/s at a distance of 200 μm from the swarm. The side view of simulation results (Fig. 10.27d, e) shows that the swarm can also induce out-of-plane 3D convection (yellow arrows), exerting effects on blood clots above the swarm. This means that the swarm-enhanced mass transfer can exert an effect on the entire clot-fluid interface even if the thickness of the clot is greater than that of the swarm pattern. Furthermore, boundary conditions affect the flow distribution near the clot-fluid interface. Figure 10.27f shows the average flow velocity distribution for one actuation cycle (0.1 s) near the interface (red line in Fig. 10.27a). The comparison of the results of the two simulations shows that deforming the swarm to fit the blood clot (case II) can further enhance the induced convection and also make the entire clot-fluid interface subject to convection.

The movement of RBCs is used to investigate the mass transport near the clot-fluid interface. In the simulation, the diameter and density of simulated red blood cells (SRBCs) are set to 6 μm and 1.09×10^3 kg/m^3, respectively. SRBCs are released from five different locations in the blood (all at a distance of 45 μm from the swarm), with 50 SRBCs released at each location (Fig. 10.28a). Under the influence of the swarm, the SRBC moves back and forth along the boundary at the same frequency as the oscillation of the chain, and its velocity reaches a maximum at $t = 4.95$ s and $t = 5$ s. Figure 10.28b simulates the situation in which SRBCs originate from dissolved clots (in PBS), and their release site is set at a distance of 95 μm from the swarm. The results show that the SRBCs are gradually moved out of the region between the clot and the swarm pattern under the influence of the flow, while new fluid will flow to this region. The simulation results in Fig. 10.28a, b indicate that mass transfer at the clot-fluid interface is enhanced by the swarm, and dissolved clot fragments will be removed from this region. Then, new tPA molecules will be transported by the flow to the clot-fluid interface to continuously dissolve the clot, contributing to the improvement of thrombolysis efficiency. Figure 10.28c presents the velocity of the induced flow between the clot and the cluster. In the PBS solution, the higher flow rate and larger fluctuation are attributed to the longer nanoparticle chain lengths. In addition to enhanced treatment mass transfer near the clot-fluid interface, rapidly changing flow profiles may also have an impact on thrombolytic efficacy. The first order of the velocity gradient is the shear rate, and the local shear stress near the interface can be expressed as the product of the shear rate and the fluid viscosity (Chong et al. 2013). Figure 10.28d shows the shear stress near the clot-fluid interface.

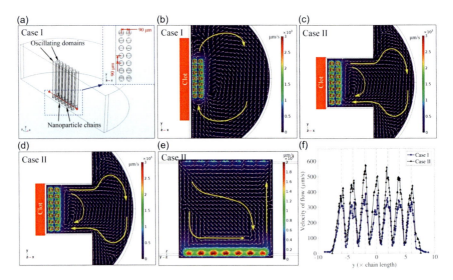

Fig. 10.27 Simulations of swarm-induced flow near the blood clot. **a** Schematic illustration of the model (case I). **b, c** The side view of the simulation results shows the swarm-induced on the cross section of the x–y plane ($z = 30$ μm). **d, e** The top view of the simulation results shows the distribution of flow velocity on the x–z and y–z planes, respectively. **f** The average flow velocity along the red line in **a** during an actuation cycle (0.1 s). Each dashed line indicates the position of the nanoparticle chain center. In **b–e**, the yellow arrows indicate the convection induced by the swarm and the boundaries, the color profile indicates the flow velocity as marked by the color legend, and the white arrows indicate the flow direction

Although the velocity gradient in blood is lower than that in PBS, the shear stress is of the same order of magnitude in both cases due to the higher viscosity of blood. The mean absolute values of shear stress in the simulations of blood and PBS are 401.580 and 458.982 mPa, respectively, indicating that the swarm can enhance shear stress on the thrombus. In thrombolysis, the tPA molecule first activates plasminogen to plasmin that degrades the fibrin network to fibrin. Enhanced shear stress accelerates the removal of reaction products, allowing the continuous exposure of new fibrin networks to tPA molecules. Notably, the shear stress induced by the swarm is less than that in normal physiological situations (e.g., 1–7 Pa in arteries) and is therefore considered biologically safe (Paszkowiak and Dardik 2003). The above simulation results suggest that microswarms can enhance fluid flow, mass transport, and shear stress near the clot-fluid interface, which is beneficial to obtain a higher lysis rate.

10.4 Accelerate Thrombolysis

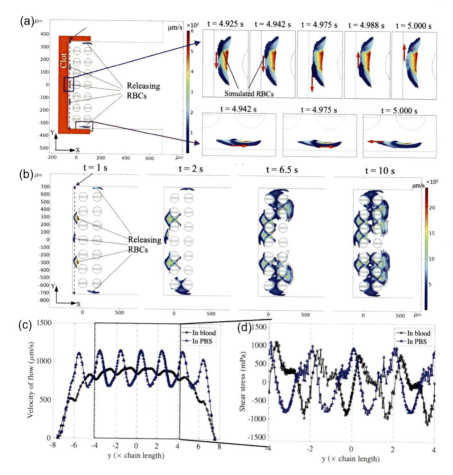

Fig. 10.28 Simulations of swarm-induced RBCs motion and shear stress. Simulation result of the motion of simulated red blood cells in **a** blood and **b** in PBS. **c** Distribution of the flow velocity between the clot-fluid interface and the swarm at $t = 5$ s. The cut lines are parallel to the Y-axis and are at the positions of $x = 45$ μm (in the blood) and $x = 95$ μm (in PBS), as indicated by purple dashed lines in (**a**) and (**b**). **d** Distribution of the shear stress along the two cut lines. Each dashed line in (**c**) and (**d**) indicates the position of the nanoparticle chain center

10.4.3 Accelerating TPA-Mediated Thrombolysis Using the Microswarm

In the experiments, an ultrasound system with a linear array probe is integrated into the magnetic actuation system (three-axis Helmholtz electromagnetic coil) for microswarm imaging. A heating plate is placed in the center of the coil, and a thermal imager is used to calibrate the temperature to ensure that the blood in the workspace is always around 37 °C. The Fe_3O_4 nanoparticle suspension is added to a PDMS tank filled with porcine whole blood. After applying the oscillating field, the nanoparticles can form a ribbon-like swarm (validated in Chap. 7), and the formation can be observed by ultrasound imaging (Fig. 10.29a). Compared with nanoparticles in the dispersed state, the ultrasound imaging contrast of the swarm is significantly enhanced, which is caused by the increased area density of the nanoparticles (from ~5.9 to ~8.2 µg/mm^2) due to the swarming effect (Wang et al. 2020). The ribbon-like swarm is able to successfully perform reversible pattern transformation in blood (Fig. 10.29a), although the relatively high viscosity of blood slows the transformation rate. The relationships between the aspect ratio and length of the swarm in blood and the input field ratio are shown in Fig. 10.29b, c, respectively. It can be found that the aspect ratio can reach around 15, and the lengths can vary from ~1 mm to ~4.5 mm (the particle dose is about 15 µg). The flexible pattern transformation capability in blood enables the ribbon-like swarm to adapt to clots of various sizes, which is critical for thrombolysis in blood vessels of different diameters. In addition, the ultrasound has no obvious influence on pattern formation and transformation of the ribbon-like swarm.

As discussed in Chap. 3, the ribbon-like swarm has two locomotion modes, i.e., longitudinal motion and lateral motion, which are realized by adding pitch angles to the alternating and static-field components, respectively (see Fig. 3.25). The swarm moving in both modes can be observed in the blood by ultrasound imaging, as shown in Fig. 10.30a. The moving velocity of the two modes increases with the pitch angle, and the maximum velocity in the blood is 35.80 ± 3.86 µm/s (longitudinal motion) and 23.97 ± 3.07 µm/s (lateral motion), respectively (Fig. 10.30b). When γ is too large, the increased fluid drag will lead to instability or even pattern collapse of the swarm, which reduces the moving velocity. The velocity of swarms performing the longitudinal motion and the lateral motion decrease when $\gamma > 10°$ and $\gamma > 7°$, respectively. The fluid drag to the swarm can be expressed as $F_D = 0.5 C_D \rho u2A$, where C_D is a dimensionless coefficient, ρ is the fluid density, u is the relative velocity, and A is the reference area. When the moving speed is the same, the swarm performing the lateral motion is subject to greater resistance than that performing the longitudinal motion. Therefore, the longitudinal motion is conducive to the swarm to obtain a more stable pattern, better motion performance, and higher efficiency for navigating to the clot area. After approaching the clot, the lateral motion is adopted to make the long edge of the pattern fit the clot, so that the entire clot-fluid interface is affected by the swarm after pattern transformation. During thrombolysis, the swarm should continue to perform lateral motion toward the clot, which makes the clot-fluid interface always subject to swarm-induced strong convection and high shear stress.

Figure 10.31a shows the schematic diagram of a PDMS channel with three branches. The channel is filled with porcine whole blood, and a blood clot with a width of 1.5 mm is formed in one of the branches by injection of an aqueous calcium chloride solution. As shown in Fig. 10.31b, after the swarm is formed, it first performs longitudinal motion to approach the clot ($t = 0$–80 s). After reaching the branch filled with blood clot, direction of the swarm is changed to make the long side of the swarm parallel to the clot-fluid interface. The swarm then moves laterally to get closer to the clot by adjusting the magnetic field. Since the swarm

10.4 Accelerate Thrombolysis

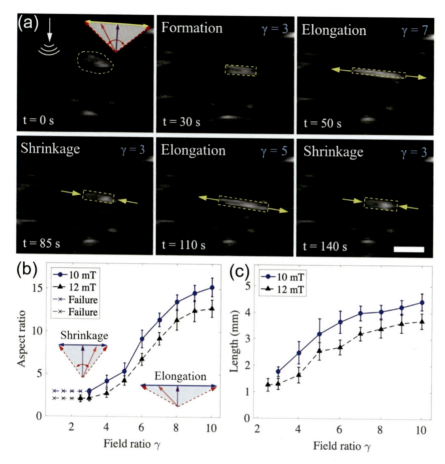

Fig. 10.29 Formation and pattern transformation of the ribbon-like swarm in porcine whole blood. **a** Ultrasound images showing formation and pattern transformation processes of the ribbon-like swarm. The scale bar is 2 mm. **b** The change in the aspect ratio of the ribbon-like swarm with the field ratio γ. **C** The change in the length of the ribbon-like swarm with the field ratio γ. Each error bar denotes the standard deviation from three trials

length is longer than the width of the channel, pattern shrinkage is conducted to adapt the swarm to the channel by reducing the field ratio. During the whole process, no obvious disturbances in the swarm locomotion induced by ultrasound are observed.

Here, commercial tPA injection is added into the tank for microswarm-assisted thrombolysis (the drug concentration is 3 mg/mL). To verify the accuracy of the ultrasound imaging, the blood in the tank is replaced with PBS, which enables the thrombolysis process be directly observed by the camera. The microswarm-assisted lysis process of blood clots in the tPA-PBS solution is shown in Fig. 10.32a. After the ribbon-like swarm is adapted to the clot area by turning direction and pattern transformation, the pitch angle is adjusted to 1° to actuate the swarm toward the clot.

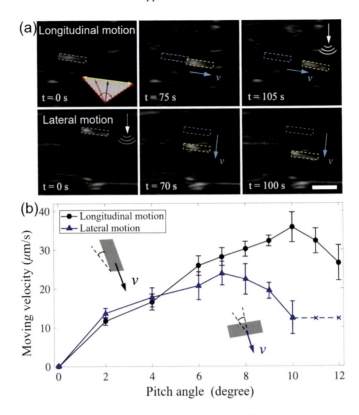

Fig. 10.30 Longitudinal and lateral motion modes of the ribbon-like swarm in porcine whole blood. **a** Ultrasound images of the ribbon-like swarm performing the longitudinal motion and the lateral motion. The scale bar is 2 mm. **b** The relationship between the moving velocity and the field ratio γ. The blue crosses indicate the failure of locomotion. Each error bar denotes the standard deviation from three trials

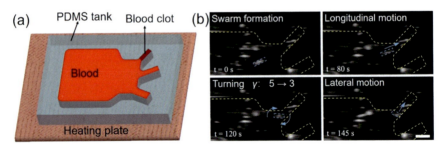

Fig. 10.31 Magnetic navigation of the ribbon-like swarm toward a blood clot. **a** Schematic diagram of a PDMS channel with a blot clot. **b** Ultrasound images of the navigation process. The scale bar is 3 mm

10.4 Accelerate Thrombolysis

At the beginning ($t = 10$ min), the locomotion of the swarm is hindered by the blood clot. Under the influence of the swarm-induced fluid flow and shear stress, the clot is continuously dissolved by tPA, resulting in a translational movement of the swarm into the channel ($t = 40$–60 min). It is worth noting that the bottom clot area dissolves faster due to the presence of the swarm (Fig. 10.32a). However, due to the effect of 3D induced flow (Fig. 10.27), the top region of the clot also gradually dissolves, and the lysis rate is higher than with the addition of tPA alone. According to the optical image of the camera and the ultrasound image, the remaining volume of the clot over time can be calculated, as shown in Fig. 10.32b. Due to the low echogenicity of blood clots, ultrasound images cannot directly present the size of the clot. Therefore, the computed results based on the ultrasound images are derived by taking the positions of the swarm as the boundary of the blood clot. The mean error of the camera-based and ultrasound-based calculation results is 5.32%, indicating the effectiveness of evaluating thrombolytic efficiency by ultrasound images. The microswarm-assisted thrombolysis is further performed with different concentrations of tPA in channels of different widths. The results show that the average lysis rate increases with drug dose, and the rate is highest in the 2.5 mm wide channel due to the larger area of the clot-fluid interface.

The ribbon-like swarm is used to accelerate tPA-mediated thrombolysis in blood (the drug concentration is 3 mg/mL), the process is shown in the schematic diagram of Fig. 10.33a. Blood clots are generated in channels of 1.5-, 2.0-, and 2.5-mm-width, with initial volumes of 11.25, 15, and 18.75 mm^3, respectively (Fig. 10.33b). Under the navigation of ultrasound imaging, the microswarm moves laterally into the channel. Thrombolysis is complete when the swarm reaches the end of the channel. The blood in the tank is subsequently drawn and filtered through a 100-mesh filter (pore size: 150 μm), and no small blood clots (thrombus fragments) are observed in the blood nor on the inner walls of the tank. This suggests that the non-contact interaction between the microswarm and the clot via fluid flow reduces the likelihood of the clot disintegrating into smaller fragments. In general thrombolysis methods using mechanical force, clot fragments that may cause secondary vascular occlusion are prone to be generated. The lysis rate of the clot as a function of time is shown in Fig. 10.33c. The lysis rates in all three cases increase with time, with the highest rates at 50–60 min. Blood clots in 1.5-, 2.0- and 2.5-mm wide channels are unclogged after 86, 80, and 69 min, with mean lysis rates of 0.1310 ± 0.0324 (1.5 mm), 0.1895 ± 0.0227 (2.0 mm), and 0.2669 ± 0.0372 mm^3/min (2.5 mm), respectively (Fig. 10.33d). The clot lysis rate in these three channels using tPA alone with the same concentration are 0.0418 ± 0.0041, 0.0710 ± 0.0175, and 0.1013 ± 0.0214 mm^3/min, respectively. The assistant effect of the swarm improves the thrombolytic efficiency of tPA significantly.

According to the previous theoretical analysis, the accelerated thrombolysis using the microswarm can be attributed to the following three reasons. First, the swarm-induced flow enhances the convection of tPA molecules. Mass transfer near the clot-blood interface is accelerated, and new tPA molecules can be continuously and actively transported to the interface, inducing dissolution of the fibrin network into degraded fibrin. Second, the convection generated by the swarm also enhances the

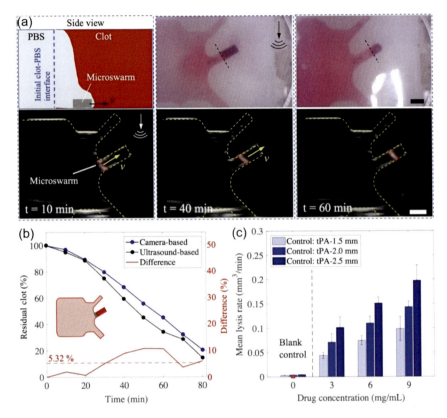

Fig. 10.32 Microswarm-assisted thrombolysis in PBS. **a** Optical (top) and ultrasound (bottom) images of the thrombolysis process. Yellow dashed curves in the ultrasound images indicate the boundary of the channel. The scale bars are 3 mm. The schematic diagram in the upper left corner shows side view of microswarm-assisted thrombolysis. **b** The relationship between residual clot and time (calculated from optical and ultrasound images). **c** The mean lysis rate varies with the tPA concentration and the width of the clot. Each error bar denotes the standard deviation from three trials

shear stress at the clot-blood interface, accelerating the removal of reaction products. Third, the swarm first lyses the bottom of the clot so that the area of the clot-blood interface becomes larger (Fig. 10.32a), resulting in an increase in the reaction area and the thrombolysis rate. Although tPA is an FDA-approved thrombolytic drug, the use of high drug doses for thrombus removal may lead to side effects such as intracranial hemorrhage. The results in Fig. 10.33d show that a higher thrombolysis rate (9 mg/mL) can be obtained with the microswarm at low tPA concentrations (3 mg/mL) than with high concentrations (9 mg/mL) of tPA alone. Before applying this microswarm-assisted thrombolysis method to the clinic, the following issues need to be considered. First, the influence of blood flow on the location of the microswarm should be minimized, and the swarm should always be close to the thrombus and maintain its pattern stability. For example, the use of gradient magnetic fields (described in

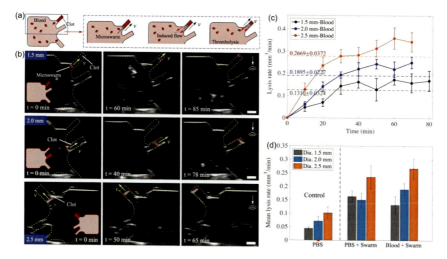

Fig. 10.33 Microswarm-assisted thrombolysis in blood. **a** Schematic diagram of the microswarm-assisted thrombolysis process. **b** Ultrasound images of microswarm-assisted thrombolysis in three clot regions with a width of 1.5, 2.0, and 2.5 mm. The red tangles indicate the position of the swarm. The scale bars are 3 mm. **c** The change of lysis rate in the three clot regions with time during thrombolysis process. **d** The mean lysis rates in the three clot regions in PBS and blood. The concentration of tPA is 3 mg/mL. Each error bar denotes the standard deviation from three trials

Chap. 9) can enhance the aggregation of nanoparticles and provide the swarm with gradient forces against flow. The use of some clinical treatments can also help the swarm resist blood flow and deliver them over long distances, such as previously introduced catheterization (Nguyen et al. 2020; Jeon et al. 2019). Blood flow can also be blocked by inflating a balloon catheter, and then nanoparticles along with the tPA drug are rapidly released to the target clot area. After thrombolysis is complete, the nanoparticles can be recovered using a magnetic guidewire through the catheter. Second, the interactions between the microswarm and blood vessels need to be further evaluated, including the generation and locomotion of microswarm on curved vessel surfaces and the effect of endothelial cells on nanoparticles. Third, the toxicity of nanoparticles also needs to be further tested and adjusted before actual thrombosis treatment. For example, the biocompatibility of magnetic particles can be improved by coating them with polydopamine.

References

Banerjee I, Pangule RC, Kane RS (2011) Antifouling coatings: Recent developments in the design of surfaces that prevent fouling by proteins, bacteria, and marine organisms. Adv Mater 23(6):690–718

Behlau I, Gilmore MS (2008) Microbial biofilms in ophthalmology and infectious disease. Arch Ophthalmol 126(11):1572–1581

Chalasani R, Vasudevan S (2011) Form, content, and magnetism in iron oxide nanocrystals. The Journal of Physical Chemistry C 115(37):18088–18093

Cheeseman S, Christofferson AJ, Kariuki R, Cozzolino D, Daeneke T, Crawford RJ, Truong VK, Chapman J, Elbourne A (2020a) Antimicrobial metal nanomaterials: From passive to stimuli-activated applications. Advanced Science 7(10):1902913

Cheeseman S, Elbourne A, Kariuki R, Ramarao AV, Zavabeti A, Syed N, Christofferson AJ, Kwon KY, Jung W, Dickey MD (2020b) Broad-spectrum treatment of bacterial biofilms using magneto-responsive liquid metal particles. Journal of Materials Chemistry B 8(47):10776–10787

Chen Z, Wang Z, Ren J, Qu X (2018) Enzyme mimicry for combating bacteria and biofilms. Acc Chem Res 51(3):789–799

Chen Z, Xia T, Zhang Z, Xie S, Wang T, Li X (2019) Enzyme-powered janus nanomotors launched from intratumoral depots to address drug delivery barriers. Chem Eng J 375:122109

Chen H, Zhang H, Xu T, Yu J (2021) An overview of micronanoswarms for biomedical applications. ACS Nano 15(10):15625–15644

Chen H, Wang Y, Liu Y, Zou Q, Yu J (2022) Sensing of fluidic features using colloidal microswarms. ACS Nano 16(10):16281–16291

Chong K, Kelly SD, Smith S, Eldredge JD (2013) Inertial particle trapping in viscous streaming. Phys Fluids 25(3):033602

Chung KK, Schumacher JF, Sampson EM, Burne RA, Antonelli PJ, Brennan AB (2007) Impact of engineered surface microtopography on biofilm formation of staphylococcus aureus. Biointerphases 2(2):89–94

Dahl O, Overgaard J (1995) A century with hyperthermic oncology in scandinavia. Acta Oncol 34(8):1075–1083

Deng H, Li X, Peng Q, Wang X, Chen J, Li Y (2005) Monodisperse magnetic single-crystal ferrite microspheres. Angew Chem 117(18):2842–2845

Deng Y-H, Ricciardulli T, Won J, Wade MA, Rogers SA, Boppart SA, Flaherty DW, Kong H (2022) Self-locomotive, antimicrobial microrobot (slam) swarm for enhanced biofilm elimination. Biomaterials 11(6):121610

Dong Y, Wang L, Yuan K, Ji F, Gao J, Zhang Z, Du X, Tian Y, Wang Q, Zhang L (2021) Magnetic microswarm composed of porous nanocatalysts for targeted elimination of biofilm occlusion. ACS Nano 15(3):5056–5067

Elbourne A, Cheeseman S, Atkin P, Truong NP, Syed N, Zavabeti A, Mohiuddin M, Esrafilzadeh D, Cozzolino D, McConville CF (2020) Antibacterial liquid metals: Biofilm treatment via magnetic activation. ACS Nano 14(1):802–817

Feliu N, Docter D, Heine M, Del Pino P, Ashraf S, Kolosnjaj-Tabi J, Macchiarini P, Nielsen P, Alloyeau D, Gazeau F (2016) In vivo degeneration and the fate of inorganic nanoparticles. Chem Soc Rev 45(9):2440–2457

Flemming H-C, Wingender J (2010) The biofilm matrix. Nat Rev Microbiol 8(9):623–633

Flemming H-C, Wingender J, Szewzyk U, Steinberg P, Rice SA, Kjelleberg S (2016) Biofilms: An emergent form of bacterial life. Nat Rev Microbiol 14(9):563–575

Fu Y, Yu H, Zhang X, Malgaretti P, Kishore V, Wang W (2022) Microscopic swarms: From active matter physics to biomedical and environmental applications. Micromachines 13(2):295

Gao L, Liu Y, Kim D, Li Y, Hwang G, Naha PC, Cormode DP, Koo H (2016) Nanocatalysts promote streptococcus mutans biofilm matrix degradation and enhance bacterial killing to suppress dental caries in vivo. Biomaterials 101:272–284

Heit JA (2015) Epidemiology of venous thromboembolism. Nat Rev Cardiol 12(8):464–474

Huang C, Yang G, Ha Q, Meng J, Wang S (2015) Multifunctional "smart" particles engineered from live immunocytes: Toward capture and release of cancer cells. Adv Mater 27(2):310–313

Huang J, Li Y, Orza A, Lu Q, Guo P, Wang L, Yang L, Mao H (2016) Magnetic nanoparticle facilitated drug delivery for cancer therapy with targeted and image-guided approaches. Adv Func Mater 26(22):3818–3836

References

Hwang, G., Paula, A.J., Hunter, E.E., Liu, Y., Babeer, A., Karabucak, B., Stebe, K., Kumar, V., Steager, E., & Koo, H. (2019). Catalytic antimicrobial robots for biofilm eradication. *Science robotics*, 4(29), eaaw2388.

Jeon S, Hoshiar AK, Kim K, Lee S, Kim E, Lee S, Kim J-Y, Nelson BJ, Cha H-J, Yi B-J (2019) A magnetically controlled soft microrobot steering a guidewire in a three-dimensional phantom vascular network. Soft Rob 6(1):54–68

Ji F, Wu Y, Pumera M, Zhang L (2022) Collective behaviors of active matter learning from natural taxes across scales. Adv Mater. https://doi.org/10.1002/adma.202203959

Jin D, Zhang L (2022) Collective behaviors of magnetic active matter: recent progress toward reconfigurable, adaptive, and multifunctional swarming micro/nanorobots. Acc Chem Res 55(1):98–109

Juenet M, Aid-Launais R, Li B, Berger A, Aerts J, Ollivier V, Nicoletti A, Letourneur D, Chauvierre C (2018) Thrombolytic therapy based on fucoidan-functionalized polymer nanoparticles targeting p-selectin. Biomaterials 156:204–216

Kälicke T, Schierholz J, Schlegel U, Frangen TM, Köller M, Printzen G, Seybold D, Klöckner S, Muhr G, Arens S (2006) Effect on infection resistance of a local antiseptic and antibiotic coating on osteosynthesis implants: An in vitro and in vivo study. J Orthop Res 24(8):1622–1640

Kaplan JÁ (2010) Biofilm dispersal: Mechanisms, clinical implications, and potential therapeutic uses. J Dental Res 89(3):205–218

Karimi M, Ghasemi A, Zangabad PS, Rahighi R, Basri SMM, Mirshekari H, Amiri M, Pishabad ZS, Aslani A, Bozorgomid M (2016) Smart micro/nanoparticles in stimulus-responsive drug/gene delivery systems. Chem Soc Rev 45(5):1457–1501

Khalil IS, Mahdy D, El Sharkawy A, Moustafa RR, Tabak AF, Mitwally ME, Hesham S, Hamdi N, Klingner A, Mohamed A (2018) Mechanical rubbing of blood clots using helical robots under ultrasound guidance. IEEE Robot Autom Lett 3(2):1112–1119

Kiristi M, Singh VV, Esteban-Fernández de Ávila B, Uygun M, Soto F, Aktaş Uygun D, Wang J (2015) Lysozyme-based antibacterial nanomotors. ACS Nano 9(9):9252–9259

Koo H, Allan RN, Howlin RP, Stoodley P, Hall-Stoodley L (2017) Targeting microbial biofilms: Current and prospective therapeutic strategies. Nat Rev Microbiol 15(12):740–755

Korin N, Kanapathipillai M, Matthews BD, Crescente M, Brill A, Mammoto T, Ghosh K, Jurek S, Bencherif SA, Bhatta D (2012) Shear-activated nanotherapeutics for drug targeting to obstructed blood vessels. Science 337(6095):738–742

Kostakioti M, Hadjifrangiskou M, Hultgren SJ (2013) Bacterial biofilms: Development, dispersal, and therapeutic strategies in the dawn of the postantibiotic era. Cold Spring Harb Perspect Med 3(4):a010306

Law J, Wang X, Luo M, Xin L, Du X, Dou W, Wang T, Shan G, Wang Y, Song P (2022) Microrobotic swarms for selective embolization. Sci Adv 8(29):eabm5752

Leclerc J, Zhao H, Bao D, Becker AT (2020) In vitro design investigation of a rotating helical magnetic swimmer for combined 3-d navigation and blood clot removal. IEEE Trans Rob 36(3):975–982

Lee J-H, Jang J-T, Choi J-S, Moon SH, Noh S-H, Kim J-W, Kim J-G, Kim I-S, Park KI, Cheon J (2011) Exchange-coupled magnetic nanoparticles for efficient heat induction. Nat Nanotechnol 6(7):418–422

Leung J, Ling T, Kung J, Vallance-Owen J (1988) The role of bacteria in the blockage of biliary stents. Gastrointest Endosc 34(1):19–22

Leung JW, Liu Y-L, Chan RC, Tang Y, Mina Y, Cheng AF, Silva J Jr (2000) Early attachment of anaerobic bacteria may play an important role in biliary stent blockage. Gastrointest Endosc 52(6):725–729

Li Y, Lu Q, Liu H, Wang J, Zhang P, Liang H, Jiang L, Wang S (2015) Antibody-modified reduced graphene oxide films with extreme sensitivity to circulating tumor cells. Adv Mater 27(43):6848–6854

Li L, Chang H, Yong N, Li M, Hou Y, Rao W (2021) Superior antibacterial activity of gallium based liquid metals due to ga 3+ induced intracellular ros generation. J Mater Chem B 9(1):85–93

Lin Y, Genzer J, Dickey MD (2020) Attributes, fabrication, and applications of gallium-based liquid metal particles. Adv Sci 7(12):2000192

Lin Z, Gao C, Wang D, He Q (2021) Bubble-propelled janus gallium/zinc micromotors for the active treatment of bacterial infections. Angew Chem Int Ed 60(16):8750–8754

Liu Y, Naha PC, Hwang G, Kim D, Huang Y, Simon-Soro A, Jung H-I, Ren Z, Li Y, Gubara S (2018) Topical ferumoxytol nanoparticles disrupt biofilms and prevent tooth decay in vivo via intrinsic catalytic activity. Nat Commun 9:2920

Liu Y, Shi L, Su L, van der Mei HC, Jutte PC, Ren Y, Busscher HJ (2019a) Nanotechnology-based antimicrobials and delivery systems for biofilm-infection control. Chem Soc Rev 48(2):428–446

Liu X, Gao Y, Chandrawati R, Hosta-Rigau L (2019b) Therapeutic applications of multifunctional nanozymes. Nanoscale 11(44):21046–21060

Maldonado RA, LaMothe RA, Ferrari JD, Zhang A-H, Rossi RJ, Kolte PN, Griset AP, O'Neil C, Altreuter DH, Browning E (2015) Polymeric synthetic nanoparticles for the induction of antigen-specific immunological tolerance. Proc Natl Acad Sci 112(2):E156–E165

Mayorga-Martinez CC, Zelenka J, Grmela J, Michalkova H, Ruml T, Mareš J, Pumera M (2021) Swarming aqua sperm micromotors for active bacterial biofilms removal in confined spaces. Adv Sci 8(19):2101301

Miller KP, Wang L, Benicewicz BC, Decho AW (2015) Inorganic nanoparticles engineered to attack bacteria. Chem Soc Rev 44(21):7787–7807

Monopoli MP, Åberg C, Salvati A, Dawson KA (2012) Biomolecular coronas provide the biological identity of nanosized materials. Nat Nanotechnol 7(12):779–786

Naha PC, Liu Y, Hwang G, Huang Y, Gubara S, Jonnakuti V, Simon-Soro A, Kim D, Gao L, Koo H (2019) Dextran-coated iron oxide nanoparticles as biomimetic catalysts for localized and ph-activated biofilm disruption. ACS Nano 13(5):4960–4971

Nguyen KT, Kim S-J, Min H-K, Hoang MC, Go G, Kang B, Kim J, Choi E, Hong A, Park J-O (2020) Guide-wired helical microrobot for percutaneous revascularization in chronic total occlusion in-vivo validation. IEEE Trans Biomed Eng 68(8):2490–2498

Oh MJ, Babeer A, Liu Y, Ren Z, Wu J, Issadore DA, Stebe KJ, Lee D, Steager E, Koo H (2022) Surface topography-adaptive robotic superstructures for biofilm removal and pathogen detection on human teeth. ACS Nano 16(8):11998–12012

Paszkowiak JJ, Dardik A (2003) Arterial wall shear stress: Observations from the bench to the bedside. Vascular Endovascular Surg 37(1):47–57

Périgo EA, Hemery G, Sandre O, Ortega D, Garaio E, Plazaola F, Teran FJ (2015) Fundamentals and advances in magnetic hyperthermia. Appl Phys Rev 2(4):041302

Pogodin S, Hasan J, Baulin VA, Webb HK, Truong VK, Nguyen THP, Boshkovikj V, Fluke CJ, Watson GS, Watson JA (2013) Biophysical model of bacterial cell interactions with nanopatterned cicada wing surfaces. Biophys J 104(4):835–840

Roussakow S (2013) The history of hyperthermia rise and decline. In: Conference papers in science. Hindawi, 428027

Saei AA, Yazdani M, Lohse SE, Bakhtiary Z, Serpooshan V, Ghavami M, Asadian M, Mashaghi S, Dreaden EC, Mashaghi A (2017) Nanoparticle surface functionality dictates cellular and systemic toxicity. Chem Mater 29(16):6578–6595

Servant A, Qiu F, Mazza M, Kostarelos K, Nelson BJ (2015) Controlled in vivo swimming of a swarm of bacteria-like microrobotic flagella. Adv Mater 27(19):2981–2988

Soto F, Kupor D, Lopez-Ramirez MA, Wei F, Karshalev E, Tang S, Tehrani F, Wang J (2020) Onion-like multifunctional microtrap vehicles for attraction–trapping–destruction of biological threats. Angew Chem 132(9):3508–3513

Sun M, Chan KF, Zhang Z, Wang L, Wang Q, Yang S, Chan SM, Chiu PWY, Sung JJY, Zhang L (2022) Magnetic microswarm and fluoroscopy-guided platform for biofilm eradication in biliary stents. Adv Mater 34(34):2201888

Sung J, Leung J, Shaffer E, Lam K, Costerton J (1993) Bacterial biofilm, brown pigment stone and blockage of biliary stents. J Gastroenterol Hepatol 8(1):28–34

Tasci TO, Disharoon D, Schoeman RM, Rana K, Herson PS, Marr DW, Neeves KB (2017) Enhanced fibrinolysis with magnetically powered colloidal microwheels. Small 13(36):1700954

Thomalla GT, Schwark C, Sobesky J, Bluhmki E, Fiebach JB, Fiehler J, Zaro Weber O, Kucinski T, Juettler E, Ringleb PA (2006) Outcome and symptomatic bleeding complications of intravenous thrombolysis within 6 hours in mri-selected stroke patients: comparison of a german multicenter study with the pooled data of atlantis, ecass, and ninds tpa trials. Stroke 37(3):852–858

Ussia M, Urso M, Dolezelikova K, Michalkova H, Adam V, Pumera M (2021) Active light-powered antibiofilm zno micromotors with chemically programmable properties. Adv Func Mater 31(27):2101178

Vaishnavi C, Samanta J, Kochhar R (2018) Characterization of biofilms in biliary stents and potential factors involved in occlusion. World J Gastroenterol 24(1):112

Van Berkel A, Van Marle J, Groen A, Bruno M (2005) Mechanisms of biliary stent clogging: Confocal laser scanning and scanning electron microscopy. Endoscopy 37(08):729–734

Vilela D, Cossío U, Parmar J, Martínez-Villacorta AM, Gómez-Vallejo V, Llop J, Sánchez S (2018) Medical imaging for the tracking of micromotors. ACS Nano 12(2):1220–1227

Villa K, Sopha H, Zelenka J, Motola M, Dekanovsky L, Beketova DC, Macak JM, Ruml T, Pumera M (2022) Enzyme-photocatalyst tandem microrobot powered by urea for escherichia coli biofilm eradication. Small 18(36):2106612

Wan M, Wang Q, Wang R, Wu R, Li T, Fang D, Huang Y, Yu Y, Fang L, Wang X (2020) Platelet-derived porous nanomotor for thrombus therapy. Sci Adv 6(22):eaaz9014

Wang H, Pumera M (2020) Coordinated behaviors of artificial micro/nanomachines: from mutual interactions to interactions with the environment. Chem Soc Rev 49(10):3211–3230

Wang Q, Zhang L (2018) External power-driven microrobotic swarm: from fundamental understanding to imaging-guided delivery. ACS Nano 15(1):149–174

Wang L, Min Y, Xu D, Yu F, Zhou W, Cuschieri A (2014) Membrane lipid peroxidation by the peroxidase-like activity of magnetite nanoparticles. Chem Commun 50(76):11147–11150

Wang L, Yuan Y, Lin S, Huang J, Dai J, Jiang Q, Cheng D, Shuai X (2016a) Photothermo-chemotherapy of cancer employing drug leakage-free gold nanoshells. Biomaterials 78:40–49

Wang L, Liu H, Zhang F, Li G, Wang S (2016b) Smart thin hydrogel coatings harnessing hydrophobicity and topography to capture and release cancer cells. Small 12(34):4697–4701

Wang B, Chan KF, Yu J, Wang Q, Yang L, Chiu PWY, Zhang L (2018) Reconfigurable swarms of ferromagnetic colloids for enhanced local hyperthermia. Adv Func Mater 28(25):1705701

Wang Q, Yang L, Yu J, Chiu PW, Zheng YP, Zhang L (2020) Real-time magnetic navigation of a rotating colloidal microswarm under ultrasound guidance. IEEE Trans Biomed Eng 67(12):3403–3412

Wang B, Kostarelos K, Nelson BJ, Zhang L (2021) Trends in micro-/nanorobotics: Materials development, actuation, localization, and system integration for biomedical applications. Adv Mater 33(4):2002047

Wang Q, Yang S, Zhang L (2022) Magnetic actuation of a dynamically reconfigurable microswarm for enhanced ultrasound imaging contrast. IEEE/ASME Trans Mechatron 27(6):4235–4245. https://doi.org/10.1109/TMECH.2022.3151983

Wechsler LR (2011) Intravenous thrombolytic therapy for acute ischemic stroke. N Engl J Med 364(22):2138–2146

Whitchurch CB, Tolker-Nielsen T, Ragas PC, Mattick JS (2002) Extracellular DNA required for bacterial biofilm formation. Science 295(5559):1487–1487

Wu Z, Chen Y, Mukasa D, Pak OS, Gao W (2020) Medical micro/nanorobots in complex media. Chem Soc Rev 49(22):8088–8112

Xie L, Pang X, Yan X, Dai Q, Lin H, Ye J, Cheng Y, Zhao Q, Ma X, Zhang X (2020) Photoacoustic imaging-trackable magnetic microswimmers for pathogenic bacterial infection treatment. ACS Nano 14(3):2880–2893

Xu T, Luo Y, Liu C, Zhang X, Wang S (2020) Integrated ultrasonic aggregation-induced enrichment with Raman enhancement for ultrasensitive and rapid biosensing. Anal Chem 92(11):7816–7821

Yan X, Zhou Q, Vincent M, Deng Y, Yu J, Xu J, Xu T, Tang T, Bian L, Wang Y-XJ (2017) Multifunctional biohybrid magnetite microrobots for imaging-guided therapy. Sci Robot 2(12):eaaq1155

Yang L, Yu J, Yang S, Wang B, Nelson BJ, Zhang L (2022) A survey on swarm microrobotics. IEEE Trans Rob 38(3):1531–1551

Yu J, Xu T, Lu Z, Vong CI, Zhang L (2017) On-demand disassembly of paramagnetic nanoparticle chains for microrobotic cargo delivery. IEEE Trans Rob 33(5):1213–1225

Yu J, Yang L, Zhang L (2018) Pattern generation and motion control of a vortex-like paramagnetic nanoparticle swarm. Int J Robot Res 37(8):912–930

Yu J, Wang Q, Li M, Liu C, Wang L, Xu T, Zhang L (2019) Characterizing nanoparticle swarms with tuneable concentrations for enhanced imaging contrast. IEEE Robot Autom Lett 4(3):2942–2949

Yu Q, Deng T, Lin F-C, Zhang B, Zink JI (2020) Supramolecular assemblies of heterogeneous mesoporous silica nanoparticles to co-deliver antimicrobial peptides and antibiotics for synergistic eradication of pathogenic biofilms. ACS Nano 14(5):5926–5937

Zenych A, Fournier L, Chauvierre C (2020) Nanomedicine progress in thrombolytic therapy. Biomaterials 258:120297

Zhang P, Chen L, Xu T, Liu H, Liu X, Meng J, Yang G, Jiang L, Wang S (2013) Programmable fractal nanostructured interfaces for specific recognition and electrochemical release of cancer cells. Adv Mater 25(26):3566–3570

Zhang Z, Wang L, Chan TK, Chen Z, Ip M, Chan PK, Sung JJ, Zhang L (2022) Micro-/nanorobots in antimicrobial applications: Recent progress, challenges, and opportunities. Adv Healthcare Mater 11(6):2101991

Chapter 11
Biohybrid Microswarm for the Removal of Toxic Heavy Metals

Abstract In addition to biomedical applications, micro/nanorobot swarms can also be applied to other fields, such as decontamination of pollutants. This chapter describes using biohybrid microrobot swarms to enhance the removal of toxic heavy metals. The biohybrid microrobots are fabricated by encapsulating *G. lucidum* spores with Fe_3O_4 nanoparticle layers. Benefiting from the synergistic effect of magnetic components and porous spores, the biohybrid microrobot swarms show excellent removal efficiency of heavy metal ions in contaminated water. The microrobots can be collected and recycled after use.

Keywords Biohybrid microrobot · Spore · Surface functionalization · Heavy metal · Magnetic microswarm

11.1 Introduction

Nowadays, as heavy metals are extensively used in industrial, agricultural, and anthropogenic activities, their accumulation in water sources and the food chain has become a severe threat to ecology and public health (Azizullah et al. 2011). Many heavy metals have been proven to be toxic to living organisms, even at trace levels of exposure, including copper, nickel, cobalt, mercury, lead, and so on (Jaishankar et al. 2014; Li et al. 2014). Therefore, the development of environmental remediation strategies capable of removing heavy metals in an efficient and green manner is urgently needed. To date, several kinds of methods have been proposed (Santhosh et al. 2016; Ersahin et al. 2012; Khin et al. 2012), such as chemical precipitation, membrane filtration, physical adsorption, ion exchange, and electrochemical treatment techniques, among which adsorption emerges as an economical and effective candidate (Adeleye et al. 2016; Li et al. 2015). There exist a variety of functional materials to be used as adsorbents (Lu and Astruc 2018; Aragay et al. 2011; Fu and Wang 2011), while the natural counterparts originating from biological organisms are highly preferred, mainly due to their abundant quantities and inherent adsorptive property. For example, various microorganisms (e.g., bacteria, fungi, and algae)

© The Author(s), under exclusive license to Springer Nature Singapore Pte Ltd. 2023
L. Zhang et al., *Magnetic Micro and Nanorobot Swarms: From Fundamentals to Applications*, Springer Tracts in Electrical and Electronics Engineering,
https://doi.org/10.1007/978-981-99-3036-4_11

(Saha and Orvig 2010; Gupta et al. 2015; Bilal et al. 2013) possess a remarkable capability to uptake heavy metals, enabling them to be an ideal biosorption candidate in an effective and green way (Kavamura and Esposito 2010). However, current removal efficiency is still limited due to the passive nature of the diffusive mass transportation during the biosorption process. The operation efficiency is also restricted by the collection and reuse process of adsorbents. Therefore, it is highly necessary to develop alternative means to improve removal efficiency and enable effective recycling.

Micro/nanorobotic technologies exhibit promising potential in the field of heavy metal remediation (Wang and Chen 2009; Li et al. 2017). Self-propelled micro/nanorobots fueled by chemicals have been verified to be useful in the adsorption and/or destruction of various pollutants (Moo and Pumera 2015; Jurado-Sanchez and Wang 2018; Wang et al. 2017). In comparison to traditional adsorbent nanomaterials, the active mobility of micro/nanorobots endows better adsorption efficiency (~4–5 times) and faster treatment speed (~5 times for certain capacity) (Wang et al. 2017, 2016). However, current studies mainly utilize Pt-H_2O_2 catalytic reaction to drive the motion of self-propelled micro/nanorobots (Wang et al. 2016; Jurado-Sanchez et al. 2015), which is non-lasting and sensitive to environmental conditions (Moo et al. 2014). In this respect, magnetic field-driven micro/nanorobots have received tremendous interest due to their remote controllability, biocompatibility, and reusability (Fischer and Ghosh 2011; Chen and Johnson 1986). Owing to the availability of swarm behaviors under the guidance of magnetic fields, high convections and intermixing can be induced to improve the ability of microrobots (Yu et al. 2018; Wang et al. 2018; Yan et al. 2015, 2017). Therefore, it can be expected that preparation of existing microorganisms into magnetized biohybrid adsorbents that combine good adsorption ability from the biological matrix and microrobotic techniques would be an ideal candidate for environmental remediation.

In this chapter, a magnetically controllable swarm of biohybrid microrobots is developed for the enhanced removal of heavy metals. After simple pretreatment, magnetic Fe_3O_4 nanoparticles are functionalized on the *Ganoderma lucidum* (*G. lucidum*) spores, thus enabling magnetic controlled locomotion. Under an external magnetic field, a swarm of the obtained organic/inorganic porous spore@Fe_3O_4 biohybrid adsorbents (PSFBAs) shows excellent adsorption capacity and shorter removal time. Furthermore, the PSFBAs can be collected with ease by a permanent magnet and recycled for next utilization by simple acid treatment and ultrasonic redispersion.

11.2 Synthesis and Characterization of PSFBA

As shown in Fig. 11.1, the fabrication process of PSFBA includes hydrothermal treatment of the original spores followed by chemical bath deposition (CBD) of magnetic nanoparticles. *G. lucidum* spores are initially treated to produce multiscale pores in their bodies. After functionalization with Fe_3O_4 nanoparticles, magnetic

11.2 Synthesis and Characterization of PSFBA

PSFBAs are obtained, which are able to locomote controllably in various contaminated solutions under the guidance of an external magnetic field. Considering the inherent adsorption capability of spores (Zhang et al. 2018; Zhang 2003), the prepared biohybrid microrobotic adsorbents are expected to rapidly remove heavy metals in aqueous environments in a collective manner. After a thorough adsorption process, the swarming PSFBAs are collected by a permanent magnet and recycled by acid treatment and ultrasonic redispersion.

It has already been demonstrated that biological spores with intricate 3D structures and abundant sources in nature can be employed as biotemplates and applied in biomedical fields (Mundargi et al. 2016a, b; Banerjee et al. 2014; Lu et al. 2004; Potroz et al. 2017). Therefore, *G. lucidum* spores are selected here to constitute the biohybrid microrobotic adsorbents. The scanning electron microscope (SEM) images in Fig. 11.2a show the morphology of intact spores to be droplet-like, and an average size of 6–10 μm is estimated. It can be found that the surface of the spore is rough and wrinkled, with many tiny pores (100–400 nm in diameter) distributed throughout the body (Fig. 11.2b). Figure 11.2c provides the SEM image of a broken spore, from which a hollow architecture with a double-walled layer can be confirmed. The surface area of such spore is measured as 6.7 m^2/g, which is not a satisfactory value for adsorbing process.

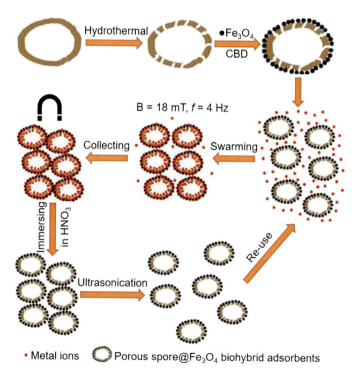

Fig. 11.1 Schematic of the fabrication process of PSFBAs and their application in heavy metal removal

310   11 Biohybrid Microswarm for the Removal of Toxic Heavy Metals

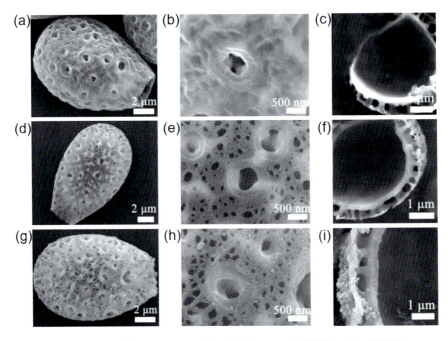

Fig. 11.2 SEM images of the product: **a–c** the original spores, **d–f** PHSs and **g–i** PSFBAs

Besides, the double wall of the spore is reported to contain a high content of nitrogen-containing chitin (about 60%) (Ma et al. 2007; Kumar 2000; Dursun et al. 2003). From the Fourier transform infrared (FTIR) spectra and non-crystalline peak in XRD patterns (Fig. 11.3a, b), the existence of carboxyl, hydroxyl, and nitrogen-based groups can be confirmed. The aforementioned structure and composition, together with the considerable negative surface charge of −29.0 mV (Fig. 11.3c), not only make the spores suitable for physical adsorption but also provide the possibility for further functionalization with magnetic nanomaterials.

However, the adsorption ability is still limited by the inherent low specific surface area of the spore. To address this problem, a hydrothermal method is used to treat the

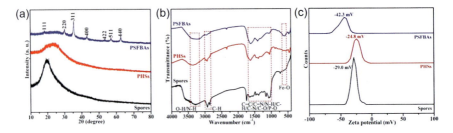

Fig. 11.3 **a** XRD patterns, **b** FTIR spectra, and **c** zeta potentials of the original spores, PHSs, and PSFBAs

spores, which is expected to increase the surface area. Figure 11.2d–f indicate that the droplet-like morphology of the spore is kept well after being heated at 180 °C for 24 h, while many new pores with different dimensions emerge, leading to the enhanced surface area up to 146.6 m^2/g. Besides, it can be found that the hydrothermally treated spores (PHSs) possess similar FTIR peaks with original spores, demonstrating the remaining abundant functional groups from G. lucidum spores (Bailey et al. 1999). Together with the stable negative surface charge, PHSs with porous structure and favorable functional groups would possess high adsorption capacity.

Subsequently, the magnetic property is endowed into PHSs via the immobilization of Fe$_3$O$_4$ nanoparticles, aiming to realize feasible operation in confined space and reutilization. As shown in Fig. 11.2g–l, the droplet-like morphology and porous structure remain after growing Fe$_3$O$_4$ nanoparticles on the surface. These magnetic nanoparticles with a diameter of 20–100 nm are uniformly coated on the wall of PHS samples. Finally, a magnetic layer with a thickness of ~140 nm is formed on the external surface. Characteristic peaks in the XRD pattern (Fig. 11.3a) confirm the existence of Fe$_3$O$_4$ phase in the biological spore. Besides, it can be found that the functional groups are well retained by comparing their FTIR spectra. It is worth noting that PSFBAs exhibit increased specific surface area and higher negative surface charge owing to the deprotonation of amino groups and protonation of carboxyl groups in alkaline ammonium solution. These features, together with the reported adsorption ability of Fe$_3$O$_4$ (Zhang 2003; Kumari et al. 2015), should endow the synthesized PSFBAs with excellent adsorption capability.

11.3 Controlled Locomotion of PSFBA Under the Guidance of Magnetic Field

Due to the good superparamagnetic property enabled by the functionalization of magnetic Fe$_3$O$_4$ nanoparticles (Fig. 11.4a), the obtained magnetic PSFBAs can controllably move in the fluid under the guidance of external magnetic fields. Such behavior can promote the mass diffusion process, improve the adsorption efficiency, as well as enable the PSFBAs to access hard-to-reach regions to perform tasks remotely. Figure 11.4b shows the moving trajectory of a single PSFBA over 13 s in contaminated water, indicating that a regular, continuous, and controllable motion is achieved via dynamic regulation of the magnetic field. Figure 11.4c presents the quantitative evaluation of average moving speed at different times in water and contaminated water. PSFBA shows a stable speed of ~28.2 m/s in the contaminated water as time goes on. Compared to that in pure water, a slight decrease by ~10% in speed occurs in the contaminated water, indicating the good stability of the motile microrobotic adsorbent. The magnetically driven motion also possesses excellent controllability achieved by adjusting the direction angle of the external rotating magnetic field. Therefore, the employment of magnetic actuation endows the PSFBA with the precise, controlled active motion for long-term applications, which is significantly

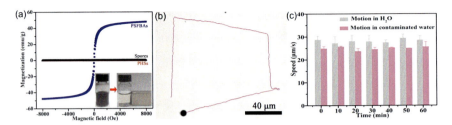

Fig. 11.4 **a** Magnetization curve of the original spores, PHSs, and PSFBAs at 300 K. **b** Controlled navigation of PSFBA in contaminated water. **c** Speed of PSFBA in water and contaminated water when moving for different times

superior to the environment-sensitive self-propelled microrobots (Xu et al. 2017; Zhao et al. 2013). When compared to the passive absorbents, the mobility of PSFBA also exhibits great advantages in enhanced diffusive performance, easy operation in confined surroundings, and facile recycling for environmental remediation.

Generally, a large amount of PSFBAs is needed to perform the actual remediation applications for massive removal due to the limited adsorption capability of individuals. The locomotion of individual motile PSFBA is not sufficient, and it is necessary to investigate the collective behavior control of PSFBAs. It has been previously reported that magnetic micro/nanorobots with swarming behaviors, like some living microorganisms (Liu et al. 2017), exhibit enhanced tasking performances than individual microrobots (Yu et al. 2017; Hong et al. 2018). In this respect, the collective behaviors of magnetic PSFBAs are studied for practical heavy metal removal applications. Figure 11.5a shows that thousands of magnetic PSFBAs are well gathered and move as an entity, verifying the feasibility of collective motion under magnetic control. In Fig. 11.5b, it can be found that the effective area of PSFBAs swarm changes under different magnetic parameters as time goes on. Higher magnetic field strength leads to a smaller area of collective microrobotic biohybrid adsorbents, while the rotating frequency plays a minor role in the swarm pattern. Besides, the translational speed of PSFBAs keeps stable with the time going or the area decreasing, indicating that aggregation does not affect the mobility of the swarm. In contrast, the magnetic field parameters determine the moving speed of swarming PSFBAs (Fig. 11.5c). Initially, the higher strength of the magnetic field induces higher locomotion velocity of collective PSFBAs with a maximal value of up to 266.3 m/s, which then keeps stable even if continuously increasing field strength. After finishing the manipulation, shrunken and aggregated PSFBAs swarm patterns can be redispersed via ultrasonication. Based on the aforementioned collective motion capability compared to the individual motile adsorbent, the PSFBAs swarm with a controllable motion may possess better adsorption and removal ability for heavy metal pollutants.

11.4 Enhanced Heavy Metal Removal Efficiency Enabled Flow Field Around ... 313

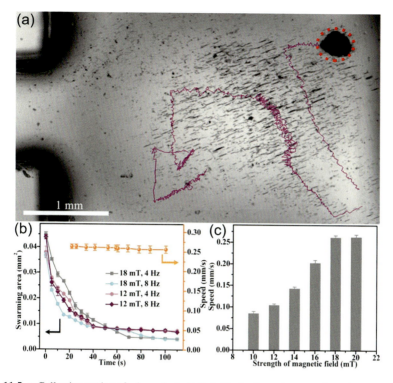

Fig. 11.5 **a** Collective motion of microrobotic biohybrid adsorbents in contaminated water under a rotating magnetic field. **b** Temporal change in shrinking area and speed of motile collective PSFBAs under different magnetic field parameters. **C** Effect of the magnetic field strength on the collective speed of PSFBAs

11.4 Enhanced Heavy Metal Removal Efficiency Enabled Flow Field Around PSFBAs Swarm

In this part, the fluid flow fields around motile PSFBAs are analyzed and simulated using the modules of rotating machinery and deformed geometry in COMSOL Multiphysics software. The motion of both the individual adsorbent and collective swarms is able to induce the directional flow and intermixing of surrounding fluids. Figure 11.6a presents the surrounding directional flow (indicated by white arrows) induced by the motion of individual PSFBA. The velocity can be as high as 900 μm/s, yet the flow field is limited to the small region near the microrobot. If the number of PSFBAs is increased (e.g., 9 in Fig. 11.6b), no obvious change in the maximum flow speed can be detected. In contrast, when these dispersed PSFBAs are gathered and controlled in a swarming mode, higher directional flow speed and larger rotating fluid flow area can be achieved, as shown in Fig. 11.6c. Furthermore, when thousands of PSFBAs are employed to constitute the swarm system as an entity, Fig. 11.6d indicates the flow speed induced by collective motion can be as high as 3500 μm/s,

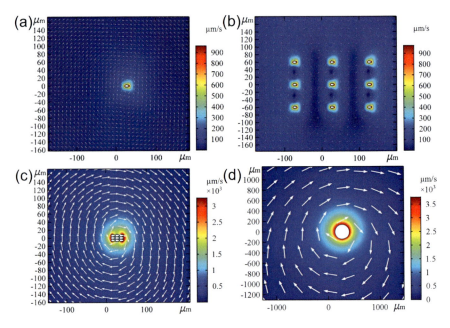

Fig. 11.6 Simulation results showing the fluid flow field near different amounts of motile PSFBAs. **a** Single motile PSFBA, **b** 9 motile PSFBAs, **c** Swarm of aggregated 9 PSFBAs, and **d** Swarm of thousands PSFBAs are used

which is far beyond those achieved by individual PSFBA and/or dispersed PSFBAs. Such high-flow speed and large flow field facilitate the fast diffusion and exchange of solutes near the swarm system, which may enable an efficient adsorption process of pollutants.

With the capability of controlled motion and generating surrounding flow, magnetic PSFBAs can be employed as swarming absorbents to remove heavy metals in the contaminated water, and the related proof-of-concept experiments are demonstrated in this part. The aqueous solution containing 5 ppm heavy metal ions without any additives is used as the model in all the removal experiments. Given the utilization of collective motion, the densities of the tested samples are set as 0.2 mg/mL in the contaminated solution. Figure 11.7a presents the removal ability in 60 min of static and swarming PSFBAs for various kinds of heavy metal ions, including Pb(II), Cd(II), Hg(II), Mn(II), Co(II), Ni(II), and ferric ions. It can be found that certain removal efficiency can be realized by magnetic PSFBAs in static or swarming modes, ranging from 27.3% to 48.3% (static) and from 29.2% to 85.2% (swarming), respectively. So, it is obvious that the collective motion of PSFBAs can enhance the removal efficiency by ~2–3 times for heavy metal ions except for ferric ions. The exception of ferric ions should be due to the adsorption saturation during the preparation process. To further investigate the improved removal performance of motile PSFBAs, a detailed experiment on the removal of Pb (II) ions is conducted. Figure 11.7b shows the results after navigating PSFBAs in the contaminated solution

11.4 Enhanced Heavy Metal Removal Efficiency Enabled Flow Field Around …

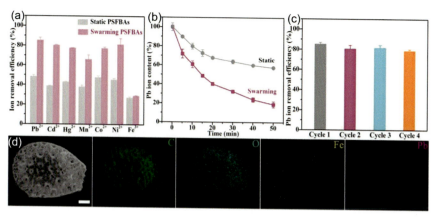

Fig. 11.7 Heavy metal removal efficiency of static and mobile PSFBAs. **a** Removal efficiency for different heavy metal ions. **b** Temporal lead removal efficiency. **c** Recycle collective PSFBAs after post-treatments. The error bars represent the standard deviation ($n = 3$). **d** SEM and EDX mapping of a single PSFBA after performing remediation, illustrating the elemental distribution of C, O, Fe, and Pb. The scale bar is 1 μm

for different times. If the collective PSFBAs are static in the contaminated solution, the concentration of lead decreases slowly, and an efficiency of 57.4% is obtained at 50 min. In comparison, more than 48.7% of pollutant is removed using swarming PSFBAs after 15 min. If the remediation time is further prolonged to 50 min, the lead concentration decreases to only 18.9% of the initial value, demonstrating a significantly improved removal efficiency due to the accelerated diffusion process and enhanced contacting chances (Vilela et al. 2016; Parmar et al. 2018).

Figure 11.7d presents the energy-dispersive X-ray (EDX) mapping of a single PSFBA and shows the distribution of various elements on the surface of the used PSFBA. Compared to the pristine PSFBA, the emergence of lead elements provides direct evidence for the adsorption of lead ions. Such adsorption ability should be ascribed to the inherited functional groups (i.e., carboxyl and amino) on the surface of spores (Yan and Viraraghavan 2003; Hu et al. 2005; Zhao et al. 2011). Besides, the morphology and structure of PSFBA are not damaged after performing decontamination tasks, which makes it possible to recycle the PSFBA by acid treatment for reutilization. The resultant removal efficiency in four removal cycles is shown in Fig. 11.7c. All cycles are conducted under the same conditions for 50 min, and the PSFBAs are refreshed by simple treatment with diluted nitric acid for 1 h. It is found that throughout four consecutive cycles, the reused PSFBA swarm always shows removal efficiency higher than 78%, which is comparable with the initial removal efficiency, suggesting the practical potential for heavy metal removal.

11.5 Experimental Section

The first step for preparing the building blocks is G. lucidum spore pretreatment, which is performed by a facile hydrothermal treatment of the original spores. 1 g of spores is first dispersed in 40 mL deionized (DI) water with ultrasonication. The resultant mixture is transferred to a 50 mL Teflon-lined stainless-steel autoclave, followed by heating at 180 °C for certain periods. After naturally cooled to room temperature, the product (i.e., PHS) is collected, washed with ethanol and DI water times by vacuum filtration, and then freezing-dried for further use.

The spores are then magnetized. A chemical deposition method is used to synthesize the PSFBAs. 0.2 g PHSs are dispersed in 60 mL DI water via ultrasonication and mechanical stirring for 5 min. Then, 0.27 g $FeSO_4$ is added and further stirred for 20 min, followed by a dropwise injection of 20 mL ammonia (25–27 wt%) in 10 min. The mixture is sealed and undergoes further stirring for 2 h. Finally, the product (i.e., PSFBAs) is collected, washed with ethanol and DI water using a permanent magnet, and freezing-dried for further use.

In the experiments of heavy metal ions removal using the as-prepared adsorbents, original spores, PHSs, or PSFBAs are added with the same concentration of 0.2 mg/mL into the aqueous solution contaminated by lead (Pb, II), cadmium (Cd, II), mercury (Hg, II), manganese (Mn, II), cobalt (Co, II), nickel (Ni, II), and ferric ions (5 ppm), respectively. The motile and static swarming PSFBAs are used to navigate in the contaminated solution for different times. Then the samples are separated using a filter membrane (220 nm in pore size), and the rest solution is subjected to ICP-MS to measure the concentration of heavy metal ions. Recycle of the PSFBAs is conducted by soaking the used samples in 0.5 M HNO_3 for 1 h.

11.6 Conclusion

In summary, a magnetic biohybrid adsorbent-based swarm system is introduced in this chapter. Through utilizing the porous structure and functional groups of *G. lucidum* spore, magnetic PSFBA is synthesized by hydrothermal treatment and chemical deposition for heavy metal adsorption. The obtained PSFBAs demonstrate improved removal efficiency than the original counterpart due to the introduction of multiscale pores, high-adsorbing components, and nanoparticles during the fabrication process. Besides, the controlled motion of collective formations disturbs the surrounding fluid and then accelerates the mass transfer process. The swarming PSFBAs are able to achieve rapid removal of Pb(II) in contaminated water within 15 min, and almost all the heavy metal ions can be removed if the remediation time is set to 50 min. Moreover, the used PSFBAs can be effectively recycled with ease by simple acid treatments. The motile biohybrid adsorbent introduced here provides a microrobotic platform for motion-enhanced heavy metal removal, holding promising potential in the field of environmental remediation.

References

Adeleye AS, Conway JR, Garner K, Huang YX, Su YM, Keller AA (2016) Engineered nanomaterials for water treatment and remediation: costs, benefits, and applicability. Chem Eng J 286:640–662

Aragay G, Pons J, Merkoci A (2011) Recent trends in macro-, micro-, and nanomaterial-based tools and strategies for heavy-metal detection. Chem Rev 111(5):3433–3458

Azizullah A, Khattak MNK, Richter P, Häder D-P (2011) Water pollution in Pakistan and its impact on public health—a review. Environ Int 37(2):479–497

Bailey SE, Olin TJ, Bricka RM, Adrian DD (1999) A review of potentially low-cost sorbents for heavy metals. Water Res 33(11):2469–2479

Banerjee J, Biswas S, Madhu NR, Karmakar SR, Biswas SJ (2014) A better understanding of pharmacological activities and uses of phytochemicals of lycopodium clavatum: a review. J Pharmacognosy Phytochem 3(1):207–210

Bilal M, Shah JA, Ashfaq T, Gardazi SMH, Tahir AA, Pervez A, Haroon H, Mahmood Q (2013) Waste biomass adsorbents for copper removal from industrial wastewater—a review. J Hazard Mater 263:322–333

Chen KS, Johnson DW (1986) Neutralization kinetics of bovine viral diarrhea virus by hyperimmune serum: one or multi-hit mechanism. Comp Immunol Microbiol Infect Dis 9(1):37–45

Dursun AY, Uslu G, Tepe O, Cuci Y, Ekiz HI (2003) A comparative investigation on the bioaccumulation of heavy metal ions by growing rhizopus arrhizus and aspergillus niger. Biochem Eng J 15(2):87–92

Ersahin ME, Ozgun H, Dereli RK, Ozturk I, Roest K, van Lier JB (2012) A review on dynamic membrane filtration: materials, applications and future perspectives. Biores Technol 122:196–206

Fischer P, Ghosh A (2011) Magnetically actuated propulsion at low reynolds numbers: Towards nanoscale control. Nanoscale 3(2):557–563

Fu FL, Wang Q (2011) Removal of heavy metal ions from wastewaters: a review. J Environ Manage 92(3):407–418

Gupta VK, Nayak A, Agarwal S (2015) Bioadsorbents for remediation of heavy metals: current status and their future prospects. Environ Eng Res 20:1–18

Hong SH, Li Y, Eom JB, Choi Y (2018) Responsive alginate-cisplatin nanogels for selective imaging and combined chemo/radio therapy of proliferating macrophages. Quant Imaging Med Surg 8(8):733–742

Hu J, Chen GH, Lo IMC (2005) Removal and recovery of Cr(VI) from wastewater by maghemite nanoparticles. Water Res 39(18):4528–4536

Jaishankar M, Tseten T, Anbalagan N, Mathew BB, Beeregowda KN (2014) Toxicity, mechanism and health effects of some heavy metals. Interdiscip Toxicol 7(2):60

Jurado-Sanchez B, Wang J (2018) Micromotors for environmental applications: a review. Environ Sci-Nano 5(7):1530–1544

Jurado-Sanchez B, Sattayasamitsathit S, Gao W, Santos L, Fedorak Y, Singh VV, Orozco J, Galarnyk M, Wang J (2015) Self-propelled activated carbon janus micromotors for efficient water purification. Small 11(4):499–506

Kavamura VN, Esposito E (2010) Biotechnological strategies applied to the decontamination of soils polluted with heavy metals. Biotechnol Adv 28:61–69

Khin MM, Nair AS, Babu VJ, Murugan R, Ramakrishna S (2012) A review on nanomaterials for environmental remediation. Energy Environ Sci 5(8):8075–8109

Kumar M (2000) A review of chitin and chitosan applications. React Funct Polym 46(1):1–27

Kumari M, Pittman CU, Mohan D (2015) Heavy metals [chromium (VI) and lead (II)] removal from water using mesoporous magnetite (Fe_3O_4) nanospheres. J Colloid Interface Sci 442:120–132

Li ZY, Ma ZW, van der Kuijp TJ, Yuan ZW, Huang L (2014) A review of soil heavy metal pollution from mines in China: pollution and health risk assessment. Sci Total Environ 468:843–853

Li RY, Zhang LB, Wang P (2015) Rational design of nanomaterials for water treatment. Nanoscale 7(41):17167–17194

Li JX, de Avila BEF, Gao W, Zhang LF, Wang J (2017) Micro/nanorobots for biomedicine: delivery, surgery, sensing, and detoxification. Sci Robot 2(4):eaam6431

Liu C, Xu T, Xu L-P, Zhang X (2017) Controllable swarming and assembly of micro/nanomachines. Micromachines 9(1):10

Lu F, Astruc D (2018) Nanomaterials for removal of toxic elements from water. Coord Chem Rev 356:147–164

Lu QY, Sartippour MR, Brooks MN, Zhang QF, Hardy M, Go VL, Li FP, Heber D (2004) Ganoderma lucidum spore extract inhibits endothelial and breast cancer cells in vitro. Oncol Rep 12(3):659–662

Ma JJ, Fu ZY, Ma PY, Su YL, Zhang QJ (2007) Breaking and characteristics of ganoderma lucidum spores by high speed entrifugal shearing pulverizer. J Wuhan Univ Technol-Mater Sci Edition 22(4):617–621

Moo JGS, Pumera M (2015) Chemical energy powered nano/micro/macromotors and the environment. Chem Euro J 21(1):58–72

Moo JGS, Wang H, Zhao GJ, Pumera M (2014) Biomimetic artificial inorganic enzyme-free self-propelled microfish robot for selective detection of Pb^{2+} in water. Chem-A Euro J 20(15):4292–4296

Mundargi RC, Potroz MG, Park S, Park JH, Shirahama H, Lee JH, Seo J, Cho NJ (2016a) Lycopodium spores: a naturally manufactured, superrobust biomaterial for drug delivery. Adv Func Mater 26(4):487–497

Mundargi RC, Potroz MG, Park S, Shirahama H, Lee JH, Seo J, Cho NJ (2016b) Natural sunflower pollen as a drug delivery vehicle. Small 12(9):1167–1173

Parmar J, Vilela D, Villa K, Wang J, Sanchez S (2018) Micro- and nanomotors as active environmental microcleaners and sensors. J Am Chem Soc 140(30):9317–9331

Potroz MG, Mundargi RC, Gillissen JJ, Tan EL, Meker S, Park JH, Jung H, Park S, Cho D, Bang SI, Cho NJ (2017) Plant-based hollow microcapsules for oral delivery applications: Toward optimized loading and controlled release. Adv Func Mater 27(31):1700270

Saha B, Orvig C (2010) Biosorbents for hexavalent chromium elimination from industrial and municipal effluents. Coord Chem Rev 254(23–24):2959–2972

Santhosh C, Velmurugan V, Jacob G, Jeong SK, Grace AN, Bhatnagar A (2016) Role of nanomaterials in water treatment applications: a review. Chem Eng J 306:1116–1137

Vilela D, Parmar J, Zeng YF, Zhao YL, Sanchez S (2016) Graphene-based microbots for toxic heavy metal removal and recovery from water. Nano Lett 16(4):2860–2866

Wang JL, Chen C (2009) Biosorbents for heavy metals removal and their future. Biotechnol Adv 27(2):195–226

Wang H, Khezri B, Pumera M (2016) Catalytic DNA-functionalized self-propelled micromachines for environmental remediation. Chem 1(3):473–481

Wang H, Potroz MG, Jackman JA, Khezri B, Maric T, Cho NJ, Pumera M (2017) Bioinspired spiky micromotors based on sporopollenin exine capsules. Adv Func Mater 27(32):1702338

Wang B, Chan KF, Yu JF, Wang QQ, Yang LD, Chiu PWY, Zhang L (2018) Reconfigurable swarms of ferromagnetic colloids forenhanced local hyperthermia. Adv Func Mater 28(25):1705701

Xu TL, Gao W, Xu LP, Zhang XJ, Wang ST (2017) Fuel-free synthetic micro-/nanomachines. Adv Mater 29(9):1603250

Yan GY, Viraraghavan T (2003) Heavy-metal removal from aqueous solution by fungus mucor rouxii. Water Res 37(18):4486–4496

Yan XH, Zhou Q, Yu JF, Xu TT, Deng Y, Tang T, Feng Q, Bian LM, Zhang Y, Ferreira A, Zhang L (2015) Magnetite nanostructured porous hollow helical microswimmers for targeted delivery. Adv Func Mater 25(33):5333–5342

Yan XH, Zhou Q, Vincent M, Deng Y, Yu JF, Xu JB, Xu TT, Tang T, Bian LM, Wang YXJ, Kostarelos K, Zhang L (2017) Multifunctional biohybrid magnetite microrobots for imaging-guided therapy. Sci Robot 2(12):eaaq1155

References

Yu JF, Xu TT, Lu ZY, Vong CI, Zhang L (2017) On-demand disassembly of paramagnetic nanoparticle chains for microrobotic cargo delivery. IEEE Trans Rob 33(5):1213–1225

Yu JF, Wang B, Du XZ, Wang QQ, Zhang L (2018) Ultra-extensible ribbon-like magnetic microswarm. Nat Commun 9:3260

Zhang WX (2003) Nanoscale iron particles for environmental remediation: an overview. J Nanopart Res 5(3–4):323–332

Zhang YB, Chan KF, Wang B, Chiu PWY, Zhang L (2018) Spore-derived color-tunable multi-doped carbon nanodots as sensitive nanosensors and intracellular imaging agents. Sens Actuators B-Chem 271:128–136

Zhao GX, Ren XM, Gao X, Tan XL, Li JX, Chen CL, Huang YY, Wang XK (2011) Removal of Pb(II) ions from aqueous solutions on few-layered graphene oxide nanosheets. Dalton Trans 40(41):10945–10952

Zhao GJ, Viehrig M, Pumera M (2013) Challenges of the movement of catalytic micromotors in blood. Lab Chip 13(10):1930–1936

Chapter 12
Ant Bridge-Mimicked Reconfigurable Microswarm for Electronic Application

Abstract This chapter describes mimicking both the structure and function of ant bridges by using the magnetic ribbon-like microswarm for electronic applications. Fe_3O_4 nanoparticles coated with dense gold surface layers form ribbon-like microswarms under the actuation of input oscillating magnetic fields. The reversible ultra-extensible ability of ribbon-like microswarm allows them to be applied to the electronics field, such as connecting two isolated electrodes, serving as microswitches, and repairing broken microcircuits.

Keywords Collective behaviors · Ant bridge · Magnetic microswarm · Surface functionalization · Microelectronics

12.1 Introduction

Swarm behaviors widely exist in natural insects and are indispensable for their living activities, such as foraging, feeding, constructing nests, struggling with predators and harsh environments, and so on (Hölldobler and Wilson 1990; Mlot et al. 2011). Through local communications, insect colonies (i.e., ants, bees, etc.) can self-organize into a variety of sophisticated structures in a collective manner. For example, when encountering a gap that cannot be crossed by a single ant, a group of ants would form a flexible and robust bridge-like structure by gripping the body of each other to march across difficult terrains. Similarly, ants are capable of self-assembling into a ball-like living structure through body connections, aiming to protect themselves in response to fire and flood (Anderson et al. 2002). Such natural swarm behaviors exhibit group-level functionality that cannot be achieved by individuals and thus are fascinating to scientists, especially researchers in the robotic field.

To date, taking inspiration from social insects, various artificial swarm robots with collective abilities have been created. Through local interactions between building blocks, swam robots cooperate with each other to coordinatively accomplish complex tasks far beyond the capability of a single building block, including task allocation and sequencing (Garattoni and Birattari 2018; Krieger et al. 2000), building construction

(Werfel et al. 2014; Gauci et al. 2014), on-demand shape transformation (Rubenstein et al. 2014; Mathews et al. 2017) and so on (Yang et al. 2018; O'Grady et al. 2010; Halloy et al. 2007). If the building blocks are downscaled to micro/nano size, the corresponding microswarms may possess the capability of entering the narrow and closed environments inside the human body, thus exhibiting great potential in various engineering fields (Felfoul et al. 2016; Martel and Mohammadi 2010; Gracias et al. 2000). However, as the functions of building blocks are severely limited by their small size, new controlling and application strategies of microswarm different from traditional robotic swarms are required. As mentioned in previous chapters, the colloid is an ideal model to investigate the underlying principles of collective behaviors of microswarms (Zhang et al. 2017). Through modulating the fabrication process, size, surface properties, as well as mutual interactions of colloidal particles (e.g., dipolar interactions (Lu and Tang 2016; Yan et al. 2012; Yu et al. 2018a; Snezhko and Aranson 2011), capillary forces (Breen et al. 1999), van der Waals forces (Gobre and Tkatchenko 2013), etc. (Miele et al. 2017)), a variety of collective systems have been established via static or dynamic self-assembly (Palacci et al. 2013; Cademartiri and Bishop 2015; Whitesides and Grzybowski 2002; Mao et al. 2013; Chen et al. 2011). Currently, the basic principles are still under investigation, making it challenging to emulate the group-level swarm behaviors of social insects.

In this chapter, using the ribbon-like magnetic colloid microswarm introduced in Chap. 3 (Yu et al. 2018b), we have further designed a functionalized microswarm that emulates the collective behaviors of ant bridges for electronic applications (Fig. 12.1a). The building blocks are Fe_3O_4 nanoparticles coated with a compact gold surface layer so that they are both paramagnetic and electrically conductive. When an external oscillating magnetic field is applied, the functionalized building blocks can be assembled into a microswarm and navigated between two disconnected electrodes. As shown in Fig. 12.1b, by transforming the microswarm into a chain-like structure with an ultrahigh aspect ratio, an electrically conductive pathway is constructed between electrodes using the bodies of building blocks, just resembling the ant bridge composed of physically connected ants for march and transportation. We demonstrate that such microswarm can be used in electronic fields, such as microswitch, microcircuit repairing, and flexible electronics, which possess the advantages of excellent controllability and high precision.

12.2 Design and Functionalization of Building Blocks

To realize the electronic applications of microswarm, the components of building blocks should be both magnetic and electrically conductive. To achieve this goal, Fe_3O_4 magnetic nanoparticles are selected and coated with a continuous and compact gold surface layer. The detailed synthesis procedures are shown in Fig. 12.2. We first use the solvothermal method (Deng et al. 2005) to prepare spherical Fe_3O_4 nanoparticles as the core of building blocks. The nanoparticles are subsequently coated with

12.2 Design and Functionalization of Building Blocks 323

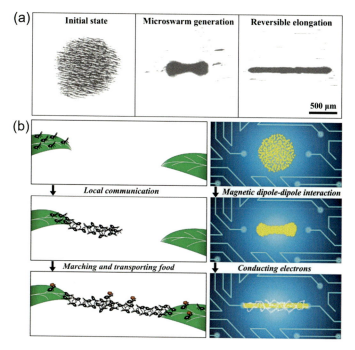

Fig. 12.1 Schematics of ant bridge-mimicked magnetic colloid microswarm for electronic applications. **a** Reconfigurable microswarm under the actuation of an oscillating magnetic field. **b** The formation procedures of ant bridge and magnetic microswarm for ants marching and electrons conduction, respectively

a polydopamine (PDA) layer on their surface (Fe$_3$O$_4$@PDA) using *in-situ* polymerization (Lee et al. 2007), followed by electrostatic adsorption of oppositely-charged gold nanoparticles (Fe$_3$O$_4$@PDA@Seeds). Then, employing the immobilized Au nanoparticles as seeds, a compact gold layer is grown on the nanoparticles (Fe$_3$O$_4$@PDA@Au) via electroless plating (Choi et al. 2017), leading to the fabrication of desired building blocks.

Scanning electron microscope (SEM) and transmission electron microscope (TEM) are first used to characterize the morphology of obtained building blocks. As shown in Fig. 12.3a, all of the nanoparticles at each fabrication process exhibit regular spherical shapes. Quantitative statistics are conducted and show that Fe$_3$O$_4$,

Fig. 12.2 Schematics of the fabrication process of building blocks

Fe$_3$O$_4$@PDA, Fe$_3$O$_4$@PDA@Seeds, and Fe$_3$O$_4$@PDA@Au nanoparticles have narrow size distribution with average diameters of ~660, 675, 670, and 675 nm, respectively, indicating that the variation in diameters is not significant during surface functionalization. In contrast, an obvious change in surface morphology is found in Fig. 12.3b. In comparison to pristine Fe$_3$O$_4$, a smooth PDA layer as thin as 8 nm is uniformly coated for Fe$_3$O$_4$@PDA, which has a positive charge of 19.3 mV in an acidic condition (pH value ~2.0), as shown in Fig. 12.4a. When mixed with negatively-charges Au seeds with a negative zeta potential of −36.4 mV (Fig. 12.4b), the Fe$_3$O$_4$@PDA nanoparticles would absorb Au seeds via efficient electrostatic self-assembly, which is verified by the rough surface morphology of Fe$_3$O$_4$@PDA@Seeds. Finally, for the Fe$_3$O$_4$@PDA@Au (i.e., the building block), incubating them in gold electroless plating solution with the appropriate amount of reductive agent facilitates the growth of a compact gold layer on the surface, thus making the TEM images totally opaque.

Besides, energy dispersive X-ray (EDX) analysis is conducted to demonstrate the elemental distribution of Au, Fe, and O in the obtained building blocks (Fig. 12.5a), which confirms the distribution of gold as a uniform layer on the surface of Fe$_3$O$_4$ nanoparticle. Then vibrating sample magnetometer (VSM) test is further performed to characterize the magnetic property of building blocks at each fabrication process, as shown in Fig. 12.5b. The synthesized nanoparticles are paramagnetic as the procedure goes on, but their saturation magnetization gradually decreases from 84.3 emu/g for

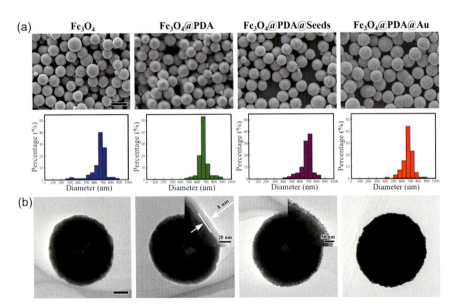

Fig. 12.3 **a** SEM images of building blocks at each fabrication stage. The statistic results are obtained by measuring the diameters of 100 nanoparticles in the corresponding SEM images. The scale bar is 1 μm. **B** TEM images of building blocks at each fabrication stage with magnified surface morphology shown in the insets. The scale bar is 200 nm

12.2 Design and Functionalization of Building Blocks

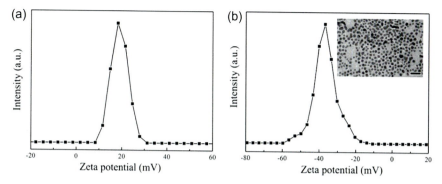

Fig. 12.4 **a** Zeta potential of Fe$_3$O$_4$@PDA when pH = 2. **b** Zeta potential of Au seeds with their TEM images shown in the insets. The scale bar is 20 nm

pristine Fe$_3$O$_4$ to 80.1, 75.0, and 57.1 emu/g for Fe$_3$O$_4$@PDA, Fe$_3$O$_4$@PDA@Seeds, and Fe$_3$O$_4$@PDA@Au, respectively. The main reason for such value reduction is due to the decreased mass ratio of magnetic material in nanoparticles. However, as shown in the inset, the obtained building blocks still possess a strong response to a permanent magnet, making them feasible to be controlled by the external magnetic field.

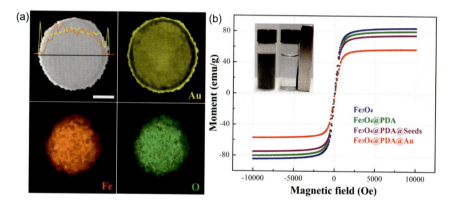

Fig. 12.5 **a** EDX analysis of building block with the yellow and orange curves indicating the lateral composition variation of Au and Fe elements, respectively. The scale bar is 200 nm. **b** Magnetic hysteresis loop of building blocks during each fabrication process. The inset shows that building blocks can be attracted by a permanent magnet

12.3 Swarm Control of the Magnetic and Electrically Conductive Building Blocks

An oscillating magnetic field is applied to wirelessly actuate the collective behaviors of building blocks. The magnetic field **B** is composed of two components, i.e., a constant field vector with the field strength of C, and a perpendicular sinusoidally alternating field vector $A\sin(2\pi ft)$, where A is amplitude (fixed as 10 mT), f is the oscillating frequency, and t is time (see Fig. 3.1). The amplitude ratio between two vectors is denoted as $\gamma = A/C$. The mechanisms have already been introduced in Chap. 3, and the generation effects of magnetic microswarm depend on the input magnetic field parameters, which are depicted as a phase diagram in Fig. 12.6. Before the application of a magnetic field, the building blocks are initially dispersed on the substrate. When the magnetic field is on, if γ is too low (region I), a large number of small chain-like structures would be generated, which oscillate with the variation of the external magnetic field. If γ is too high (region III), several irregular ribbon-like long patterns would be formed, which however, cannot be well controlled. When γ and f with optimized collocations in region II are imposed, the ribbon-like microswarm with a dynamically stable pattern can be successfully generated.

The morphology of the microswarm can be further adjusted by changing the values of γ and f in region II. As shown in Fig. 12.7, when γ is 4 and f is 10 Hz, a relatively thick and short pattern with an aspect ratio of 3.2 is formed. Elevating the values of γ and f to 9 and 30, respectively, makes the pattern of the microswarm become thinner and longer, and an aspect ratio as high as 45.3 can be achieved. Then decreasing γ and f to their initial values once again would make the microswarm recover to its original state, indicating that the elongating and shrinking pattern transformation of the microswarm is reversible and can be fully controlled by the magnetic field.

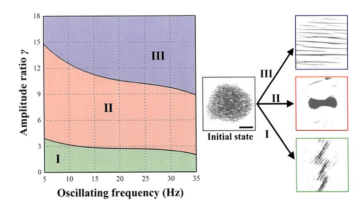

Fig. 12.6 Phase diagram of the magnetic ribbon-like microswarm under oscillating magnetic fields with different oscillating frequencies and amplitude ratios. The scale bar is 500 μm

12.3 Swarm Control of the Magnetic and Electrically Conductive Building ...

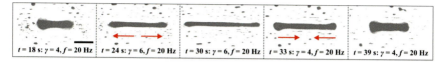

Fig. 12.7 Reversible transformation of microswarm under different amplitude ratios. The red arrows indicate the transformation direction. The scale bar is 500 μm

Besides, the states of building blocks, including concentration and diameter, also play important effects on the collective behaviors of microswarm. As shown in Fig. 12.8, 3 μL nanoparticle solution with different concentrations (1, 2, and 4 mg/mL) is added to generate the ribbon-like microswarm. Microswarm with more building blocks is obviously longer than that with fewer building blocks when the same magnetic fields are applied. In contrast, the width of the microswarm does not change obviously when the concentration of building blocks is increased. For example, if $\gamma = 6$ and $f = 20$ Hz, the length of microswarms is ~1200, 2050, and 3300 μm, respectively, when the concentrations are 1, 2, and 4 mg/mL. At the same time, the corresponding width is 130, 140, and 140 μm, respectively. Therefore, it could be concluded that increasing the concentration of building blocks would elongate the microswarm while keeping its width.

Three kinds of magnetic nanoparticles with different diameters are prepared by modifying the process of the solvothermal method and are employed as the building blocks to investigate the size effect on collective behaviors. From the statistical

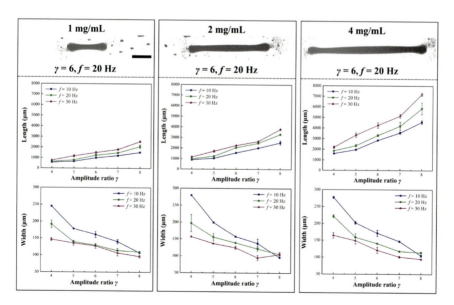

Fig. 12.8 Effect of the building block concentrations on the dimensions of microswarm. The error bars represent the standard deviation ($n = 3$). The scale bar is 500 μm

analysis in Fig. 12.9, the average diameters of small, medium, and large nanoparticles are 135, 400, and 660 nm, respectively. Then the programmed oscillating magnetic field is applied, and their phase diagrams are provided. It can be found that with the increase in the size of building blocks, the generation condition of microswarm does not change too much, and the effective region II only exhibits a slight global shift toward the area with a higher amplitude ratio. The influences of the input field parameters on the morphology of microswarm for all kinds of building blocks are evaluated in Fig. 12.10a. The aspect ratio of microswarms exhibits the same variation tendency when changing the magnetic field, even though the diameters of building blocks are different. However, the absolute values of the aspect ratio are different. As shown in Fig. 12.10b, it can be found that if the parameters of the magnetic field are the same, increasing the size of building blocks will decrease the aspect ratio of the microswarm, which means making the microswarm relatively shorter and thicker.

After generating a stable ribbon-like pattern, the locomotion of microswarms can be actuated by tuning the pitch angle of the magnetic field. Figure 12.11a exhibits the relationship between the moving velocity v of microswarm and pitch angle β. Microswarm remains at the same position when β is 0°. Its motion would be initiated and then accelerated by gradually increasing the value of β. When β and f are set as 4° and 30 Hz, respectively, the microswarm achieves a translational velocity as

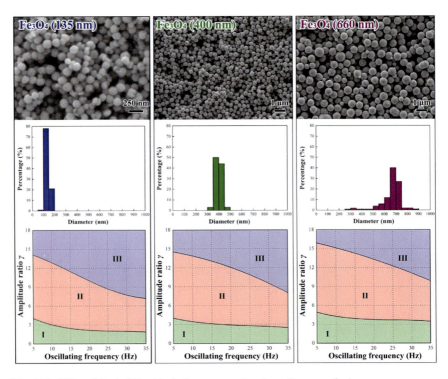

Fig. 12.9 Effect of the building block diameters on the phase diagrams of microswarms

12.3 Swarm Control of the Magnetic and Electrically Conductive Building ...

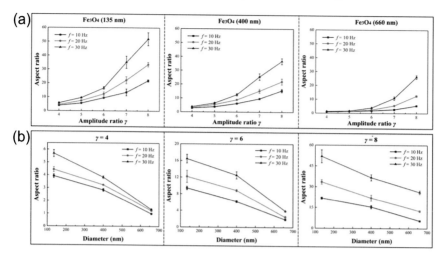

Fig. 12.10 Effect of the building block diameters on the aspect ratio of microswarm. The error bars represent the standard deviation ($n = 3$)

high as ~180 μm/s, indicating effective movement. In contrast, decreasing the pitch angle from 0° to −4° would trigger the locomotion in the opposite direction, but the speed is comparable. When $|\beta|$ is beyond 4°, the microswarm cannot maintain its pattern anymore. Therefore, by adjusting the direction angle and pitch angle of the magnetic field, the magnetic microswarm can be guided to navigate along any desired trajectories, such as the square route shown in Fig. 12.11b, indicating excellent motion controllability.

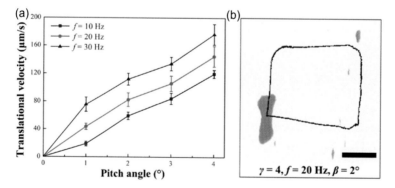

Fig. 12.11 a Effect of pitch angle and oscillating frequency on the velocity of ribbon-like microswarm when amplitude ratio γ is 4. **b** Navigation of microswarm along a square trajectory with the blue line indicating the motion path. The error bars represent the standard deviation ($n = 3$), and the scale bar is 500 μm

12.4 Construction of Ant Bridge-Mimicked Magnetic Microswarm for Electrical Connection

After establishing the magnetic control strategy of the electrically conductive building blocks, the ant bridge-mimicked microswarm is applied to self-organize a conductive pathway for electrons between two disconnected electrodes (Fig. 12.12a). The building blocks with polyvinylpyrrolidone solution are added around the electrodes, and then an oscillating magnetic field with appropriate parameters ($\gamma = 4, f = 20$ Hz) is imposed to trigger the generation of ribbon-like microswarm. Through dynamically adjusting the direction angle α and pitch angle β of the magnetic field, the microswarm is guided to the middle position between two disconnected electrodes and then aligned with the horizontal direction. Subsequently, the amplitude ratio γ is gradually increased to trigger the elongation of the microswarm until its two terminals come into contact with the electrodes. Finally, excess polyvinylpyrrolidone solution is removed, followed by drying in the air. In this manner, a chain-like pattern assembled by building blocks for electron transportation is formed between two disconnected electrodes, which mimics both the structure and function of an ant bridge.

Figure 12.12b shows the temporal variation of electrical resistance between two disconnected electrodes during the construction of an ant bridge-mimicked microswarm. The resistance is too large to be measured initially, and it changes to ~110 kΩ once suspension of building blocks is added. After the ribbon-like microswarm is successfully constructed to connect electrodes, a slight decrease in the electrical resistance to ~90 kΩ occurs, mainly due to the existence of poorly conductive water droplets between the building blocks (Li et al. 2015). Therefore, after water is completely evaporated during the natural drying process (~20 min), the resistance dramatically decreases to a minimum value as low as 50 Ω, indicating the successful electrical connection between electrodes. In this respect, the densely packed building blocks in microswarm provide sufficient and direct physical contact with each other and thus form a conductive network for electrons based on their gold surface layer (Fig. 12.13).

Besides, control experiments are also performed by using Fe_3O_4, Fe_3O_4@PDA, and Fe_3O_4@PDA@Seeds nanoparticles as the building blocks. In Fig. 12.12c, two electrodes are not electrically conductive enough with a resistance of 0.7 MΩ when using Fe_3O_4 and Fe_3O_4@PDA to constitute microswarm, while the resistance decreased to ~10kΩ for Fe_3O_4@PDA@Seeds, suggesting that the immobilization of dispersed gold nanoparticles on iron oxide nanoparticle is insufficient to improve the inferior conductivity of pristine building blocks to an acceptable level. In contrast, after a continuous and compact gold layer is coated, the resistance of the microswarm can be as low as 20Ω, demonstrating the crucial role of gold coating functionalization. Moreover, from the inset figure, it can be found that the electrical resistance of the microswarm is approximately linearly related to its aspect ratio, which suggests that we are able to not only achieve an electrical connection between two electrodes but also customize the corresponding resistance by tuning the magnetic field parameters.

12.5 Applications of Microswarm in Electronic Field

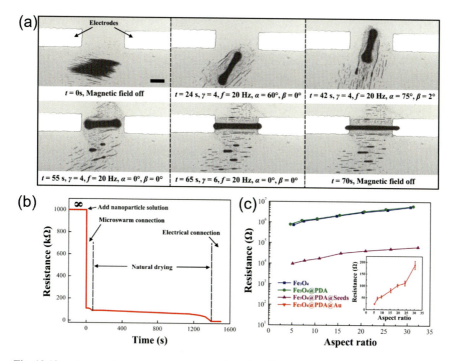

Fig. 12.12 **a** Establishment process of the ribbon-like microswarm between two isolated electrodes with the input magnetic field parameters. The scale bar is 500 μm. **b** Resistance variation between electrodes as a function of time. **c** Effect of microswarm aspect ratio on the resistance when using Fe_3O_4, Fe_3O_4@PDA, Fe_3O_4@PDA@Seeds, and Fe_3O_4@PDA@Au as building blocks, respectively. The error bars represent the standard deviation ($n = 3$)

12.5 Applications of Microswarm in Electronic Field

The excellent collective behavior controllability and capability in electrical connection facilitate microswarm to be a promising tool for various microelectronic devices. For example, the ribbon-like microswarm is able to act as a microswitch for a four-way microcircuit. At first, all circuits are disconnected, and the "C", "U", "H" and "K" LED arrays are off. Then an oscillating magnetic field is applied to generate the ribbon-like microswarm on the substrate. Through tuning the magnetic field parameters, the microswarm can be navigated to connect two specific isolated electrodes in an intuitive manner, leading to the lightening of corresponding LED arrays, as shown in Fig. 12.14. Besides, after erasing the microswarm by sonication in ethanol, the microswitch can be reset to an open state, which verifies the ability of repeatable switching on and off.

Besides, the ribbon-like microswarm can be used to repair broken microcircuits, as shown in Fig. 12.15a. Compared to the traditional manual soldering method, which suffers from relatively low precision and high risk in short connection, the

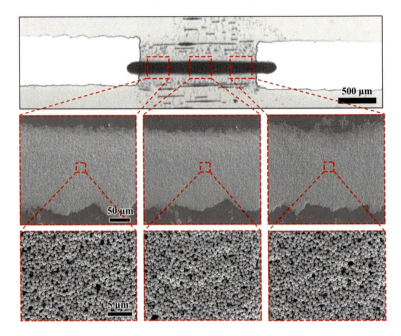

Fig. 12.13 Microstructures of the ribbon-like swarm. After drying, the microswarm-based wire still has a uniform chain-like structure. The average diameter of building blocks is ~675 nm and thus three orders of magnitude lower than the distance between two electrodes (~1.5 mm), indicating electrical connection across a gap far beyond the reach of individuals has been realized by swarm behaviors

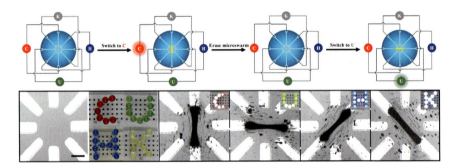

Fig. 12.14 Schematic and experimental demonstration of the microswarm-based microswitch. The scale bar is 500 μm

mending process using magnetic microswarm is composed of three steps, i.e., dropping building block suspension, generating and navigating microswarm by applying the programmed magnetic field, and removing and drying excess solution. Such a strategy is quite straightforward yet effective, and a broken wire array with an interval distance as low as ~200 μm can be well connected. Figure 12.15b–d present

12.6 Experimental Section

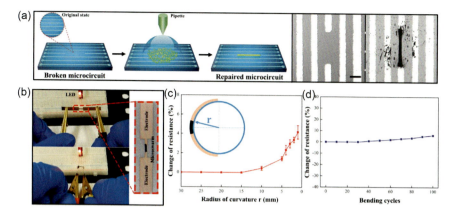

Fig. 12.15 **a** Schematic and experimental demonstration of microswarm for broken microcircuit repair. The scale bar is 250 μm. **b** Application of microswarm for flexible electronics. **c** Effect of bending curvature on the electrical resistance of microswarm with the inset showing the schematic of radius r. The error bars represent the standard deviation ($n = 3$). **d** The stability of the microswarm in electrical conductivity after being bent 100 times

the effectiveness of microswarm in flexible devices. A polyethylene terephthalate film is selected as the flexible circuit substrate, and two disconnected copper-based electrodes are patterned on it. Then microswarm is generated on the film using the same method described above, followed by natural drying. In this manner, the ribbon-like microswarm can stick firmly to the film, and even manually bending the substrate would not destroy the formed chain-like structure. The effects of bending degree on the electrical resistance of the microswarm are quantitively evaluated. Although increasing the deformation of the substrate deteriorates the conductivity, the microswarm still shows resistance as low as 5.9 Ω when the bending radius is 1.0 mm. Moreover, from the anti-fatigue investigation, only 5.9% increase in resistance is caused by bending the film to a radius of 1.0 mm for 100 times. It is worth noting that the circuit connected by the microswarm can maintain its function for at least one month without significant degradation, demonstrating the excellent stability of the electrical connection enabled by the microswarm. Such performance is especially important for flexible electronic devices, endowing the ant bridge mimicked microswarm with great potential in wearable devices, electronic skin, and implantable electrodes (Wang et al. 2018).

12.6 Experimental Section

The preparation of building blocks starts with fabricating Fe_3O_4 nanoparticles with a solvothermal method and then employing an *in-situ* polymerization process to coat a thin and smooth PDA layer on the surface of Fe_3O_4 nanoparticles, which is introduced in Sect. 4.6.1 in detail. After Fe_3O_4@PDA nanoparticles are obtained, an electrostatic self-assembly process is performed to immobilize gold seeds on the surface of Fe_3O_4@PDA nanoparticles. 0.158 g $C_6H_8O_7 \cdot H_2O$ is dissolved in 50 mL DI water under sonication, and the pH is adjusted to 2.0 using 1 M HCl. 10 mg Fe_3O_4@PDA nanoparticle is then added, followed by mechanical stirring and sonication for 15 min, aiming to completely disperse the nanoparticles and tune the surface charge of Fe_3O_4@PDA to positive. Then 600 μL $HAuCl_4 \cdot 3H_2O$ solution (1.0 wt.%) and 600 μL $C_6H_5Na_3O_7 \cdot 2H_2O$ solution (25 mM) are mixed with 58.8 mL DI water, and stirred for 10 min, followed by the addition of 900 μL freshly prepared $NaBH_4$ solution (200 mM) to prepare gold nanoparticles with a negative charge. Next, the suspensions of gold and Fe_3O_4@PDA nanoparticles are quickly mixed, which further undergo mechanical stirring and sonication for 30 min. The functionalized nanoparticles are washed with DI water and collected by a permanent magnet. Finally, the nanoparticles are dispersed in DI water with a concentration of 10 mg/mL for further use.

In the last step, a modified electroless plating process is used to grow a continuous gold surface layer on nanoparticles. Typically, 0.0125 g K_2CO_3 and 750 μL $HAuCl_4 \cdot 3H_2O$ solution (1.0 wt.%) are dissolved in 50 mL DI water to obtain the gold plating solution. 10 mg Fe_3O_4@PDA@Seeds nanoparticle is then added, followed by mechanical stirring and sonication for 15 min, aiming to completely disperse the nanoparticles. Afterwards, 250 μL formaldehyde is added dropwise to trigger the growth of gold seeds, followed by mechanical stirring and sonication for 10 min. At this time, 500 μL mercaptosuccinic acid solution (25 mM) is added to avoid the aggregation of nanoparticles. The functionalized nanoparticles are washed with DI water and collected by a permanent magnet. Finally, the nanoparticles are dispersed in DI water with a concentration of 2 mg/mL for generating the ribbon-like swarms.

In the Magnetic actuation experiments, 3 μL suspension of building blocks (Fe_3O_4@PDA@Au nanoparticles) is added into an open tank filled with 1.5 mL 0.1 wt.% polyvinylpyrrolidone ($M_w = 1,300,000$) solution. Then a permanent magnet is used to concentrate the building blocks, followed by the insertion of tank into a customized 3-axis Helmholtz electromagnetic coil setup. For the electrical connection experiments, building blocks with polyvinylpyrrolidone solution are directly dropped on the substrate patterned with electrodes. Finally, an oscillating magnetic field is applied to generate and control the swarm behaviors.

A mask-assisted physical vapor deposition process is used to pattern copper-based electrodes on substrate for Electrode fabrication. A commercial laser cutting platform (Universal Laser Systems PLS6.75, USA) is first utilized to engrave the mask with desired geometry in an A4 paper. After adhering the mask to substrate, an e-beam evaporator (EB-600, Innovative Vacuum Solution Co., Ltd., Taiwan, China) is employed to sequentially deposit 20 nm Ti and 100 nm Cu. Finally, the mask is torn off, and various substrates with patterned electrodes are obtained.

12.7 Conclusion

In conclusion, a microswarm system capable of forming an electrically conductive pathway across a gap far beyond the reach of individuals is developed in this chapter to emulate the structure and function of ant bridges. Core–shell structured building block with a magnetite nanoparticle core and a continuous gold shell is designed and synthesized to possess both magnetic and electrically conductive properties. Under the wireless actuation of a programmed oscillating magnetic field, a ribbon-like magnetic microswarm is generated using the building blocks, which shows controllable shape transformation and navigation. Such artificial miniature swarms are demonstrated to be able to precisely connect two isolated electrodes on demand, thus exhibiting promising potentials in the electronic field, such as microswitch, microcircuit repairing, and flexible electronics. Future efforts may be devoted to the development of building blocks with more customized functionalities, which may endow the microswarm with diverse group-level functions, and facilitate to better understand the natural complex swarm behaviors.

References

Anderson C, Theraulaz G, Deneubourg J-L (2002) Self-assemblages in insect societies. Insectes Soc 49(2):99–110
Breen TL, Tien J, Scott R, Hadzic T, Whitesides GM (1999) Design and self-assembly of open, regular, 3D mesostructures. Science 284(5416):948–951
Cademartiri L, Bishop KJ (2015) Programmable self-assembly. Nat Mater 14:2–9
Chen Q, Bae SC, Granick S (2011) Directed self-assembly of a colloidal kagome lattice. Nature 469(7330):381–384
Choi CKK, Zhuo X, Chiu YTE, Yang H, Wang J, Choi CHJ (2017) Polydopamine-based concentric nanoshells with programmable architectures and plasmonic properties. Nanoscale 9(43):16968–16980
Deng H, Li X, Peng Q, Wang X, Chen J, Li Y (2005) Monodisperse magnetic single-crystal ferrite microspheres. Angew Chem 117(18):2842–2845
Felfoul O, Mohammadi M, Taherkhani S, De Lanauze D, Xu YZ, Loghin D, Essa S, Jancik S, Houle D, Lafleur M (2016) Magneto-aerotactic bacteria deliver drug-containing nanoliposomes to tumour hypoxic regions. Nat Nanotechnol 11(11):941
Garattoni L, Birattari M (2018) Autonomous task sequencing in a robot swarm. Sci Robot 3(20):eaat0430
Gauci M, Chen J, Li W, Dodd TJ, Gross R (2014) Clustering objects with robots that do not compute. In: Proceedings of the 2014 international conference on autonomous agents and multi-agent systems. international foundation for autonomous agents and multiagent systems, 421–428
Gobre VV, Tkatchenko A (2013) Scaling laws for van der Waals interactions in nanostructured materials. Nat Commun 4:2341
Gracias DH, Tien J, Breen TL, Hsu C, Whitesides GM (2000) Forming electrical networks in three dimensions by self-assembly. Science 289(5482):1170–1172
Halloy J, Sempo G, Caprari G, Rivault C, Asadpour M, Tâche F, Saïd I, Durier V, Canonge S, Amé J-M (2007) Social integration of robots into groups of cockroaches to control self-organized choices. Science 318(5853):1155–1158
Hölldobler B, Wilson EO (1990) The ants. Harvard University Press

Krieger MJ, Billeter J-B, Keller L (2000) Ant-like task allocation and recruitment in cooperative robots. Nature 406(6799):992–995

Lee H, Dellatore SM, Miller WM, Messersmith PB (2007) Mussel-inspired surface chemistry for multifunctional coatings. Science 318(5849):426–430

Li J, Shklyaev OE, Li T, Liu W, Shum H, Rozen I, Balazs AC, Wang J (2015) Self-propelled nanomotors autonomously seek and repair cracks. Nano Lett 15(10):7077–7085

Lu C, Tang Z (2016) Advanced inorganic nanoarchitectures from oriented self-assembly. Adv Mater 28(6):1096–1108

Mao X, Chen Q, Granick S (2013) Entropy favours open colloidal lattices. Nat Mater 12(3):217–222

Martel S, Mohammadi M (2010) Using a swarm of self-propelled natural microrobots in the form of flagellated bacteria to perform complex micro-assembly tasks. In: Robotics and automation (ICRA), 2010 IEEE international conference on IEEE, 500–505

Mathews N, Christensen AL, O'Grady R, Mondada F, Dorigo M (2017) Mergeable nervous systems for robots. Nat Commun 8:439

Miele E, Raj S, Baraissov Z, Král P, Mirsaidov U (2017) Dynamics of templated assembly of nanoparticle filaments within nanochannels. Adv Mater 29(37):1702682

Mlot NJ, Tovey CA, Hu DL (2011) Fire ants self-assemble into waterproof rafts to survive floods. Proc Natl Acad Sci 108(19):7669–7673

O'Grady R, Groß R, Christensen AL, Dorigo M (2010) Self-assembly strategies in a group of autonomous mobile robots. Auton Robot 28(4):439–455

Palacci J, Sacanna S, Steinberg AP, Pine DJ, Chaikin PM (2013) Living crystals of light-activated colloidal surfers. Science 1230020

Rubenstein M, Cornejo A, Nagpal R (2014) Programmable self-assembly in a thousand-robot swarm. Science 345(6198):795–799

Snezhko A, Aranson IS (2011) Magnetic manipulation of self-assembled colloidal asters. Nat Mater 10(9):698

Wang C, Wang C, Huang Z, Xu S (2018) Materials and structures toward soft electronics. Adv Mater 30(50):1801368

Werfel J, Petersen K, Nagpal R (2014) Designing collective behavior in a termite-inspired robot construction team. Science 343(6172):754–758

Whitesides GM, Grzybowski B (2002) Self-assembly at all scales. Science 295(5564):2418–2421

Yan J, Bloom M, Bae SC, Luijten E, Granick S (2012) Linking synchronization to self-assembly using magnetic Janus colloids. Nature 491(7425):578–581

Yang G-Z, Bellingham J, Dupont PE, Fischer P, Floridi L, Full R, Jacobstein N, Kumar V, McNutt M, Merrifield R (2018) The grand challenges of science robotics. Sci Robot 3(14):eaar7650

Yu J, Yang L, Zhang L (2018a) Pattern generation and motion control of a vortex-like paramagnetic nanoparticle swarm. Int J Robot Res 37(8):912–930

Yu J, Wang B, Du X, Wang Q, Zhang L (2018b) Ultra-extensible ribbon-like magnetic microswarm. Nat Commun 9:3260

Zhang J, Luijten E, Grzybowski BA, Granick S (2017) Active colloids with collective mobility status and research opportunities. Chem Soc Rev 46(18):5551–5569

Chapter 13
Summary and Outlook

Abstract In this chapter, we summarize the contents of this book and provide perspectives on the development of micro/nanorobot swarms. We summarize and look forward to the fundamental research of microswarms, including swarm formation, pattern transformation, locomotion, heterogeneous and 3D swarm, and swarm control. In terms of applications of microswarms, we briefly review progress on biomedical applications of microswarms and provide insights into the challenges of achieving clinical applications of microswarms.

Keywords Micro/nanorobot swarms · Fundamental · Application · Challenge

As a hot research topic in robotics and active matter, micro/nanorobot swarms have aroused extensive attention in academia and society (Kaspar et al. 2021; Jin and Zhang 2022; Ji et al. 2022; Yang et al. 2018). In 2018, experts in robotics grouped robot swarms as one of the ten grand challenges in robotics, which may have breakthroughs with significant research and/or socioeconomic impact in the next 5 to 10 years (Yang et al. 2018). The strategies for building robot swarms at the micro/nanoscale are quite different from that at the macro scale, and many problems need to be addressed. Five years later, we look back on developments in this field and find that plenty of major advances have been made in understanding collective behaviors of micro/nanorobots and constructing different types of microswarms (Yang et al. 2022a; Wang and Zhang 2018; Wang and Pumera 2020; Fu et al. 2022). In this book, we take influential key efforts carried out by our group as examples to introduce the outstanding achievements in the field of magnetic micro/nanorobot swarm, ranging from fundamental research to specific applications (Fig. 13.1). For fundamental research, we introduce the formation, transformation, and locomotion of different types of microswarms, and the underlying mechanisms are revealed by building theoretical models and using finite element simulations. The experimental environments for fundamental research are generally ideal and differ greatly from practical application environments. Therefore, preliminary studies oriented to applications (especially biomedical applications) are carried out, such as the influence of the physiological environment on the microswarm and the imaging strategies in the

338   13 Summary and Outlook

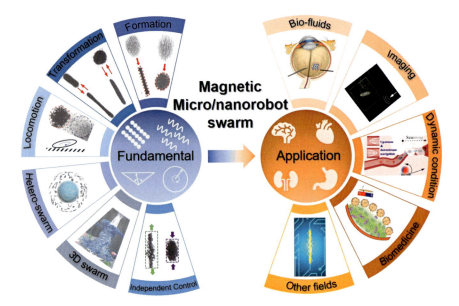

Fig. 13.1 Schematic illustration of key research topics in magnetic micro/nanorobot swarms covered in this book

human body. Microswarms with different properties and functions are used in various specific applications, and we present some representative examples in biomedical and engineering applications. In this chapter, we summarize the key research topics of microswarms and provide perspectives on related future challenges.

13.1 Fundamentals

13.1.1 Formation

The formation of magnetic micro/nanorobot swarms originates from the response of magnetic agents to the applied magnetic field. The microswarm needs to constantly dissipate energy to maintain its swarm pattern, which is different from static equilibrium assembled structures (Chen et al. 2011; Ma et al. 2015; Yan et al. 2012). Therefore, the swarm pattern disappears accordingly once the magnetic field is turned off. We summarize the magnetic microswarms into three types, i.e., the medium-induced swarm, magnetic field-induced swarm, and weakly-interacted swarm (Yu et al. 2019a). In addition to the weakly-interacted swarm, the agent-agent interactions within the microswarm play a crucial role in the formation of the microswarm. During the formation process, the force and moment exerted on each individual constantly change until a dynamic-equilibrium state are reached. For example, the two representative microswarms introduced in this book, the vortex-like microswarm (Chap. 2) and the ribbon-like microswarm (Chap. 3), are dominated by hydrodynamic and magnetic interactions, respectively (Yu et al. 2021, 2018a, b). Based on the generation strategies of these two microswarms, we have developed a series of other types of microswarms. By combining the light and the rotating magnetic field, a tornado-like microswarm with a three-dimensional structure is generated (Chap. 5) (Ji et al. 2020). A rotating magnetic field with gradients is used to generate microswarms with enhanced gathering effects (Chap. 9) (Wang et al. 2021a). The building blocks of the microswarm can be replaced with other magnetic agents, such as functionalized Fe_3O_4 nanoparticles (Chap. 4) (Jin et al. 2021) and Ni nanorods (Chap. 5) (Du et al. 2021). We hope that more novel micro/nanorobot swarms can be developed in the future, which may be realized by designing and programming magnetic fields, using new magnetic materials, and combining them with other physical fields. The generation method based on new interaction mechanisms is able to greatly enrich the types of microswarms, which is conducive to further research and applications. It is worth mentioning that the research on micro/nanorobot swarms is initially inspired by various collective behaviors in nature. Plenty of microswarm reported so far have structures or functions similar to swarms in nature (Sun et al. 2021; Jin et al. 2019; Ahmed et al. 2017). We believe that inspiration from fascinating natural collective behaviors will continuously promote the development of microswarms. On the other hand, the study of microswarms is also expected to provide a new perspective for understanding the principle of swarming behaviors in nature. The investigation of fundamental mechanisms of microswarms, such as the way individuals in the swarm communicate with others and the mechanism of how the swarm exerts influence on individuals, may help to address problems in the study of biological swarms.

13.1.2 Transformation

The micro/nanorobot swarms generally have transformable morphologies similar to natural swarms (Snezhko and Aranson 2011; Xie et al. 2019). Compared with individual micro/nanorobots (e.g., soft micro/nanorobots) capable of performing deformation, the pattern transformation of microswarms is easier to trigger and can have a larger degree. The pattern transformation of microswarms is achieved in a passive or active manner. Passive transformation refers to the shape of the swarm varying due to changes in the surrounding environments instead of the input field. For example, the light-driven microswarm adaptively changes its patterns when passing through a narrow channel (Mou et al. 2019a). The active transformation is achieved by adjusting the external field. We introduce active pattern transformation behaviors of the vortex-like and ribbon-like microswarms, including reversible expansion, elongation, shrinkage, and so on (Yu et al. 2021, 2018a, b). Both active and passive transformations of the microswarm are attributed to tunable interaction forces between the building blocks. That is, the morphology of the microswarm is a reflection of the equilibrium state of the building blocks, and the transformation process makes the microswarm change from one dynamic-equilibrium state to another. Research on pattern transformation of microswarms may deepen the understanding of natural collective transformation behaviors. In addition, the pattern transformation ability plays an important role in enhancing the environmental adaptability of microswarms. Further progress may be made in three aspects: achieving more types, greater degrees, and faster rates of pattern transformation. First, pattern transformation of most microswarms is relatively simple. Increasing the modalities and realizing the multimode transformation can endow the microswarm with rich functionalities to adapt to different environments and abilities to perform complex tasks. Second, the limit of microswarm pattern transformation theoretically depends on the size of the building blocks. Individuals at the micro/nanoscale provide infinite possibilities for pattern transformation of microswarms. How to achieve a greater degree of deformation while maintaining the stability of the swarm pattern is a question worthy of consideration. Third, the controllable pattern transformation rate can further enhance the environmental adaptability and task execution efficiency of the microswarm. We propose a strategy to control the pattern transformation rate of two microswarms by tuning the magnetic field parameters to coordinate the inner interactions (Chap. 6) (Yang et al. 2022b). For other types of microswarms, the pattern transformation rate control is expected to be realized by using a similar method, which requires further investigation of the relationship between the inner interaction force and the external inputs. The magnetic field is easily programmed and thus has unique advantages in realizing active pattern transformation. The key principle of microswarm pattern transformation is adjusting the inner agent-agent interactions through carefully designing the magnetic field and the building blocks.

13.1.3 Locomotion

In a low Reynolds number environment where the inertia can be ignored, the micro/nanorobots produce effective displacement through non-reciprocal motion (Purcell 1977). The locomotion of magnetic microswarms needs to break the motion symmetry, which is usually achieved by special designs of the building blocks (e.g., the helical structure) or by using boundaries. For a microswarm driven by a uniform magnetic field, a simple way to achieve locomotion is tilting the magnetic field to the horizontal plane by adding a pitch angle. Magnetic agents in the microswarm, such as nanoparticle chains, generate a net displacement when subjected to asymmetric forces as they move near the boundary. This locomotion strategy is widely used in magnetic microswarms, including the vortex-like and ribbon-like microswarm. For microswarms that use boundaries to perform locomotion, the influence of boundary conditions (e.g., roughness and hydrophilicity) needs to be considered in further research. A single form of locomotion may not be able to cope with complex and changeable external conditions. Therefore, more locomotion strategies of microswarms are needed, which may be inspired by the movement of animal swarms in nature. A practical strategy is to combine magnetic fields with other physical fields to drive the movement of the swarm (Ahmed et al. 2017; Lin et al. 2017; Palacci et al. 2013; Xu et al. 2015). In addition, from the perspective of practical application, it is hoped that the moving velocities of microswarms are as fast as possible to improve the efficiency of performing tasks. However, moving faster often means reducing the pattern stability because there always be differences in shapes or properties of the building blocks. These differences are easily amplified by excessive moving velocities, resulting in deformation or even collapse of the swarm pattern. Enhancing interaction forces between building blocks can significantly improve the stability of the swarm, but meanwhile, it will also lead to a reduction in the pattern transformation ability and even make the swarm unable to be reconfigured. Therefore, the trade-off among the moving velocity, pattern stability, and transformation ability according to specific scenarios is needed for microswarms actuated by any method.

13.1.4 Heterogeneous Microswarm

Most of the current magnetic micro/nanorobot swarms are composed of the same building blocks and hence have simple functionalities. One strategy for obtaining microswarms capable of performing multiple tasks is to endow each building block with rich functions. However, it requires elegant design and complicated fabrication, which is difficult to achieve at the micro- and nanoscale. Another approach is constructing heterogeneous microswarms. Different functionalized Fe_3O_4 nanoparticles are used to form microswarms with heterogeneous structures, which can realize synergistic therapy through domino reactions (Jin et al. 2021). In addition to functionalized nanoparticles, building blocks with different magnetic properties, shapes, and structures are also expected to be adopted to form heterogeneous microswarms. The different types of building blocks can not only enhance the functionality of the microswarm but also hold the potential for generating novel collective behaviors. For example, microparticles with different sizes and dielectric properties form hierarchical leader–follower-like microswarm in the AC electric field (Liang et al. 2020). Two species of passive microparticles exhibit predator–prey swarming behavior (Mou et al. 2019b). Similar biomimetic collective behaviors are also expected to be realized in magnetic field-driven microswarms. It should be noted that heterogeneous microswarms do not refer to the simple mixing of two or more types of building blocks. Careful studies on properties of different building blocks and interactions among them are required in the design. In addition, the functions of each component in a heterogeneous microswarm should be clarified, such as sensing, actuation, loading, and treatment. Overall, the heterogeneous microswarm is an effective strategy for building multifunctional robotic systems at small scales without the need for sophisticated equipment or high costs.

13.1.5 3D Microswarms

Natural swarms, such as bird flocks, fish schools, and honeybee colonies, generally have three-dimensional structures and can move throughout the whole space. However, most of the existing microswarms can only exist near the interface. Their formation, transformation, and locomotion are all restricted to a two-dimensional plane and hence are considered 2D microswarms. Constructing microswarms with 3D patterns against gravity has been a great challenge in the research of micro/nanorobot swarms. The key point is to enable the building blocks to produce vertical displacement, which requires the force provided by external fields or inner interactions to be greater than or equal to gravity. We introduce a method for generating a tornado-like magnetic microswarm via light-induced convection (Chap. 5). This tornado-like microswarm has a 3D structure and is capable of performing vertical mass transport, but is deficient in locomotion. The position of the swarm is basically fixed, which is because chain-chain interactions are easily destroyed during locomotion under the influence of strong light-induced convection. Recently, several magnetic microswarms capable of "standing up" have been reported (Sun et al. 2021; Yue et al. 2022; Law et al. 2022; Li et al. 2022). Despite having thin sheet-like patterns, these microswarms can stand up and roll or translate on the substrate like a wheel. The particles mainly rely on strong magnetic forces to overcome gravity, so the swarms are relatively stable. Although much progress has been made in building 3D microswarms, there are still huge challenges in this area. To date, 3D microswarms still need to depend on the substrates or boundaries. We expect microswarms with stable patterns can break free from boundaries and move as an entity in three-dimensional space like their natural counterparts. It may be a feasible strategy to realize 3D motion of microswarms using additional external fields (e.g., light and acoustic fields) or magnetic field gradients. In addition, the inner agent-agent interactions need to be strong enough to maintain the stability of the swarm as it moves in three dimensions.

13.1.6 Independent Control

The independent control of micro/nanorobot swarms is a worthy research topic in this field. The independent control of multiple microswarms under the same input can endow the entire swarm system with the ability to perform multiple tasks in parallel and achieve enhanced functions through cooperation between different microswarms. An effective independent control strategy for microswarms is to use building blocks with different structures or properties to obtain different responses to the same input field. Chapter 5 introduces Ni nanorods and Fe_3O_4 nanoparticles form two ribbon-like microswarms, which exhibit different pattern transformation behaviors in specific magnetic field parameter intervals (Du et al. 2021). Although only independent control of pattern transformation is achieved, this study illustrates that microswarms consisting of different building blocks can exhibit different collective behaviors under the same input. Further research can leverage this approach to realize independent motion control of microswarms, which has more significant potential in practical applications. Greater diversity in building blocks can make independent control of microswarms easier to achieve. In addition, the hybrid field actuation is more conducive to realizing independent control because the difference in the response of building blocks to different applied fields is usually larger.

13.2 Applications

Although the development of micro/nanorobot swarms is still in its infancy, microswarms with various functions have been designed and have great potential for applications in many fields. Among them, applications in the biomedical field present the most challenges but are also the most exciting and anticipated, which we would like to highlight and discuss here (Wu et al. 2020; Chen et al. 2021). Fundamental research on microswarms is carried out in relatively ideal environments, including pure deionized water, smooth horizontal substrates, and stagnant fluids. The actual application environment, such as the inside of the human body, is much more complicated than these ideal environments. Therefore, before using microswarms in specific biomedical applications, it is necessary to study the influence of the actual environments on microswarms.

First, we discuss the influence of bio-fluids on microswarms in Chap. 7. Since most magnetic microswarms are in a liquid environment and rely on hydrodynamic forces, the properties of bio-fluids, including viscosity, ionic strength, mesh-like network, colloidal jamming by blood cells, the formation of protein corona on nanoparticles, may have a significant impact on the microswarm. Taking the vortex-like and ribbon-like microswarms as examples, we propose a strategy for selecting optimized microswarms for bio-fluids with different properties, which has a guiding significance for the application of microswarms in bio-fluids. However, there are differences between different microswarms, so experimental validation for specific bio-fluids and microswarms is an indispensable step in designing in vivo applications.

Imaging microswarms in the human body is also critical for biomedical applications. Microswarms in open containers can be clearly observed by general optical imaging methods. However, this approach is not suitable for imaging microswarms in the human body where the light has difficulty penetrating deep tissues. Microswarms require real-time imaging feedback for performing tasks such as targeted drug delivery. Various medical imaging methods can be used to localize microswarms in opaque tissues or organs. We give examples of ultrasound imaging, fluorescence imaging, and photoacoustic imaging in Chap. 8 (Yu et al. 2019b; Wang et al. 2022). In addition, magnetic resonance imaging (MRI), positron emission tomography (PET), optical coherence tomography (OCT), etc., have also been applied to the imaging of microswarms (Wu et al. 2018; Yan et al. 2017; Vilela et al. 2018). These imaging methods have both advantages and limitations. For example, ultrasound imaging has the advantages of being able to penetrate deep tissues, being harmless to the human body, and low cost, but it is difficult to obtain sufficient imaging contrast in certain parts of the human body (e.g., gas-filled lungs). In addition to enhancing the contrast of the microswarm, the design of the imaging system can also play a key role. There may be interference between the actuation device (e.g., electromagnetic coils) and the imaging device (e.g., MRI system) (Yang and Zhang 2021). Integrating the actuation system with the imaging system has proved to be a promising strategy for optimizing the imaging of microswarms (Erin et al. 2019). In addition, the combination of multiple imaging methods is also conducive to the imaging of microswarms in different parts of the human body, which can expand the application range of microswarms (Yan et al. 2017; Servant et al. 2015). Overall, the real-time feedback image of the microswarm is crucial to realize the imaging-guide therapy of microswarms in the human body.

Another major challenge is the dynamic fluid environment represented by blood vessels. We introduce using the rotating permanent magnet system to navigate microswarms in dynamic environments in Chap. 9. Ultrasound Doppler imaging is used for the localization of microswarms in this case. This research mainly uses magnetic field gradients to make the microswarm resist the disturbance of the flowing fluid (Wang et al. 2021a). The microswarm clings to the boundaries under the influence of magnetic field gradients, where the flow velocity is at a minimum. Other studies have also demonstrated the strategy that actuates microswarms close to boundaries to reduce the effects of fluid flow (Fonseca et al. 2022; Ahmed et al. 2021). In addition, the blood flow can be effectively restrained using a balloon catheter for generating microswarms (Wang et al. 2021a). To date, studies on generating and navigating microswarms in dynamic environments are relatively few. The research should focus on how to generate microswarms and maintain pattern stability in dynamic environments, and further enabling microswarms to move upstream, which requires the designed microswarm to have a well-formed pattern and strong propulsion.

In addition to the above-mentioned problems, many challenges still remain to be solved for the in vivo application of microswarms. For example, the inside of the human body is not a 2D plane. Microswarms with three-dimensional structures and the ability to move in 3D space are required. Microswarms performing long-distance locomotion have problems with low access rates and efficiency. Traditional catheters (or endoscopes) are able to deliver microswarms near the target area, and following navigation into hard-to-reach sites is conducted using the magnetic field (Fig. 13.2). Catheters equipped with tiny cameras/sensors can also be used as tools for real-time monitoring of microswarms. This method solves the long-distance delivery problem of microswarms and greatly expands the area where microswarms can be applied. In fundamental research, features of microswarms are explored independently under ideal conditions with simple equipment. Recently, we developed a system combining endoscopy, ultrasound imaging, and electromagnetic coils to deliver and navigate microswarms (Fig. 13.3) (Wang et al. 2021b). This system has a human-scale workspace and enables fast and accurate delivery and control of microswarms. Furthermore, we hope to realize intelligence of microswarms in applications based on such well-developed clinical platforms. The image processing

13.2 Applications

method based on deep learning can be combined with medical images to extract information about the microswarms and the surrounding environment, thereby realizing the automatic navigation and distribution planning of the microswarms in the biological environment (Yang et al. 2022c).

Before the clinical application, the safety of microswarms needs to be considered. Microswarms consisting of magnetic materials harmful to the human body cannot be used in clinical medicine. Therefore, biocompatibility of materials should be evaluated for in vivo applications such as targeted drug delivery using microswarms (Gao et al. 2021). Micro/nanorobot swarms based on biocompatible materials (e.g., hydrogels) or natural biological templates (e.g., spore, pollen, cells) are considered to have good application prospects. In addition, microswarms are composed of numerous

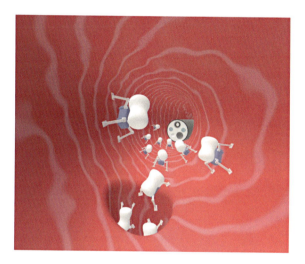

Fig. 13.2 Conceptual diagram of catheter-assisted in vivo application of microswarms

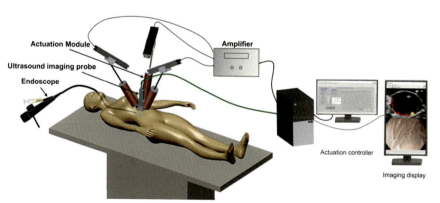

Fig. 13.3 Conceptual diagram of the integrated clinical platform for endoscope and ultrasound imaging-guided delivery of microswarms and microrobots

building blocks, some of which may remain in human organs or tissues after treatment. Therefore, the biodegradability of microswarms also needs to be considered (Li et al. 2019).

In order to realize the clinical application of micro/nanorobot swarms, an advanced medical platform with full functions and comprehensive features is required. Great efforts have been made to promote translational research and the productization of medical microswarms. The hybrid operating room at the Multiscale Medical Robotics Center (MRC) in Hong Kong Science Park enables the development and preclinical evaluation of novel medical devices, including micro/nanorobot swarms, through live animal and cadaver studies (Fig. 13.4). Advanced medical machines are equipped in the hybrid operating room, such as robotic-assisted C-arm X-ray imaging system (Siemens Artis Zeego), GE Artist MRI scanner, and Olympus endoscopic and laparoscopic system, which greatly promotes the clinical research of medical robots. In addition, the Multiscale Medical Robotics Center, supported by the Chinese University of Hong Kong in collaboration with ETH Zurich, Imperial College London, and Johns Hopkins University, provides a platform for researchers in different fields, including clinicians, engineers, and roboticists, to contribute their efforts to realize the clinical application of micro/nanorobot swarms through interdisciplinary cooperation, and to ultimately benefit patients around the world (The Multiscale Medical Robotics Center at the Hong Kong Science Park, n.d.).

In summary, micro/nanorobot swarms are in their infancy in fundamental research and practical applications. The progress made in past decades is gratifying and

Fig. 13.4 Hybrid operating room of the MRC laboratory. The device shown in the figure is the Siemens Artis Zeego C-arm X-ray system

exciting, but the future development of microswarms still faces many severe challenges, which just shows the huge potential of this field. In-depth understanding of the nature of collective behaviors at the micro/nanoscale and realizing clinical applications of microswarms are inseparable from interdisciplinary communication and close cooperation. It is foreseeable that exciting results will continue to emerge in this field. We look forward to the vigorous development of micro/nanorobot swarms and hope that the content introduced in this book can provide a reference or inspiration for researchers in this field.

References

Ahmed D, Baasch T, Blondel N, Laubli N, Dual J, Nelson BJ (2017) Neutrophil-inspired propulsion in a combined acoustic and magnetic field. Nat Commun 8:770

Ahmed D, Sukhov A, Hauri D, Rodrigue D, Maranta G, Harting J, Nelson BJ (2021) Bioinspired acousto-magnetic microswarm robots with upstream motility. Nat Mach Intell 3(2):116–124

Chen Q, Bae SC, Granick S (2011) Directed self-assembly of a colloidal kagome lattice. Nature 469(7330):381–384

Chen H, Zhang H, Xu T, Yu J (2021) An overview of micronanoswarms for biomedical applications. ACS Nano 15(10):15625–15644

Du X, Yu J, Jin D, Chiu PWY, Zhang L (2021) Independent pattern formation of nanorod and nanoparticle swarms under an oscillating field. ACS Nano 15(4):4429–4439

Erin O, Gilbert HB, Tabak AF, Sitti M (2019) Elevation and azimuth rotational actuation of an untethered millirobot by mri gradient coils. IEEE Trans Rob 35(6):1323–1337

Fonseca ADC, Kohler T, Ahmed D (2022) Ultrasound-controlled swarmbots under physiological flow conditions. Adv Mater Interfaces 9(26):2200877

Fu Y, Yu H, Zhang X, Malgaretti P, Kishore V, Wang W (2022) Microscopic swarms: from active matter physics to biomedical and environmental applications. Micromachines 13(2):295

Gao C, Wang Y, Ye Z, Lin Z, Ma X, He Q (2021) Biomedical micro-/nanomotors: from overcoming biological barriers to in vivo imaging. Adv Mater 33(6):2000512

Ji F, Jin D, Wang B, Zhang L (2020) Light-driven hovering of a magnetic microswarm in fluid. ACS Nano 14(6):6990–6998

Ji F, Wu Y, Pumera M, Zhang L (2022) Collective behaviors of active matter learning from natural taxes across scales. Adv Mater 35(8):2203959. https://doi.org/10.1002/adma.202203959

Jin D, Zhang L (2022) Collective behaviors of magnetic active matter: recent progress toward reconfigurable, adaptive, and multifunctional swarming micro/nanorobots. Acc Chem Res 55(1):98–109

Jin D, Yu J, Yuan K, Zhang L (2019) Mimicking the structure and function of ant bridges in a reconfigurable microswarm for electronic applications. ACS Nano 13(5):5999–6007

Jin D, Yuan K, Du X, Wang Q, Wang S, Zhang L (2021) Domino reaction encoded heterogeneous colloidal microswarm with on-demand morphological adaptability. Adv Mater 33(37):2100070

Kaspar C, Ravoo B, van der Wiel W, Wegner S, Pernice W (2021) The rise of intelligent matter. Nature 594(7863):345–355

Law R, Chen H, Wang Y, Yu J, Sun Y (2022) Gravity-resisting colloidal collectives. Sci Adv 8(46):eade3161

Li X, Liu L, Li S, Wan Y, Chen J-X, Tian S, Huang Z, Xiao Y-F, Cui X, Xiang C (2019) Biodegradable π-conjugated oligomer nanoparticles with high photothermal conversion efficiency for cancer theranostics. ACS Nano 13(11):12901–12911

Li M, Zhang T, Zhang X, Mu J, Zhang W (2022) Vector-controlled wheel-like magnetic swarms with multimodal locomotion and reconfigurable capabilities. Frontiers Bioeng Biotechnol 10:877964

Liang X, Mou F, Huang Z, Zhang J, You M, Xu L, Luo M, Guan J (2020) Hierarchical microswarms with leader–follower-like structures: electrohydrodynamic self-organization and multimode collective photoresponses. Adv Func Mater 30(16):1908602

Lin Z, Si T, Wu Z, Gao C, Lin X, He Q (2017) Light-activated active colloid ribbons. Angew Chem 129(43):13702–13705

Ma F, Wang S, Wu DT, Wu N (2015) Electric-field-induced assembly and propulsion of chiral colloidal clusters. Proc Natl Acad Sci 112(20):6307–6312

Mou F, Zhang J, Wu Z, Du S, Zhang Z, Xu L, Guan J (2019a) Phototactic flocking of photochemical micromotors. Iscience 19:415–424

Mou F, Li X, Xie Q, Zhang J, Xiong K, Xu L, Guan J (2019b) Active micromotor systems built from passive particles with biomimetic predator-prey interactions. ACS Nano 14(1):406–414

Palacci J, Sacanna S, Steinberg AP, Pine DJ, Chaikin PM (2013) Living crystals of light-activated colloidal surfers. Science 339(6122):936–940

Purcell EM (1977) Life at low Reynolds number. Am J Phys 45(1):3–11

Servant A, Qiu F, Mazza M, Kostarelos K, Nelson BJ (2015) Controlled in vivo swimming of a swarm of bacteria-like microrobotic flagella. Adv Mater 27(19):2981–2988

Snezhko A, Aranson IS (2011) Magnetic manipulation of self-assembled colloidal asters. Nat Mater 10(9):698–703

Sun M, Fan X, Tian C, Yang M, Sun L, Xie H (2021) Swarming microdroplets to a dexterous micromanipulator. Adv Func Mater 31(19):2011193

The Multiscale Medical Robotics Center at the Hong Kong Science Park. https://www.mrc-cuhk.com/

Vilela D, Cossío U, Parmar J, Martínez- AM, Gómez-Vallejo V, Llop J, Sánchez S (2018) Medical imaging for the tracking of micromotors. ACS Nano 12(2):1220–1227

Wang H, Pumera M (2020) Coordinated behaviors of artificial micro/nanomachines: from mutual interactions to interactions with the environment. Chem Soc Rev 49(10):3211–3230

Wang Q, Zhang L (2018) External power-driven microrobotic swarm: from fundamental understanding to imaging-guided delivery. ACS Nano 15(1):149–174

Wang Q, Chan KF, Schweizer K, Du X, Jin D, Yu SCH, Nelson BJ, Zhang L (2021) Ultrasound doppler-guided real-time navigation of a magnetic microswarm for active endovascular delivery. Sci Adv 7(9):eabe5914

Wang B, Chan KF, Yuan K, Wang Q, Xia X, Yang L, Ko H, Wang Y-XJ, Sung JJY, Chiu PWY (2021) Endoscopy-assisted magnetic navigation of biohybrid soft microrobots with rapid endoluminal delivery and imaging. Sci Robot 6(52):eabd2813

Wang Q, Yang S, Zhang L (2022) Magnetic actuation of a dynamically reconfigurable microswarm for enhanced ultrasound imaging contrast. IEEE/ASME Trans Mechatron 27(6):4235–4245. https://doi.org/10.1109/TMECH.2022.3151983

Wu Z, Troll J, Jeong H-H, Wei Q, Stang M, Ziemssen F, Wang Z, Dong M, Schnichels S, Qiu T, Fischer P (2018) A swarm of slippery micropropellers penetrates the vitreous body of the eye. Sci Adv 4(11):eaat4388

Wu Z, Chen Y, Mukasa D, Pak OS, Gao W (2020) Medical micro/nanorobots in complex media. Chem Soc Rev 49(22):8088–8112

Xie H, Sun M, Fan X, Lin Z, Chen W, Wang L, Dong L, He Q (2019) Reconfigurable magnetic microrobot swarm: Multimode transformation, locomotion, and manipulation. Sci. Robot 4(28):eaav8006

Xu T, Soto F, Gao W, Dong R, Garcia- V, Magana E, Zhang X, Wang J (2015) Reversible swarming and separation of self-propelled chemically powered nanomotors under acoustic fields. J Am Chem Soc 137(6):2163–2166

Yan J, Bloom M, Bae SC, Luijten E, Granick S (2012) Linking synchronization to self-assembly using magnetic Janus colloids. Nature 491(7425):578–581

Yan X, Zhou Q, Vincent M, Deng Y, Yu J, Xu J, Xu T, Tang T, Bian L, Wang Y-XJ (2017) Multifunctional biohybrid magnetite microrobots for imaging-guided therapy. Sci Robot 2(12):eaaq1155

Yang L, Zhang L (2021) Motion control in magnetic microrobotics: From individual and multiple robots to swarms. Annu Rev Control Robot Auton Syst 4:509–534

Yang G-Z, Bellingham J, Dupont PE, Fischer P, Floridi L, Full R, Jacobstein N, Kumar V, McNutt M, Merrifield R (2018) The grand challenges of science robotics. Science robotics 3(14):eaar7650

Yang L, Yu J, Yang S, Wang B, Nelson BJ, Zhang L (2022a) A survey on swarm microrobotics. IEEE Trans Rob 38(3):1531–1551

Yang S, Wang Q, Jin D, Du X, Zhang L (2022b) Probing fast transformation of magnetic colloidal microswarms in complex fluids. ACS Nano 16(11):19025–19037

Yang L, Jiang J, Gao X, Wang Q, Dou Q, Zhang L (2022c) Autonomous environment-adaptive microrobot swarm navigation enabled by deep learning-based real-time distribution planning. Nat Mach Intell 4(5):480–493

Yu J, Yang L, Zhang L (2018a) Pattern generation and motion control of a vortex-like paramagnetic nanoparticle swarm. Int J Robot Res 37(8):912–930

Yu J, Wang B, Du X, Wang Q, Zhang L (2018b) Ultra-extensible ribbon-like magnetic microswarm. Nat Commun 9:3260

Yu J, Jin D, Chan KF, Wang Q, Yuan K, Zhang L (2019a) Active generation and magnetic actuation of microrobotic swarms in bio-fluids. Nat Commun 10:5631

Yu J, Wang Q, Li M, Liu C, Wang L, Xu T, Zhang L (2019b) Characterizing nanoparticle swarms with tuneable concentrations for enhanced imaging contrast. IEEE Robot Autom Lett 4(3):2942–2949

Yu J, Yang L, Du X, Chen H, Xu T, Zhang L (2021) Adaptive pattern and motion control of magnetic microrobotic swarms. IEEE Trans Rob 38(3):1552–1570

Yue H, Chang X, Liu J, Zhou D, Li L (2022) Wheel-like magnetic-driven microswarm with a band-aid imitation for patching up microscale intestinal perforation. ACS Appl Mater Interfaces 14(7):8743–8752

Printed in the United States
by Baker & Taylor Publisher Services